Shaping Smart for Better Cities

Shaping Smart for Better Cities

Rethinking and Shaping Relationships between Urban Space and Digital Technologies

Edited by

Alessandro Aurigi
Professor of Urban Design and Associate Dean: Research,
University of Plymouth, UK

Nancy Odendaal
Associate Professor, City and Regional Planning,
University of Cape Town, South Africa

ELSEVIER

ACADEMIC PRESS
An imprint of Elsevier

Academic Press is an imprint of Elsevier
125 London Wall, London EC2Y 5AS, United Kingdom
525 B Street, Suite 1650, San Diego, CA 92101, United States
50 Hampshire Street, 5th Floor, Cambridge, MA 02139, United States
The Boulevard, Langford Lane, Kidlington, Oxford OX5 1GB, United Kingdom

Library of Congress Cataloging-in-Publication Data
A catalog record for this book is available from the Library of Congress

British Library Cataloguing-in-Publication Data
A catalogue record for this book is available from the British Library

ISBN: 978-0-12-818636-7

For information on all Academic Press publications
visit our website at https://www.elsevier.com/books-and-journals

Publisher: Brian Romer
Acquisitions Editor: Graham Nisbet
Editorial Project Manager: Alice Grant
Production Project Manager: Kiruthika Govindaraju
Cover Designer: Matthew Limbert

Typeset by SPi Global, India

Contents

3. Responsive public spaces: Five mechanisms for the design of public space in the era of networked urbanism

Martijn de Waal, Frank Suurenbroek, and Ivan Nio

4. Smart plays

Ben van Berkel, Filippo Lodi, and Wael Sami Batal

5. Snowfall on Piazza Castello: Stubborn dispositions and multiple publics in a (temporarily smart) Milanese square

Yulya Besplemmenova and Andrea Pollio

6. Designing for hyperlocal: The use of locative media to augment place narratives

Efstathia Kostopoulou and Ava Fatah gen Schieck

7. Place-based design as method of accessing memories and meanings: Historical augmentation in the harbor promenade of Lahti

Aale Luusua, Henrika Pihlajaniemi, Mika Hakkarainen, Petri Honkamaa, Eveliina Juntunen, and Sami Huuskonen

Contributors

Numbers in parenthesis indicate the pages on which the authors' contributions begin.

María José Castelazo André (239), Infrarrojo Producciones, Guadalajara, México

Alessandro Aurigi (1, 11), University of Plymouth, Plymouth, United Kingdom

María José Sánchez Varela Barajas (239), Instituto Tecnológico y de Estudios Superiores de Occidente (ITESO) and SarapeLab, Guadalajara, México

Ben van Berkel (55), UNStudio, Amsterdam; UNSense, Amsterdam; Rietveld Academy, Amsterdam, Netherlands; Architectural Association, London, England; Harvard University Graduate School of Design, Cambridge, MA, United States

Yulya Besplemmenova (71), Oblo, Service Design and Research, Milan, Italy

Ana Bilandzic (283), QUT Design Lab, Queensland University of Technology, Brisbane, QLD, Australia

Annemiek van Boeijen (123), Faculty of Industrial Design Engineering, Delft University of Technology, Delft, The Netherlands

Cat Button (357), School of Architecture, Planning and Landscape, Newcastle University, Newcastle upon Tyne, United Kingdom

Glenda Amayo Caldwell (239), Queensland University of Technology, Brisbane, QLD, Australia

Marianella Chamorro-Koc (239), Queensland University of Technology, Brisbane, QLD, Australia

Paul Cureton (267), Imagination Lancaster, Lancaster University, Lancaster, United Kingdom

Rudi Darson (123), Shop owners' association of Middelland, Rotterdam, The Netherlands

Nick Dunn (267), Imagination Lancaster, Lancaster University, Lancaster, United Kingdom

Kathryn Ewing (339), Department of Architecture, Planning and Geomatics, University of Cape Town, Cape Town, South Africa

Ava Fatah gen Schieck (87), The Bartlett School of Architecture, University College London, London, United Kingdom

Marcus Foth (283), QUT Design Lab, Queensland University of Technology, Brisbane, QLD, Australia

Joel Fredericks (177, 239), Design Lab, School of Architecture, Design and Planning, The University of Sydney, Sydney, NSW, Australia

Amalia De Götzen (373), Aalborg University, Copenhagen, Denmark

Carlos Granell-Canut (157), Universitat Jaume I, Castellón de la Plana, Spain

Mirko Guaralda (283), QUT Design Lab, Queensland University of Technology, Brisbane, QLD, Australia

Prince K. Guma (321), British Institute in Eastern Africa, Nairobi, Kenya

Mika Hakkarainen (107), VTT Technical Research Centre of Finland, Otaniemi, Espoo

Iliana Hernández-García (307), Department of Aesthetics, School of Architecture and Design, Pontificia Universidad Javeriana, Bogotá, Colombia

Jaime Hernández-García (307), Department of Aesthetics, School of Architecture and Design, Pontificia Universidad Javeriana, Bogotá, Colombia

Luke Hespanhol (239), Design Lab, School of Architecture, Design and Planning, The University of Sydney, Sydney, NSW, Australia

Petri Honkamaa (107), VTT Technical Research Centre of Finland, Otaniemi, Espoo

Sami Huuskonen (107), Tehomet Oy, Kangasniemi, Finland

Eveliina Juntunen (107), VTT Technical Research Centre of Finland, Otaniemi, Espoo

Efstathia Kostopoulou (87), The Bartlett School of Architecture, University College London, London, United Kingdom

Michael Krause (339), Violence Prevention Through Urban Upgrading NPC, Cape Town, South Africa

Filippo Lodi (55), UNStudio, Amsterdam, Netherlands; MSc Civil Engineering and Architecture, Università di Bologna, Bologna, Italy; Architecture, MA Advanced Architectural Design, HfBK Städelschule, Frankfurt am Main, Germany; MBA, Universiteit van Amsterdam, Amsterdam, Netherlands

Aale Luusua (107), INTERACT Research Group and Oulu School of Architecture, University of Oulu, Oulu, Finland

Nicola Morelli (373), Aalborg University, Copenhagen, Denmark

Ingrid Mulder (123), Faculty of Industrial Design Engineering, Delft University of Technology, Delft, The Netherlands

Ivan Nio (33), Amsterdam University of Applied Sciences, Amsterdam, The Netherlands

Nancy Odendaal (1, 203), University of Cape Town, Cape Town, South Africa

Callum Parker (177), Design Lab, School of Architecture, Design and Planning, The University of Sydney, Sydney, NSW, Australia

Henrika Pihlajaniemi (107), Oulu School of Architecture, University of Oulu, Oulu, Finland

Andrea Pollio (71), Institute for Culture and Society, Western Sydney University, Sydney, NSW, Australia

Manuel Portela (157), Universitat Pompeu Fabra, Barcelona, Spain

Gillian Rose (221), School of Geography and the Environment, University of Oxford, Oxford, United Kingdom

Ummu Sakiinah (123), Faculty of Industrial Design Engineering, Delft University of Technology, Delft, The Netherlands

Wael Sami Batal (55), UNStudio, Amsterdam, Netherlands; University of California, Los Angeles, CA; University of Colorado, Boulder, CO, United States

Luca Simeone (373), Aalborg University, Copenhagen, Denmark

Frank Suurenbroek (33), Amsterdam University of Applied Sciences, Amsterdam, The Netherlands

Martin Tomitsch (177), Design Lab, School of Architecture, Design and Planning, The University of Sydney, Sydney, NSW, Australia; CAFA Beijing Visual Art Innovation Institute, China

Alan-Miguel Valdez (221), School of Engineering and Innovation Management, The Open University, Milton Keynes, United Kingdom

Martijn de Waal (33), Amsterdam University of Applied Sciences, Amsterdam, The Netherlands

Edward Wigley (221), Department of Geography, The Open University, Milton Keynes, United Kingdom

Oliver Zanetti (221), School of Geography and the Environment, University of Oxford, Oxford, United Kingdom

Authors' biography

Alessandro Aurigi is Professor of Urban Design at the University of Plymouth, having previously worked at the University of Newcastle and University College London (UCL). His research focuses on the relationships between our digital society and the ways we conceive, design, and manage urban space, to enhance and support place quality. Alex has published *Digital and Smart Cities* and *The Routledge Companion to Smart Cities* (Routledge, with Katharine Willis), the multidisciplinary book *Augmented Urban Spaces* (Ashgate, with Fiorella De Cindio), and *Making the Digital City* (Ashgate).

Wael Batal received a master's degree in architecture from UCLA and is an architect at UNStudio where he has worked on numerous projects across scales and typologies. He believes that architecture is a cultural discipline that operates between theory and practice. As a Knowledge Platform Coordinator at UNS, he is active in the production of knowledge through writing and lecturing about the intersections between architecture, technology, and society.

Yulya Besplemennova is a product service system designer currently working with service design and research studio Oblo in Milan. The focus of her work is designing complex systems that put together human and technological components, especially when it comes to the public realm, urban services, and spaces.

Ana Bilandzic is a PhD candidate with the Urban Informatics Research Group in the QUT Design Lab, Brisbane, Australia. Her research is on social and spatial precursors to innovation in casual creative environments, e.g., coworking spaces, makerspaces, and hackerspaces. She explores the spread of these spaces in Brisbane and the strategies of users to find and navigate them. Further, her research unpacks a better understanding of innovation precursors in spaces that focus on social innovation.

Dr. Cat Button is a senior lecturer (associate professor) in the School of Architecture, Planning and Landscape at Newcastle University, United Kingdom. Water has been her research focus for over a decade, particularly looking at domestic water shortages in India. She is interested in how people deal with not having enough water (and what happens when there is too much water in the wrong place). She is currently a coinvestigator on two international and transdisciplinary UKRI GCRF research hubs. This research focuses on cultural, social, and spiritual significances of water.

Dr. Glenda Amayo Caldwell is Associate Professor in Architecture at the School of Design, Creative Industries Faculty, Queensland University of Technology (QUT). She is an architecture and design scholar with expertise in physical, digital, and robotic fabrication, leading Industry 4.0 innovation through human-centered research in design robotics and media architecture. Glenda leads the QUT Design Lab's Designing Creative and Resilient Communities Program and is the author of numerous publications in the areas of media architecture, urban informatics, and design robotics.

Majo Castelazo André was born on May 23, 1983, in Cuernavaca, Mexico. The Castelazo family migrated to Guadalajara, the city where she grew up, studied, and worked for the next few years. Seeking justice, she studied law, but following Oscar Wilde's steps, she participated in more than 30 theatrical productions. With the conviction to learn from other cultures, she engages with tech-related communities and participates all over the world in workshops specialized in smart cities as a solution that combines with the arts in efforts to make the world better.

Associate Professor Marianella Chamorro-Koc is an industrial design researcher and educator. Her work aims to contribute to the design of enhanced people's interactions with everyday technologies. Marianella investigates the experiential knowledge embedded in people's activities and the contextual aspects shaping them, and her applied research is positioned in the design of technologies applications to help people manage their health and well-being. She leads the QUT Design Lab Designing for Health Program and is a member of the Design Research Society (DRS).

Paul Cureton is Senior Lecturer in Design at ImaginationLancaster, Lancaster University, United Kingdom, and a member of the Data Science Institute (DSI). His primary research interests include future cities, geodesign, GIS, UAVs, mapping, modeling, and digital fabrication. His recent publications include the monographs, *Strategies for Landscape Representation: Digital and Analogue Techniques* (Routledge, 2016) and *Drone Futures: UAS for Landscape & Urban Design* (CRC Press, 2020). He is also working on digital twin concepts through the development of Lancaster City Information Model (LCIM) with Garsdale Design, supported by ESRI, Bluesky Ltd., and CyberCity 3D.

Rudi Darson (M.Sc.) studied commercial sciences in Suriname and the Netherlands and public administration at Leiden University. He has a degree in change management from Maastricht University and he is an independent consultant on capacity building and entrepreneurship for creating economies. For more than 10 years, he has been supervising multidisciplinary students' teams from TU Delft, doing projects in developing countries on entrepreneurship and development. Now, he is focused on creating virtual and physical places to help revitalize economies.

Amalia de Götzen is Associate Professor at Aalborg University in Denmark where she coordinates the master's program in Service Systems Design. Her research activity focuses on digital social innovation and in particular on the intersection between interaction design and service design. She is interested in tools and methods that bridge the analog and digital world of services. She participated in several European projects as investigator and work package leader and has an extensive research experience on sonic interaction design.

Martijn de Waal is a lector (professor), at the Play & Civic Media research group at the Amsterdam University of Applied Sciences. His work focuses on the experience and design of public spaces in a networked society. He is the author of *The City as Interface* (Nai010 Publishers, 2014) and *The Platform Society* (Oxford University Press, 2018—with Jose van Dijck and Thomas Poell), and edited *The Hackable City* (Springer 2018) with Michiel de Lange. He collaborated in the research project Co-creating Responsive Urban Spaces (2017–19). He is part of the executive committee and general chair of the Media Architecture Biennale 2020.

Nick Dunn is Executive Director of ImaginationLancaster, the design research lab at Lancaster University, United Kingdom, where he is Professor of Urban Design. He is also Senior Fellow at the Institute for Social Futures, examining the insights that the arts and humanities can bring to the ways we think, envision, and analyze the futures of people, places, and planets. He has published numerous books and journal articles and is lead author on various commissioned reports. Nick's expertise on cities, futures, and health has led to curated exhibitions, keynotes, and plenary presentations around the world. He is coauthor, with Paul Cureton, of *Future Cities: A Visual Guide* (Bloomsbury, 2020).

Kathryn Ewing is a senior lecturer and convener of the Urban Design Programme at the University of Cape Town. She has architectural, urban design, planning, and research experience working through a multidisciplinary approach in Southern Africa. She is one of the founding directors and currently a nonexecutive director of VPUU NPC. Her work focuses on safe cities, participative design, and public space, building on lessons from design practice for policy development and teaching and knowledge exchange in African cities.

Ava Fatah gen Schieck is an educator, architect, and researcher. She is Associate Professor in Media Architecture and Urban Digital Interaction and leads the two studios, Body as Interface and City as Interface, on the M.Sc. Architectural Computation at the Bartlett School of Architecture, UCL. Ava is also the director of "Architectural Space and Computation" PhD program, and the departmental tutor for the Architecture and Digital Theory PhD program, UCL. She is the principal investigator of "Screens in the Wild: exploring the potential of networked urban screens for communities and culture," resulting in a unique "living lab" environment of four interactive networked screens (connected through video feed) in London and Nottingham (United Kingdom).

Marcus Foth (PhD, FACS) is Professor of Urban Informatics in the QUT Design Lab at Queensland University of Technology, Australia. He is also an honorary professor in the School of Communication and Culture at Aarhus University, Denmark. His transdisciplinary research focuses on interaction design, smart cities, community engagement, media architecture, and sustainability. Foth has authored or coauthored over 200 peer-reviewed publications. He founded the Urban Informatics Research Lab in 2006 and the QUT Design Lab in 2016. He tweets via @sunday9pm.

Dr. Joel Fredericks is a Lecturer in Design at the University of Sydney's School of Architecture, Design and Planning. Joel's research is transdisciplinary and sits across the domains of digital placemaking, media architecture, urbanism, smart cities, and immersive technologies. He has investigated collaborative and creative approaches to designing and deploying urban interventions that create curiosity and encourage people to playfully interact. He has authored and coauthored many publications in journals, edited books, and conference proceedings.

Carlos Granell currently holds a 5-year Ramón y Cajal postdoctoral fellowship at the Universitat Jaume I (UJI) of Castellón, Spain. Before rejoining UJI in 2015, he worked 4 years as a postdoc in the European Commission's Joint Research Centre. His research interests lie in the multidisciplinary application of GIS (science/systems), spatial analysis and visualization of sensor- and user-generated geographic content, and reproducibility research practice.

Mirko Guaralda (PhD, MHEd, DArch) is an academic at Queensland University of Technology (QUT). His work focuses on people-place interaction, enquiring into the complex issues of urban density, place quality, and community engagement. He was a researcher at the United States Study Centre of Sydney in 2012; in 2017, he was visiting professor at the Thammasat University of Bangkok, Thailand. Since 2018, he has also been engaged in research and teaching at the Jiangxi University of Science and Technology (China).

Prince K. Guma is a research fellow and assistant country director at the British Institute in Eastern Africa, Nairobi, Kenya. His research work draws on the intersection of STS, urban studies, and postcolonial studies to examine the contingent and place-based articulations of cities and how these are mediated through the diffusion and uptake of ICT-based plans and infrastructures. His findings are hoped to provide a menu for new explorations and to enhance our understanding of urban and digital possibilities in Africa.

Mika Hakkarainen received his B.Sc. (Eng.) degree in electronics and information technology at the Savonia University of Applied Science in 1997. He joined VTT in 1997, working on real-time video coding methods and applications, and received his M.Sc. (Tech) in computer science at Helsinki University of Technology in 2005. His master's thesis "SymBall: Camera controlled virtual mobile game for Symbian OS/Series 60 mobile phone," pioneered in gesture-based games for mobile phones and it was also reviewed in MIT Tech Review. Since 2001, he has been working on augmented reality-related projects as chief system designer and end-user application developer, with the main emphasis on AEC (Architecture, Engineering, and Construction) visualizations outdoors and indoors. Several of his implementations have been the world's leading solutions, including real-time AR webcam solutions and utilizing the BIM models in the real construction site. He was the visiting academic in HIT Lab NZ for 8 months in 2007–08 for designing and developing the mobile phone-based AR assembly solution.

Iliana Hernández-Garcia is Professor at the Department of Aesthetics in School of Architecture and Design of the Pontificia Universidad Javeriana in Bogotá, Colombia. She is an architect with a master's degree in aesthetics and a PhD in arts and arts sciences from the University of Sorbonne, Paris, France. She teaches at the PhD program in social sciences and at the master's degree in artificial landscapes. Her research expertise is in virtual words, architecture/science/technology, aesthetics/arts, artificial life and built environments, and posthuman studies.

Jaime Hernández-Garcia is Professor at the Department of Aesthetics in School of Architecture and Design of the Pontificia Universidad Javeriana in Bogotá, Colombia. He is an architect with a master's degree in architecture from the University of York and a PhD from the University of Newcastle, United Kingdom. He has more than 20 years of teaching and research experience. His research interests and expertise are in informal settlements/housing, open/public space, conflict/peace studies, community participation, and local knowledge/expression. He has published widely in Spanish and English.

Dr. Luke Hespanhol is an interaction designer, lecturer, researcher, media artist, and producer. His research ranges from creative technologies, digital storytelling, digital placemaking, and technology-mediated social interactions, to urban robotics, media architecture, and smart cities. He is the lead designer and curator of the Footbridge Gallery at the University of Sydney and a member of the Media Architecture Institute.

Petri Honkamaa is a senior scientist at VTT Digital Engineering. He received his M.Sc. (Tech.) in computer science from the Lappeenranta University of Technology in 1997. After graduation, he started at VTT focusing first on video compression technologies. Later, research topics include various computer vision and optimization algorithms. Currently, his focus is on 3D tracking algorithms and augmented reality applications. For 3 years (2011–13), he leads an 8-person research team on the topic of 3D tracking technologies. He has also had a key role developing several widely used software modules in VTT, including video compression method MVQ, open-source augmented reality library ALVAR, and more advanced 3D tracking library ALVAR Mobile.

Sami Huuskonen is a Finnish industrial designer (Master of Arts) and has been working in Tehomet, a Valmont Company, since 2011. Sami is specialized in different smart city-related product solutions and concepts. Currently, he is mainly involved in publicly funded LuxTurrim5G + Ecosystem project that consists of the consortium of 26 different companies, universities, and research institutions led by Nokia Bell Labs.

Eveliina Juntunen currently works as a senior project manager at RISE Research Institutes of Sweden in its Digital Platforms unit located in Västerås. Juntunen has an IPMA C project management certificate and M.Sc. (Tech.) and D.Sc. (Tech.) degrees in electrical engineering from the University of Oulu in 2004 and 2014, respectively. Before joining RISE in 2019, she had 15 years of working experience as research scientist and senior scientist at VTT Technical Research Centre of Finland. Her technical expertise is in the area of electronics packaging and manufacturing including LED lighting, thermal management, and reliability. She is also experienced in modern manufacturing methods of printed, flexible electronics, structural electronics, and hybrid integration and has wide multidisciplinary collaborations with public sectors (municipalities) and other technology topics such as end-user acceptance, architecture design, and ICT.

Efstathia Kostopoulou is an architect whose work spans across urban design and heritage, to media architecture, infrastructure, and the public. She is currently a doctoral researcher at the Bartlett School of Architecture, University College London. Her research looks into affective digital and physical experiences that relate public spaces to local memory and culture. She further works in place management, commissioning, and facilitating a number of placemaking projects and activities in East London.

Michael Krause is a placemaker who believes in negotiating solutions to shape urban environments. An urban designer from Bauhaus-Universität Weimar, Germany, he has been working since 1995 in South Africa, mainly in informal settlements including Phoenix Settlement—Gandhi's second ashram. As CEO of Violence Prevention through Urban Upgrading NPC, Michael leads a highly dedicated and transversal team of people that continuously learns with residents in low-income areas how to improve the quality of life in the most difficult circumstances.

Filippo Lodi holds several master's degrees in engineering, architecture, art, and business, making him an all-around thinker. Filippo, as Head of Innovation and Knowledge Management, leads UNSx, UNStudio's in-house innovation think tank and experience lab, where he works, taking a human(ity)-centric approach, on the development of disruptive technologies for the built environment from a unique coating that cools down buildings, responsive lighting system—as well as providing consulting services to companies on their workflow and rebranding strategies.

Aale Luusua, D.Sc. (Tech), Architect (M.Sc.), is a postdoctoral researcher at the INTERACT research group at the University of Oulu, Finland. With a general research focus on urban environments, citizen participation, digitalization, and design theory, she currently leads the Academy of Finland postdoctoral project *Experiencing Artificial Intelligence in the Smart City: Co-creating applications for urban life (AICity)*. Her works have been published in leading international scientific books and journals published by SAGE, Routledge, and Springer, and in leading scientific conference proceedings, such as the ACM Conference on Human Factors in Computing Systems and the ACM Conference on Designing Interactive Systems.

María José S.V.B. Minski is a Mexican philosopher, dancer, entrepreneur, and researcher. She has a concentrated interdisciplinary work portfolio and publications in the field of artistic embodied experience and phenomenology grounded in the philosophy of Edmund Husserl and Merleau-Ponty. María led the successful bid of Guadalajara as UNESCO city of Media Arts, involving a unique collaboration across academia, industry, governance, and citizens. She mentors emerging entrepreneurs at several universities across Mexico. She also cofounded Culturista, a digital culture and tourism start-up. This innovative proposal continues its mission to research about the future of intelligent cities and the interrelationship of how we cohabit spaces.

Nicola Morelli is Professor at Aalborg University in Denmark. He has previously worked at RMIT University in Australia and at Politecnico di Milano, where he also completed his PhD in industrial design. He is coordinating the Service Design Lab at Aalborg University, Copenhagen, which is working on different research projects on service design, with a special focus on public services and social innovation. He has also published several articles on service design methodologies, social innovation, and sustainability.

Dr. Ingrid Mulder is an associate professor at the Faculty of Industrial Design Engineering, Delft University of Technology. She has a background in policy and organization sciences (MA, University of Tilburg) and educational science and technology (PhD, University of Twente). In her current research, she studies the city as a space for transition while experimenting with participatory techniques and systemic design approaches for scaling and infrastructure social change.

Ivan Nio is a senior researcher at the Amsterdam University of Applied Sciences (Faculty of Applied Social Sciences and Law). He obtained his PhD in social sciences at the University of Amsterdam. In his research and publications, he has explored diverse themes on the interface of urban design and urban sociology. He collaborated with Frank Suurenbroek and Martijn de Waal in the research project Co-creating Responsive Urban Spaces.

Nancy Odendaal is Associate Professor in City and Regional Planning at the University of Cape Town, in South Africa. Her research examines the intersection between urban infrastructure, human agency, and city spaces. Her work has primarily focused on the impact of digital infrastructure on cities of the Global South. She is the current chair of the Association of African Planning Schools.

Dr. Callum Parker is Lecturer in Interaction Design at the University of Sydney's School of Architecture, Design, and Planning. His research explores the potential of interactive augmented city interfaces, such as pervasive displays, {augmented, virtual, and mixed} realities, smart cities, and media architecture. He has compiled his research findings in books, journal articles, and conference papers.

Henrika Pihlajaniemi, who is an Architect (M.Sc.), D.Sc. (Tech), lighting designer, and researcher, lectures and teaches Architectural and Urban Lighting in the Oulu School of Architecture (OSA), University of Oulu, leading a continuous education program of lighting design. She worked during years 1999–2005 and 2011–13 in the Department of Architecture at the University of Oulu and as project leader and researcher in several research projects concerning daylighting, housing façades, participatory design, and adaptive lighting, respectively. In 2016, she defended her doctoral thesis "Designing and experiencing adaptive lighting: Case studies with adaptation, interaction and participation." Recently, she has led "SenCity: Intelligent lighting as a service platform for innovative cities" project in OSA, where smart lighting solutions were piloted in six cities in Finland. Currently, she works in OSA as a postdoctoral researcher funded by the Academy of Finland leading several projects. Her research is related to intelligent lighting and well-being in learning and working environments, lighting education, and the use of lighting to develop cities. She has also worked as a visiting researcher in VTT Technical Research Centre of Finland and the Lighting Design research group, Aalborg University Copenhagen. Additionally, she works as a lighting design consultant in her architectural office (M3 Architects). She and M3 Architects have received numerous prizes in national and international architectural and urban design competitions.

Andrea Pollio holds a PhD from the Institute for Culture and Society, Western Sydney University. He is interested in the relationship between cities and techno-entrepreneurial cultures. His doctoral research has focused on the performativity of economic and humanitarian knowledge in the making of Cape Town's start-up "scene." His previous publications also addressed questions of technocratic expertise around the introduction of smart city policies in Italy.

Manuel Portela is a multidisciplinary postdoctoral researcher at Universitat Pompeu Fabra. He holds a degree in graphic design, an M.Sc. degree in urban studies, and a PhD in geoinformatics from Universitat Jaume I (UJI) of Castellón, Spain. He is interested in the impact of technology in society, including topics such as artificial intelligence, citizen science, and smart cities. Currently, he researches the development of explainable and participatory approaches to artificial intelligence.

Gillian Rose is Professor of Human Geography at the University of Oxford and a Fellow of the British Academy and the Academy of Social Sciences. She is the author of *Feminism and Geography* (Polity, 1993), *Doing Family Photography* (Ashgate, 2010), and *Visual Methodologies* (Sage, fourth edition 2016), as well as many papers on images, visualizing technologies, and ways of seeing in urban, domestic, and archival spaces. Her current research interests focus on contemporary digital visual culture and the so-called smart cities.

Ummu Sakiinah (M.Sc.) is a user researcher. She has an educational background in industrial product design from the Institute Technology of Bandung. She continued her study at the Delft University of Technology in the Netherlands, majoring in strategic product design as she wished to take design on a more strategic level to have more impact. She is interested in the creative research process and in circular and regenerative economy. Currently, she is working closely with millions of small business owners who are also app users across Indonesia.

Luca Simeone is Assistant Professor and a member of the Service Design Lab at Aalborg University. He holds a PhD in interaction design from the Malmö University. He is particularly interested in the organizational, strategic, economic, and financial dimensions of design and in how design can support entrepreneurship and innovation processes.

Frank Suurenbroek is a professor (lector) of Spatial Urban Transformation at the Amsterdam University of Applied Sciences Faculty of Technology. His chair conducts multiple action-research projects on urban transformation, with a focus on the relation between the physical and the social fabric. Recent research takes a neuro-architectural approach (sensingstreetscapes.com). He collaborated in the research project Co-creating Responsive Urban Spaces (2017–19). He is part of the executive committee of the Media Architecture Biennale 2020/21.

Martin Tomitsch is Associate Professor and Head of Design at the University of Sydney's School of Architecture, Design and Planning; visiting professor at the Beijing Central Academy of Fine Arts Visual Art Innovation Institute; and founding member of the Media Architecture Institute. His research investigates the way new technologies will shape people's interactions in cities. His recent books include *Design Think Make Break Repeat: A Handbook of Methods* (BIS) and *Making Cities Smarter: Designing Interactive Urban Applications* (Jovis).

Miguel Valdez is a lecturer in technology management and innovation at the Open University. Through his research, Miguel has studied how the various public as well as civic, industrial, and governmental bodies make sense of innovative technologies and collectively negotiate the future of their city. His research draws insights from urban geography and innovation studies to examine how the experimental spaces in cities may facilitate the development of new knowledge about sustainable innovation, and on how such knowledge coalesces and travels.

Ben van Berkel studied architecture at the Rietveld Academy in Amsterdam and the Architectural Association in London, receiving the AA Diploma with Honours in 1987. In 1988, he and Caroline Bos set up the architectural practice UNStudio, and in 2018, he founded UNSense, an arch tech company that designs and integrates human-centric tech solutions for the built environment. He has lectured and taught at many architectural schools around the world and, from 2011 to 2018, held the Kenzo Tange Visiting Professor's Chair at Harvard University Graduate School of Design.

Annemiek van Boeijen (M.Sc./PhD) works full time at the Faculty of Industrial Design Engineering at the Delft University of Technology. Her research is focused on the role of culture in design processes, with the goal of designing methods geared to support designers in cultivating a culture-conscious approach. She is initiator and coeditor of the Delft Design Guide and author of *Culture Sensitive Design: A guide to Culture in Practice*, and teaches a double-blended online course with the same name.

Edward Wigley is a staff tutor at the Open University, following a research post on the ESRC-funded project Smart Cities in the Making: Learning from Milton Keynes based at the Open University. His recent journal contributions have been published in *Social and Cultural Geography* (2016; 2018; 2019), *Mobilities* (2018), *Globalizations* (2018), *Cultural Geographies* (2019), and *Geografiska Annaler* (2020).

Oliver Zanetti is a senior researcher at the innovation foundation, Nesta, where he leads research into policy and practical interventions that can be made to support a thriving and inclusive innovation landscape in the UK's cities and regions. Prior to this, he was a postdoctoral researcher in the School of Geography and Environment at the University of Oxford, which he contributed to the ESRC-funded project Smart Cities in the Making. He has a PhD in human geography from the Open University.

Chapter 1

Introducing shaping smart for better cities

Alessandro Aurigi[a] and Nancy Odendaal[b]
[a]*University of Plymouth, Plymouth, United Kingdom,* [b]*University of Cape Town, Cape Town, South Africa*

Chapter outline

Setting a challenge around smart place quality

Debates around the rather elusive and flexibly interpreted concept of "smart city" can often be framed within a polarized landscape, where the "making" side of smart is often—and to an extent necessarily—driven by a technological optimism, while the critical policy-informing analysis of it focuses on identifying inequalities, deficiencies, and flaws within the "making" side. Kitchin (2015) has interestingly noted how useful it can be bringing these two "spheres" closer and engaging with both, allowing them to inform each other. However, while closing the loop between making and critiquing can be essential toward bettering our understanding and execution of smart city initiatives, this requires a nontrivial interdisciplinary dialogue between interaction designers, spatial planners, architects and urban designers, social scientists, and computer scientists. The bringing together of disciplines and the many interpretations of space and digital agency is naturally bound to be subject to debate. This needs an arena that privileges the contamination of ideas and is not overconcerned with homogeneity and all-round coherence. The tensions that stem from different interpretations of the "smart" city and approaches to it require taking a step back from narrower questions specific to one's niche—

Shaping Smart for Better Cities. https://doi.org/10.1016/B978-0-12-818636-7.00019-6

technological or social—and ease back into a zone that is not proprietary and, as such, negotiable and enriched from different quarters and traditions. An interesting possibility is engaging with the wide, certainly interpretable but also overarching notion of place and what this means for creating opportunities for living better lives.

The notion of place is central to the theme of this book. Moreover, to encourage the aforementioned "contamination" and diversity, we invited colleagues from diverse backgrounds and traditions to get their hands "dirty" with the question: "how do we shape and design smart urbanity to actually make the city a better place, and better places in the city?" As editors, we are aware of the difficulties of assembling contributions that translate those terms in potentially different and incomparable ways, as well as of course of the rather extreme interpretive flexibility to which those very concepts are hostage. But it is also a powerful question and in a way a useful expedient. It confronts all of us—whatever our native knowledge—with a "so what" challenge that takes us out of the internal coherence of disciplines and forces us to think more broadly. This book aims at making technologists, designers, and social scientists return to the notion of "better" places, as the ultimate answer, whatever "better" means to them. In doing so, we have had the privilege of sharing this platform with architects, urban designers, planners, geographers, sociologists, human–computer interaction (HCI) specialists, and urban informatics theorists, sharing content from over 15 countries. This disciplinary and geographic diversity does not necessarily make this a "global" or in any way a comprehensive response to the central question posed by the volume. Rather, it affirms both the pervasiveness and complexity of smart urbanism. These different inputs provide a nuanced interpretation of "smart" urban environments, rather than a simplified, tamed, and self-referential one. As such, this is neither a single issue or a single-theory collection, where like-minded scholars elaborate and test a specific proposition. And it is not a technology-dominated book—despite all contributions referring to aspects of technologically based smart—but rather a book on design, collaboration, and policy processes aimed at "doing better," through which technology is inscribed. Notwithstanding the diversity in language, discourse, and vision of the contributions, some interesting themes emerge that can inform the critical making and shaping of smart places.

Grounding, and the layered complexity of place

Throughout the volume, there is an awareness of "place" as a complex and multidimensional arena and the need to face its complexity rather than withdraw into simplified stories when defining and analyzing it. Such stories can only apparently be coherent and working—as Cureton and Dunn (Chapter 14) note when referring to digital "twinning"—yet carry the huge risks of overreducing reality to some model that cannot comprehend it.

This encourages zooming into place intended not as a large territory with a title or denomination that can be scrutinized through a map's bird's view and a set of statistics but as a multilayered, inhabited, immanent, culturally and existentially experienced location where some form of "sense of place" exists, collectively and individually. Places are therefore people and event-rich and meant to be embedded with history and multiple experiences. Capturing or augmenting those through digital means requires a consideration of physical place, people, and digital possibilities, as argued by Kostopolou and Fatah gen Schieck (Chapter 6). The notion of "grounding" has therefore permeated much of the content of this book. Grounding within what could be seen as a "hyperlocal" scale is what makes our contributors confront diversity and complexity and above all imagine their version—and suggestions—of "better." All chapters, in their own different ways, tackle the importance of embedding smart spatially, culturally, and socially, in the inhabited, live location it becomes a part of. Despite the capacity of smart networks to transcend space, their functioning and character are nevertheless tied to place, as explored by Krause and Ewing (Chapter 18). Many of the chapters in this volume emphasize the many nuances of what constitutes a deeply contextual approach.

In doing so a number of the volume's chapters relate to examples drawn from Africa, India, or South America. Yet, we would argue that this is dealt with in a way that represents a refreshing departure from a global North-South dichotomy. The way we see this and the reason why we have included these narratives is not to explore a specific Global South perspective per se or even provide a North-South balance or comparison. These contributions are instrumental within the volume's challenge to reinforce the importance of looking at two aspects of contextualization. The first relates to the shaping of smart within specific communities that are strongly bound to their territory and are scarcely represented by the general and highly gentrified idea of the smart city user being a professional and highly mobile—to the extreme of nomadism—civic customer with little stake in place itself. These communities inhabit, often in relatively permanent ways, their locales and continuously interact and invest in them, by choice or necessity. We believe that such communities are the majority of urban dwellers and not just in Global South metropoles. The second is the value of community-grounded appropriation in shaping smart places in lateral and original ways, beyond any initial mainstream global propositions or standardized "product" approaches. It does not have to be a "smart in a box" solution to be valuable, and its value goes well beyond a "doing with less" logic to highlight a "doing better" one. This can be a function of necessity and scarcity in certain environments, as explored by Hernández- Garcia and Hernández-Garcia, in Chapter 16, where the authors make a compelling argument for considering the parallels between innovation in informal settlements and the functioning of digital technology. We believe these to be key aspects of a more place-embedded smart design and shaping. On the one hand, it can be argued that "micro" South conditions exist everywhere, but beyond this, we feel that the opportunity of envisaging more grounded, endogenous ideas for organizing, designing, and running places is increasingly key within a global perspective on urban and social sustainability.

The grounding of smart within specific contexts requires consideration of the relationships that exist and evolve within those environments, in social, cognitive, and practical design terms. A variety of different agents/actants contribute to the making of digitally augmented places, and many of the chapters in this book explore the mechanisms through which this can be achieved. For example, through neighborhood-based consultation and careful cultural analysis, as explored by Sakiinah, Mulder, van Boeijen, and Darson (Chapter 8) in their account of how locally sensitive and connected smart design practices could offer a rich and meaningful way to revitalize a commercial street in Rotterdam. Another multifaceted perspective is introduced by Parker, Tomitsch, and Fredericks (Chapter 10) where the design of local interactive projects is framed and improved through managing the interplay between individual users, the space they inhabit and the temporal dimension. This volume collectively contributes to critical perspectives that move away from envisaging prepackaged "solutions." It does so by privileging place-based generative logics, rather than add-on views of brand-new technologies taking overall control. These logics can be enacted through "middle-out engagement" to facilitate a design-based approach that transcends more simplistic divides between top-down and bottom-up to engage a wider and more diverse range of local stakeholders effectively, as explored by Caldwell, Fredericks, Hespanhol, Chamorro-Koc, Sanchez, and Castelazo (Chapter 13). Or it can involve the local shaping of smart learning networks to favor more effective mechanisms of cocreation, as discussed by Ewing and Krause (Chapter 18). What all these approaches have in common is an emphasis on how essential it is to capture local energy in space. Guma (Chapter 17) takes this argument a step further by revealing the contingency of infrastructure in informal spaces—continuously being made and remade in accordance with temporal and place-based specificities.

Shaping smart place top-down and bottom-up

This shift to place, as well as deep contextual engagement with its qualities, is also reflected in how the volume is organized. Rather than being structured by the rather frequently encountered sections of "smart something"—where the something is "transport," "utilities," "homes," "enterprise," and so on, each showcasing systems, products, solutions, or specific methodologies to tackle that sector of activity, this book looks at how to make place and how place could become smarter. This involves two sections that embrace a complexity of views in how different categories of "actant" can deal with making places smart.

The first section is about the role of design—and designers—in this, while the second is on communities, not only seen as the local people to consult but also as coproducers, constructors, and potentially disruptors of place and—therefore—of its smart version too. In many ways, this does represent an argument, a sensibility that embraces both sides of the equation. In fact, these are not mutually exclusive. Valdez, Wigley, Zanetti, and Rose (Chapter 12) effectively engage the overlap between critical social analysis and design choices when looking at developing a toolkit for inclusion in smart city projects. Almost in a diametrically opposite guise but equally capturing the overlaps, Besplemennova and Pollio (Chapter 5) observe how a local smart urban design intervention can have agency and an impact over wider—and more persistent—debates on the publicness of urban space. Nevertheless, we divided the book into these two sections as they provide a useful and simple framework in outlining and distinguishing these various agencies.

Designing smart places

Intervening through an act of planned and proactive design under these premises involves first of all a holistic view of it, from both the viewpoint of what its elements and dimensions are and who designs it. Above all, it requires a vision that not only is functional and utilitarian but also involves a deeper and wider affirmation of values on what the "future city" is or should be. As shown by Van Berkel, Lodi, and Batal (Chapter 4), this opens up the need and opportunity to tackle urban place as a combination or recombination of physical, digital, and social dimensions. It promotes the need to see inhabitation not only as fruition or usage by a series of civic "customers" but also as deeply grounded sharing, learning, playing, and dealing with the existential aspects of being in the city. Many of the contributions are not about making some urban machine perform through some clever design, but about a set of deeper and intertwined relationships and play between physical spaces, digital layers and connections, and individual and collective living. Naturally, in the effort of tackling such a multifaceted task, designers can find themselves compelled to develop or update frameworks and toolboxes that will help them make sense of this and construct a design process to address it. When they look at tech-rich environments, these face two parallel challenges: understanding the increased complexity not only in terms of dimensions and layers of place but also in terms of a recombined design arena itself, with the need to bridge interpretations and practice. More than one chapter tries to contribute to this need, and it is interesting how different backgrounds and viewpoints produce frameworks that on the one hand are not only partial and different but also potentially complementary, proving that the complexity of the field cannot be easily reduced to a single "proprietary" formula. Aurigi (Chapter 2) focuses on how both spatial relationships and affordances are challenged and extended by smart technologies,and what this might mean for a more holistic design approach. The chapter by de Waal and Suurenbroek (Chapter 3) somehow flips the perspective to focus on bridging interpretations of smart urban design practices. It is meaningful as it stems from an effort to build a non-niched toolbox able to facilitate a dialogue and cocreation among different actors within the smart urban design arena. All contributions acknowledge how this feels at the same time as relatively uncharted territory, given the disruptive contribution of high technology, as well as an activity that needs and deserves some anchoring in what we know about the understanding, making, and designing of places. After all an aspect of place grounding has to involve not pretending that reality has simply shifted 180 degrees and all needs to be reinvented from scratch. Smart technologies do not replace, they augment and deepen, and they influence the recombination of place-based elements.

This applies not only to the "harder" elements of the built environment but also to its softer components too. We tend to think of technology as a replacement of the past, as the brave new frontier for urbanism. What this volume also shows is that history, memory, emotion, and physicality are core components of the qualities of place that cannot be dismissed and, instead, might be amplified by smart intervention. In many instances, the use of digital tools to capture and celebrate such qualities reminds us that grounding does not just mean the simple "locating" or "placing" of a piece of technology in a space that acts as its support but rather "emplacing" or embedding any technological means within the richness that already exists in the context. It involves acknowledging and dealing with the existential aspects of place, working with those meanings—individual and collective ones—that make cities places. Whether this is through foregrounding and enabling individual experiences and recent cognitive histories, as captured by Portela and Granell-Canut (Chapter 9) in their discussion on empathy and atmospheres or the celebration of place-based memory and subjective appropriation, by Luusua, Pihlajaniemi, Hakkarainen, Honkamaa, Juntunen, and Huuskonen (Chapter 7), the experiential qualities of place can indeed be deepened by digital augmentation. And this after all feels a much more thorough, concrete, and—dare we say—respectful vision of civic improvement.

Foregrounding agency and context for coshaping smart

As designers or planners, we are inevitably tempted to envision and treat augmented design processes as a simple act of consultative placemaking. What this volume captures very effectively is that the diversity of agency and stakeholders requires an approach that embraces more, not less, complexity, counterintuitively respect to any expectation of digital simplification or streamlining. Of course, this requires abandoning simplistic ideas of all-encompassing and automated algorithmic management and decision-making. Going back to place shows that agency is complex, and it is by understanding and successfully articulating its various aspects that more becomes possible toward shaping better smart places. The analyses by Caldwell, Fredericks, Hespanhol, Chamorro-Koc, Sanchez, and Castelazo (Chapter 13) and Simeone, Morelli, and De Götzen (Chapter 20) bring into focus the need to deromanticize and deidealize bottom-up approaches, toward making different types of agent and institutional approaches work together and learn from each other. The tourism-based analysis of Foth, Bilandzic, and Guaralda (Chapter 15) also points at the need to overcome simplistic narratives and easy global-local polarizations in favor of more nuanced approaches where smart platforms can be used to proactively enact synergies between diverse actors, making local character and culture work with new ideas. Odendaal (Chapter 11) and Guma (Chapter 17) challenge any inclusiveness by consultation rhetoric as also widely practiced in smart urbanism—see, for instance, Mattern's (2020) account of "Post-it Note City" on the Toronto Sidewalk project—highlighting local appropriation of technological innovation as a key function of complexity, where a story that was designed as simple and replicable faces reality by being reinterpreted and redesigned by its "users" and the inhabitants of a specific place. Button (Chapter 19) and Hernández-Garcia and Hernández-Garcia (Chapter 15) capture some of the tensions between local innovation and smart impositions, by celebrating the agency that emerges from the necessities of place, particularly in infrastructure-poor circumstances.

What the chapters—in the second half of the book in particular—capture is recognition that agency evolves from the interface and interaction between human and nonhuman actants. The evolution of smart places is dynamic and negotiable, as captured by many of the authors in this volume. This does not underplay the role of the designer, but simply what a more nuanced reading of the design process can reveal, as iterative and dynamic. We argue, through the many chapters of the book, that this is not merely constitutive of more contextually appropriate augmented design, but simply the creation of better places.

Toward a future of better smart places?

This book therefore represents a series of perspectives for understanding and shaping smart places, which rather than being tidily aligned along a theoretical trajectory or a single theme, try to assemble an holistic view of contextualized, place-based smart, informed by different frameworks and backgrounds. As such, we believe the message and contribution of the volume is a pluralistic one. While working on the book, we have felt that the suggestion that successful smart places involve recombining the physical, the digital, and the deeply personal also might require an effort to recombine our knowledge, research, and practice. This in turn needs bridging different interpretations through shareable narratives and principles. And above all, it challenges us to try and maintain a rather delicate balance between an effort not to retreat too easily into a disciplinary cocoon, with its shared but sometimes exclusive language and way of thinking, yet make sure that those very thinking and knowledge are proactively included in the mix.

All the contributions in the book suggest that making better smart places—and cities—means embracing change and trying to reinvent things, yet grounding them in knowledge of design and social processes that can be decades or centuries old, as our places could be too. What is discussed and envisaged is very much more work in progress and exploration than established practice, probably too tentative and dirty to fix any prepackaged expertise offer. In many ways the city is never a finished product but one that evolves with technology, rather than because of it. Thus beyond easy proclaims of brand-new "solutions" for our cities, we would argue that this degree of openness and enquiry is what is really needed to envisage really "smart" cities and places that are resilient, functional, and dignified. We believe that each of the chapters encapsulated in this book provides us with insights into how the many contingent smart places that evolve through design-based and coproductive measures are not only indicative of recombination and augmentation but also a function of contemporary urbanism.

References

Kitchin, R., 2015. Making sense of smart cities: addressing present shortcomings. Camb. J. Reg. Econ. Soc. 8, 131–136.

Mattern, S., 2020. Post-it note city. Places J. https://doi.org/10.22269/200211. February.

Author biography

Alessandro Aurigi is Professor of Urban Design at the University of Plymouth, having previously worked at the University of Newcastle and University College London (UCL). His research focuses on the relationships between our digital society and the ways we conceive, design, and manage urban space, to enhance and support place quality. He has published *Digital and Smart Cities* and *The Routledge Companion to Smart Cities* (Routledge, with Katharine Willis), the multidisciplinary book *Augmented Urban Spaces* (Ashgate, with Fiorella De Cindio), and *Making the Digital City* (Ashgate).

Nancy Odendaal is an Associate Professor in City and Regional Planning at the University of Cape Town, in South Africa. Her research examines the intersection between urban infrastructure, human agency, and city spaces. Her work has primarily focused on the impact of digital infrastructure on cities of the global South. She is current chair of the Association of African Planning Schools.

Section A

Designing and shaping smart places

Chapter 2

Designing smart places: Toward a holistic, recombinant approach

Alessandro Aurigi
University of Plymouth, Plymouth, United Kingdom

Chapter outline

Introduction

Much is being discussed and written about smart cities and places. Fairly recently, extensive surveys of smart city-related literature have been carried out (Mora et al., 2017) giving a much-needed overview of the field. Yet the pervasiveness of technology makes this a vast and diverse landscape. Two very different foci, both contributing toward city shaping, seem to prevail. One, mainly stemming from the disciplines of human geography, planning, and economics, looks at how at an urban or metropolitan level, high technologies are becoming part of the planning and organizing of the city, its services, and its economic and social and community relationships. Another, generated within the wide domain of the "design" disciplines to include not only industrial/product but also system, interface, and interaction design, emphasizes the role of high tech

Shaping Smart for Better Cities. https://doi.org/10.1016/B978-0-12-818636-7.00020-2

in conceiving smart urban elements such as screens, interactive urban furniture, robots, and systems. These augment the built environment and facilitate various degrees of "located" public information and participation—from surveying the public to new ways of appropriating the public sphere through collaborative appropriation of the city (de Lange and de Waal, 2019).

Reflecting on the "design" of smart places means first of all acknowledging how the field is extremely vast, diverse, and subject to competing interpretations, as can be the notion itself of "design" if—beyond the specificities of established disciplines—we look at it as a proactive and planned act of modifying place with the intent of improving it. So, diverse areas of smart city thinking—it could be argued—can all naturally contribute.

Smart urbanism perspectives on metropolitan and regional scale focused on productivity and innovation (Mora et al., 2018; Yigitcanlar and Bulu, 2018); political and social discourse, equality, and participation (Kitchin et al., 2015; Söderström et al., 2014; Vanolo, 2014; Hollands, 2008, 2015; Foth et al., 2015); urban ecologies (Caprotti, 2014); the role and politics of digital infrastructure (Luque-Ayala and Marvin, 2020); urban large-scale visions (Melhish et al., 2017; Rose and Willis, 2019); and more comprehensive reviews of the field (Willis and Aurigi, 2017), all inform the shaping of place. Almost inevitably discourses focusing on the planning aspects of the city and its socio, political, and economic implications embrace a metropolitan or regional scale. Smart urbanism and political economy look at the influence of policy over the development of civic technological ecosystems. This of course implies studying and understanding how relationships among communities, institutions, industry, and capital unfold, how technologies are socially shaped within such arena, and how they can affect civic shape and functioning. Their contribution toward moderating otherwise easily pushed technological deterministic views and the hype of smart urbanism as the new panacea for sustainable development are invaluable. However, while they are "spatial" as they can inform city shaping through large-scale infrastructural intervention and improvements in governance mechanisms, they operate less at the local scale and have less of a direct bearing on the specific design of places as inhabitable parts of it.

At the other end of the spectrum, there are approaches to both critically discuss and practice the design of those systems and (smart) objects and artifacts that constitute the key building bricks, and practical ideas, on which the smart city vision can be materialized. See, for instance, the Carlo Ratti Associati-designed prototype for a "dynamic street" modular intelligent paving proposed to the now defunct Waterfront Toronto scheme led by Sidewalk Labs (Walsh, 2018) or the work of the Media Architecture Institute (http://www.mediaarchitecture.org/) or the Urban Informatics research group (https://research.qut.edu.au/designlab/groups/urban-informatics/) or international networks and festivals of urban interactive installations, such as Bristol-born The Playable City (https://www.playablecity.com/). There are of course differences between built-environment and art-type projects, with the former focusing more closely on fulfilling a functional program and a higher degree of permanency. Yet much of the smart installation–related debates tend to be focused on an extremely hyperlocal scale often centered on the technological artifact and its positioning in its immediate surroundings, very often in a similar guise as what can be seen in the design of urban furniture. Practical smart design often focuses on the product-like shaping of "located" yet highly portable interactive objects and systems. Within such logic the prevailing perspectives tend to concentrate on the microscale of the citizen-user and their interaction with the new technological "intelligence" being added onto the urban scenery, buildings, and services. What are designed are mainly systems and objects that will be "located" in an urban scene. Much attention goes toward the creative configuration of new functions and possibilities experienced through technology and the all-crucial interfacing of such systems with their human users.

Both these prevailing approaches are not geared for looking at the dimension of the urban place—squares, streets, and neighborhoods—or more generally "transactional zones" (Smith, 1977; 180) as local environments that are inhabited and experienced, and ideally designed, as a spatial whole. The large and the hyperlocal scale views can look at civic space as a "territory" for metropolitan phenomena and networked systems or as a physical scene or support for the location and functioning of smart objects and installations. Geographers have acknowledged the advantages of connecting the social and functional aspects of these two dimensions and enriching sociopolitical analysis by building bridges with the world of practice (Kitchin, 2015; 134–135). It remains equally relevant to try and fill the spatial scale "gap" of the design of that intermediate dimension that more strongly and fully can relate to the notions of place and public space. This means looking at the design of urban habitats not as the steering of institutional and social relationship or the configuration of intelligent products, but as a strongly multidimensional task of dealing with place relationships existing in a physical/digital recombined world. This chapter tries to adopt this perspective and look at the following:

Firstly, how smart placemaking has been characterized by approaches that focus either on the technological artifact and its design or on the social implications of installations, in terms of use, codesign, and participation. This is only apparently opposite mainstream architecture's emphasis on the preeminence of physical space and place, but it shares with it an inability to look at and design place as a whole. The prevalent narratives—and projects—of digital placemaking seem to be based on linear trajectories where an aspect of place, normally technology, is the change agent.

Consequently, how overcoming partial interpretations of smart place design toward a more holistic and recombined view of it requires a critical reengagement with important design notions. Rethinking context-based relationships and multidimensional aspects of agency and affordance reconnects us with a wider field of spatial and urban design knowledge and practice, and challenges and extends our "smart" design approach and process.

The making of place as a nonlinear endeavor

When architects design "spaces," they tend to orchestrate a two-way relationship between context and proposal, often focusing on a linear process and trajectory, privileging the role of designers in determining the object/building that the client wants. This is often embedded in professional guidelines. For instance, the UK's RIBA Plan of Work (2020) is strongly informed by transactional views, where design is rather sharply separated from the social aspect of "use"—to which stage n.7 of the Plan is dedicated, and that is nevertheless still concerned with expert-owned actions such as postoccupancy evaluation and facilities management. A societal and community dimension of spatial design is mentioned in the sustainability strategy part of the document, with some generic reference to placemaking and consultation, yet still mainly framing people as end users of the final build.

Many valuable contributions have emerged on the need to move away from simplistic one-way, one-dimensional, and often one-agent perspectives on how our inhabited and social environments can be designed. For instance, the idea that the shaping of space could not be simplistically framed as something exclusive to a restricted group of experts owes much to Henri Lefevbre's work (1991), discussing how space is socially produced. More specifically, architectural culture assuming that buildings and spaces are completed by design (and by a designer) and will then become immutable agents that influence their occupants' behavior has been challenged by Stewart Brand's (1997) much-celebrated critique in *How Buildings Learn*.

Widening these perspectives however does not imply a "swap" or replacement of agency, where the designed space's role moves from dominance to irrelevance. Awan et al. highlight a mutual relationship and influence where "spatial agents are neither impotent nor all powerful: they are negotiators of existing conditions in order to partially reform them" (Awan et al., 2011; 31). Urbanist Jan Gehl, and before him Jane Jacobs (1992, original 1961) who strongly informed his work, has also aimed to shift the focus in urban design away from a built form-led perspective on civic renewal and the need to integrate everyday "life" not as an afterthought but as a generator of place design and quality (Gehl, 2006; 75). These ideas affirming a holistic and community-involved approach toward the shaping of our spaces have fuelled the theorization and practice of "placemaking," advocating the need to discuss and tackle how we can "create a place, not a design" (Project for Public Spaces, 2018; 13).

This shift from a transactional to a combined and complex view of placemaking and design marks an understanding of and converging toward mutual, nondeterministic trajectories and relationships between the various dimensions of place. By doing this, it introduces the need to rebalance and enrich our approach, rather than replace it altogether. Within this logic the rise of high technologies does not and should not make redundant the knowledge we have accumulated on good civic and spatial design (Aurigi, 2013). This remains relevant within debates involving the participation and potential disruption of high tech in the making of place, something that we can call "smart" place-making, or even smart place design, assuming this embraces the progressive views just discussed. Gehl, for example, argues that while communication and interaction technologies can offer "Abundant possibilities (...) precisely for this reason, the fact that there is still widespread criticism of the neglected public spaces is indeed thought provoking" (Gehl, 2011; 49). It would therefore be a mistake to accept technology as the new replacing and determining factor and somehow an independent variable that gets "added on" to induce change. Has the way we design "smart" benefited from these more "circular" and intertwined ways of framing the making of place? Only in part the next section argues.

Smart place design: Determinism, partial approaches, and recombination

Early debates on the implications of the emergence of "cyberspace" for architectural and urban environments saw on the one hand an obvious emphasis on the replacement power of ICTs and the fact that these could revolutionize lifestyles and conceptions of space and ecologies leading to a prevalence of the digital over the physical dimension of place (see, for instance, the various contributions in Benedikt, 1991). On the other hand, some less one-way deterministic positions on how space, technology, and people would mutually influence and challenge each other were formulated, particularly by William Mitchell with the idea of "recombinant architecture" (Mitchell, 1995; 47–105) and through a follow-up volume by Thomas Horan (2000). The core merit of Mitchell's and Horan's insight was looking at ICT-rich spaces as the product of two main phenomena. The first saw physical agency at play, where elements of the built environment could exploit ICTs to fragment and reconfigure into new typologies and spatial combinations. The second complemented the former with "soft" agency where digital elements would take part in the recombination process to generate smart elements—spaces or objects—where increasingly "function follows code" (Mitchell, 1999; 50). These two main dynamics would of course work concurrently and frame smart spatial design as a complex and circular challenge involving the mutual interaction of space and software. Horan (2000) further expanded on this approach through discussing the ideas of "fluid locations," extending the discourse on new typologies to look at how spaces could be proactively programmed, and of "meaningful places" as "the need to design digital places in a manner that respects the functional and symbolic associations that places often contain" (Horan, 2000; 15–16). He would also more punctually raise the need to combine the social dimension in the making of digital' city, through the construct of "democratic designs" (Horan, 2000; 20–21), calling for a process of codesign of smart places, and closing the loop between physical space, ICTs, and community.

In the following years however, particularly when it comes to the actual design of smart places, often associated with the prototyping of new system and devices, sometimes through practices of art installation, a deterministic prioritization of the linear effects of technology has reemerged. Only some more advanced cases have tackled issues of public participation and cocreation, on the lines of democratizing design. Yet often a deeper and more complete intertwining of spatial and digital design has been overlooked. Picon (2015), for instance, notes a disconnection between spatial and smart city design as "formal inventiveness is not the priority, and references to existing forms proliferate. For example, the promoters of Songdo set out to borrow boulevards from Paris, the Central Park principle from New York and canals from Venice" (112), fundamentally ignoring any mutual influence and opportunities between technology and space.

It is interesting to notice how the concept of a "responsive" environment has also evolved as a consequence of the emergence of digital technologies, with its new interpretation replacing, rather than actually including, complementing, and directly challenging, its previous incarnation. Exemplar of this is the existence of two popular volumes in the architecture and urban design arena, both titled *Responsive Environments*, and representative of these two different perspectives. The first, by Bentley, identifies the designed space as embedding choices and politics (Bentley, 1985; 9). The designer therefore reacts to an existing physical context in sensitive and proactive ways to give it positive embedded agency. The second volume, published about 20 years later by Lucy Bullivant, looks at how both high and low technologies, once applied to spaces, make these more "responsive" and interactive toward their users by engaging "with our wishes and bodily sensations on an existential level" (Bullivant, 2006; 7). The agency of responding is in this case shifting from space to technology—and the design of the latter. This polarized approach on "responsiveness" also suggests the need to look at digital spaces as recombined landscapes. However, the vast majority of projects presented in Buillvant's seminal book are art installations displayed in museums, art galleries, and laboratories, that is, highly controlled environments, purposefully adapted to the exhibit to the point of becoming a-spatial or at least a-contextual. Even pieces of architecture—as in the case of Diller and Scofidio's *Blur* (p. 41)—tend to focus more on generating their own environment, a disorienting one in the case of *Blur*, and encouraging visitors to interact with it, much more than having a close dialogue with the many aspects of the existent one. There are a few more "localized" exceptions, but those examples are mainly there to communicate either environmental information or specific atmospheres, often to enrich corporate showrooms and receptions.

A successive volume, also edited by Bullivant—*4D Hyperlocal*—promises a departure from the controlled, a-contextual situations of the previous cases, to add a much stronger emphasis on the social and participative dimension of the making of digital spaces. The editor herself comments that "Rather than being a generic 'tech kit,' the hyperlocal's emerging, alternative toolsets respond to specific commons" (Bullivant, 2017; 8). A similar locally applied approach permeates many of the contributions in the area of "Media Architecture," where a focus on making urban spaces content and interaction rich, through the deployment of both soft and hard infrastructure, large displays for instance, is dominant (Hespanhol et al., 2017). In a European context the public space-centered work of the Cyberparks network, though within a wider scope of approaches and methods, also tackles the relationship between high technology and place from the prevailing viewpoints of accommodating digital media in public space, increasing people–tech interactions, and enhancing participation and planning processes through digital means (Zammit and Kenna, 2017).

The British Design Council's (2020) Framework for Innovation and its associated "Double Diamond" methodology (https://www.designcouncil.org.uk/news-opinion/what-framework-innovation-design-councils-evolved-double-diamond) are also widely adopted to guide collaborative and socially involved design processes. These strongly resonate in product and service design communities, as well as in interaction design and human–computer interface (HCI) arenas, where much of the smart design for urban environments practice originates. For instance, recent work on "designing urban robots for hybrid placemaking experiences" by Hoggenmueller et al. (2020) describes the deployment of a robot drawing on urban ground with chalk "to affect people's engagement with public visualizations and to foster social interaction." Passersby interviewed by the researchers note how the device made the lane where it was positioned "a graffiti place," making it feel more inclusive and participated. It is much less clear however the way the robot creates—if at all—a more precise relationship and bespoke engagement with the spatial context itself, apart from using it as a canvas. In other words, nothing in the project implies that the robot's design—rather than its behavior—intertwines with the actual place. What it does is adding a degree of performance to it.

These approaches align well with Horan's concept of democratic designs. Yet, their ultimate focus on the "product," more than on the place as a whole, makes them less engaged with recombining all its different components and above all space. Often, even when the design of urban digital/physical hybrid interventions makes sure space is not overlooked, this is driven by principles for adapting and complementing hardware and software with the built environment. It emphasizes user interfaces, consistency and meaning of experience, the need for prototyping, and enabling interaction, with an added sensitivity toward space, but still mainly focusing on the design of technology (Tomitsch, 2018; 126–127). This is in line with a legacy of research discourses related to the design and integration of screens within the urban environment and calls for such installations to "be integrated into existing physical surroundings" (Dalsgaard and Halskov, 2010; 2280). While there is an awareness of the importance of context and of designing city apps "as part of the urban environment (…) rather than taking over" (Tomitsch, 2018; 148), these accounts do not yet go as far as seeing the shaping of place as a strategic activity combining equally and mutually active dimensions.

Toward a holistic design of recombinant places

Approaches to smart placemaking stemming from visions and processes of codesign and public participation are of extreme importance and parallel the evolution of views on architectural and urban design mentioned earlier. However, in the process of focusing on new technological possibilities and how to make them inclusive, the need to involve all dimensions of place is often overlooked. Scholars such as Shepard (2011; 20) have legitimately asked whether "software" infrastructures deserve an attention shift from the "hardware" of urban spaces. While questioning this has been pertinent, especially when faced with the conservatism of much of the architecture discipline, the risk is for the pendulum to swing too far away from the spatial dimension. Relph had already warned against new notions of digitally based communities as "an overly extreme denial of the importance of physical setting in place experience" (Relph, 1986; 33). The complex relational nature of place risks not being fully engaged as space is taken for granted and somehow marginalized by a general view of the city having become a "computer" (Mattern, 2017). We could paraphrase Gehl's concluding comments on the emergence of mediated urbanity, arguing that despite all new possibilities offered by it, "something is missing" (Gehl, 2011; 49).

A tentative framework to inform the shaping and design of smart places in a more holistic way could therefore consider the relationships among the three dimensions of space, people, and technology. These of course are in themselves complex concepts and can be easily individually expanded, multilayered, and problematized. "Space" could be seen as articulating physical and virtual aspects. "People" is clearly a coarse generalization that will include a galaxy of distinctions embracing the social and institutional spectrum. And similarly, "technology" is in many ways a complex field and indeed a socially constructed one (MacKenzie and Wajcman, 1985; Bijker and Law, 1994), so it is somehow inextricable from the other dimensions. Smart placemaking is therefore a complex activity that cannot be tackled with oversimplistic formulas. A first contribution to this is envisaging an approach that combines mutual agency and the active role of space, people, and technology. This engages place more than partial processes and in doing so multiplies possibilities and combinations in the range of placemaking actions and designs that can be conceived, particularly by reactivating space as a full actant in the overall equation (see Table 1).

TABLE 1 Typologizing approaches to designing smart places.

	Add-on smart Digital machine city	Socially shaped smart Digital community	Smart place (holistic approach)
Space	Passive: physical support/location	Passive: physical support/location	Active: relational, affording
People	Passive: end user/data source	Active: participant/ shaper	Active: participant/ shaper
Technology	Active: change agent	Active: participant agent/actant	Active: participant agent/actant

If we assume that a more holistic and comprehensive approach to smart place and its shaping involves including all three dimensions in a process of analysis and design of place, we can also argue that neither technological add-on or digital community approaches can fulfill it. These will suggest visions and actions, but more can be brought into the debate to suggest an evolution of the design approach and process. Given the discussion so far, looking at two important notions for spatial design seems a good way to start discussing smart place design challenges and possibilities. The first is context, the place that already exists and participates—willingly or not—in any new scheme. The second is affordance and how this plays out in the relationships among the various dimensions of smart place design through agency and functions. What extended narratives can we engage with, in dealing with context and affordances, and what can they mean for design?

Recombined context

Despite the growing fascination of city officials toward often self-absorbed sculptural buildings used as global landmarks, most architectural and spatial design theory will point at the need to think in terms of relationships. Nothing happens in a void, and the act of design implies not just adding to a blank (or irrelevant) canvas but above all changing and tweaking. Any design intervention changes, shifts, challenges, or complies with an already complex, preexisting set of relationships. High technology does not operate outside of this framework. Context and the notion of material place are undoubtedly disrupted by it, yet this does not mean that context becomes irrelevant and can be ignored. In fact, without knowing and understanding what already exists or how it could change, it becomes impossible to proactively, and positively, try to influence relationships within it.

Grasping the richness and potential of context

The first challenge therefore is related to grasping and interpreting context, in both its spatial and social aspects. Existing place embeds existing relationships. It suggests spatial and cultural clues, opportunities, and threats that can be harnessed in a process of place-tweaking —rather than making anew. Greenfield (2006; 22), for instance, refers to the importance for hi-tech designers to engage with the "many different 'everydays'" of local cultures that already manifest themselves through buildings and artifacts. And understanding what a place stands for, why it is there, can also be key. Technology might be able to boost the ways we use place, but place is or should also be able to boost or orientate the ways we use technology. Yet, with the exception of initiatives aiming at using ICTs for new ways of analyzing urban spaces through sensors and big data (see, for instance, Shepard, 2020), the majority of media/screen and IoT-based installations, urban apps, and other similar interventions do not stem from a design process based on a deep understanding and interpretation of any specific place. This strongly contrasts with virtually any live or educational urban design or architectural exercise, where the absence of deep site-based considerations that significantly contribute to the genesis of a scheme would be normally regarded as a major deficiency in the process. As Vande Moere and Wouters (2012; 4) have noted "While the official approval for architectural or urban interventions always involves some sort of site analysis, the same might be made applicable for media architecture, which holds the potential to have a similar, if not greater, impact on the environment than the physical building itself." Understanding the context or indeed the different hyperlocal contexts a city can embed and the constraints and possibilities they offer is the first requisite for looking at how space and place can recombine.

One place becoming many

Katharine Willis (2016; 2) provocatively questions whether when sitting in a café while accessing their Wi-Fi we are in the actual physical place or indeed somewhere else, in a "network space." The notions and feelings of being "there" or somewhere else—and maybe in multiple places at once—are one of the fundamental conundrums that electronic communication poses to a sense of space and place, as well as our behaviors in them. This problematizes context and place, making traditional analysis necessary but probably not sufficient. If already the notion of "place" has always been a highly subjective one—being our sense of place related to our own existential condition and personal memories and culture—technology disrupts this further by making one place become many not only at an existential level, but at a practical and functional one too.

This multilayering of place, where potentially a very large range of "subplaces" can appear or disappear thanks to networked connections and the individual digital customization of activities and atmospheres, seems to present a novel set of tensions. On a positive note, it has been discussed how this can challenge and contest accepted assumptions of what a place is or is for, and afford alternative and bottom-up ways of appropriating it. Shepard refers to Fujimoto's account on how Japanese young women create more protective spaces by "transforming the paternalistic communities of city streets and subway cars into private territories for women and children" (Fujimoto, 2006, quoted in Shepard, 2011; 24). Willis (2016; 43) has adopted similar examples and commentaries to discuss the "in-between" inhabited through these digital practices as an additional form of "third place" (Oldenburg, 1999). Odendaal (2018, 2019) and Rekow (2013) among others have discussed local and alternative appropriations of technology and place by low-income communities and migrants in the Global South. Such possibilities, which normally arise in spontaneous ways, can certainly be facilitated by design.

Notwithstanding these, placemakers looking to proactively tweak spatial relationships to encourage a socially diverse but collectively engaged use of public urban areas also face an opposite consideration. Should this possibility of enhanced multiplicity hamper any efforts to design a place suggesting possible collective uses and meanings? McCullough (2013; 102) notes how "as architecture arranges interpersonal distances in space, configures everyday processes, represents organizations, and shapes everyday habits within them, it also inobtrusively supports sensemaking (...) It tacitly cues what to say where, how to act in groups, and toward what goal these arrangements have been institutionalized." To what extent can a wide provision of information and functions oblivious to their physical context create the risk of producing a generalized and individualistic platform, a "click here and choose your experience" type of place, irrespective of other human and nonhuman actors present, history, character? As debates on cyberspace emerged, Shapiro (1995; 10) distinguished the potential for digital meeting places to become homogenized monocultural suburbia-like communities or to be purposefully shaped into virtual sidewalks where "people may be inconvenienced by views they don't want to hear. (…) places where bothersome, in-your-face expression flourishes and is heard." This way, he emphasized the importance of diversity coupled with maintaining a collective and open—albeit not always comfortable—experience of place. Should therefore spatial designers proactively intervene on this hybrid context keeping in mind how important it is to frame and support local social experience, through what Norberg-Schulz called a needed "stable system of places" (Norberg-Schulz, 1971; 114)? Therefore the fact that a place can become many introduces the need for a balancing act between facilitating open-source and democratized appropriations of place and overfragmented, individualized, and "neutralized" visions of it. The multilayering of smart place might therefore benefit from a proactive look at sociospatial relationships—existing and to be designed—to avoid a disconnection of people acting as detached individuals, and place itself, as some socially and existentially useful form of shared meaning and use could be lost.

Different places becoming one

Urban planners in the US West Coast area started reflecting in the second half of the 1990s on what the new possibilities offered by technologies to network places could mean for—and how could help reframe—urban sprawl. This was conceptualized as network-oriented development (NOD), seen as a strategy to "retrofit" sprawl and enhance places "by using digital networks to import from elsewhere in the city, county, and region many of the activities or functions needed in that neighborhood" (Page et al., 2003; 69).

The networking of places can therefore give a different meaning to their fragmentation. In truly recombinant logics, fragmented cities can recompose in socially purposeful and planned/designed ways, and different places become one. This can be envisaged in an additive way, where places lacking certain features or services can digitally "borrow" those from elsewhere. But it can also be imagined in an amplifying way, as an equal, two-way enhancement, where the same event, gathering, performance, or even serendipitous encounters happen synchronously or asynchronously in different physical places but combine them together into a joint one. And it can be harnessed by design. Pawley (1998; 202) had for this very reason critiqued much of "futuristic" architecture as failing to grasp these new opportunities and challenges by never questioning its "commitment to real time and unchanging space."

Among various possible implications the challenge of generating an enhanced or extended context as the sum of recombined noncontiguous fragments seems key. Working with such principle and its possibilities in mind involves a logic where urban smart technology does not act or is designed on its own. It requires understanding different recombining contexts and purposefully connecting them through an act of strategic design, with places—extended places—in mind. For example, it can mean enlivening peripheral neighborhoods by connecting them together and with the cultural center of a city, mitigating the effects of modernist physical zoning and urban edges and barriers. It can mean enhancing connections, innovation potential, encounters, and social ties among otherwise isolated small rural centers (Willis, 2017). And it can be done not by ignoring context, but by making real place, its character, and people central to such interventions.

Positively playing with hypervisibility and connectivity at an urban level, with those recombination aims in mind, goes well beyond sticking a large screen on a building façade. It requires going back to an understanding of a place's townscape (Cullen, 1961) and a purposeful articulation of its "heres" and "theres," be them physical or virtual. It can mean understanding and challenging the nature of "nodes," "landmarks," and other elements of legibility and meaning in the city (Lynch, 1960) to work with or against them proposing recombined enhancements to public space. How can two noncontiguous urban "nodes" change when they combine? How can they combine and through which hybrid (physical and digital) gateways? These are all questions and themes that aim at tackling the design of smart places as "wholes" involving the existence of an actively participating context. As such, they require reconnecting spatial and digital practice.

Programming place through extended affordances

Maier et al. (2009; 395) discuss the idea of affordance as originally stemming from the field of perceptual psychology (Gibson, 1979) where it is centered on the complementarity of an environment offering potential and facilitation for actions or conditions and a generic "animal" able to take advantage of those. They explain how the idea has become central in discourses and practice of design of human-computer interaction (HCI) and in artificial intelligence. Rietveld and Kiverstein (2014) further extend this noting that "in our account 'context' is interpreted as the rich landscape of affordances in which skillful action unfolds" (p. 346). So, in a complementary way to dealing with context, smart place design can benefit by looking at the concept of affordance. When it comes to the shaping and functioning of smart environments, conventional views of agency and roles in what makes place can be challenged by a new, more complex landscape.

In architecture the idea of affordance has probably had its main widespread merit in relativizing the relationship between designed space and its users, away from strong deterministic views of cause and effect between form and behavior. Betsky (2015) argues that "A theory of affordance lets us understand buildings not as objects, but as environments that afford us possibilities, that open and enclose, that respond and give us clues." However, while designers are expected to adopt a less demiurgic stance toward the spatial experience, this is still framed within a view of space providing affordances to people. Maier et al. (2009; 397) propose a more complex view of mutual affordances when they explain that "Examples exist of course between artifacts and users (e.g., turnability of a door-knob, readability of a sign) between multiple users (e.g., conversations, mating, fighting, etc.), and finally between multiple artifacts (e.g., walls affording support to roofs, sprinklers affording suppression of fires). We call the latter relationships 'artifact-artifact affordances' (AAA)." The already intertwining nature of the relationships and affordances that happen in the built environment can be further amplified by digital-rich lifestyles and smart designs. These however do not simply mean that agency transfers to technology. It requires envisaging a landscape where its dimensions are mutually responsive in nonlinear ways. Questioning then what responds to what—or whom—introduces some interesting issues and opportunities.

Smart places are such because within them smart elements exist that afford on the one hand the recombination of contexts and spaces and on the other the extension of agency. Smart elements extend and augment place by on the one hand affording new ways—sometimes unachievable without high technology—of perceiving and using the environment. The range of uses, functions, and even atmospheres of a place is augmented and multiplied through digital and technological means. On the other hand, they rather importantly blur and challenge any expected distinction between supporting/passive and sentient/active actors in place. Smart elements in public spaces and in our homes and offices, from large-scale intelligent facades to smart speakers, are not just objects providing a tentative environmental affordance or supporting their users' actions. They proactively participate in and further augment the place-based "ballet" (Jacobs, 1992—original 1961; 50) by embedding active intentions, rules, and behaviors in the form of code and AI-based learning. However, being participants to place means they do not just facilitate possibilities one-way, but equally require affordances from other parts of the environment and indeed from human actions generating the base of data they can feed on, and function and evolve by.

So, placemaking in a smart world is not a matter of designing spaces as a sequence of moves providing affordances to people, but the strategic shaping of a more complex set of multiagent mutual relationships, facilitations, and constraints. We have discussed how, when looking at smart design through the lens of context, "recombination" is a function of how different environments can articulate in novel ways, as a consequence of technology providing a series of affordances to spaces, allowing them to connect, and "be" more. When looking at smart design through the lens of affordance and agency, this involves a need to consider the implications of an extended and intertwined perspective on the programming of place. Mainstream architectural practice programs spaces—buildings and other environments—by articulating what the space affords. Probably the most elementary yet disturbing example is the "defensive architecture" application of spikes on buildings' surfaces to prevent homeless people sleeping on them, denying the otherwise natural affordance of those elements. Things statically embed decisions, agency, and politics through form, materiality, and low technology. Humans dynamically interact with them.

Smart design however blurs distinctions between objects and humans further and recombines their relationships. Sentient things can become proactive actors, not only offering a static embedded set of values or possibilities but also able to challenge, change, and act within place. While a person in a space can connect with and inhabit it by negotiating its affordances, a smart place plays an active role. It might want to connect or not with you, allow uses, and selectively and dynamically adapt, changing what is allowed and even how other elements of the place behave. It is self-evident then how nontrivial this can be. Any discourse suggesting a future of flexible intelligent environments has to be at the very least counterbalanced by considerations of how smart place can proactively become controlling and how hi-tech responsive automation should be part of a wider set of strategic choices on the overall program and experience of space. A paradox is one of places endowed with technologies potentially augmenting possibilities yet able to block any alternative use or interpretation, in an even worse way than the most inflexible physical setup.

Articulating affordances: Contrast, coherence, compliance, and overload

Another key tension is ending up with places where physical and digital affordances are in contradiction or contrast. Generally the challenge for designers is again one of understanding and coordinating a complex and less linear way of seeing affordances and spatial programs, where built, digital, and human factors converge and present (or deny) affordances to each other. Fatah et al. (2006) note how "In these locations the social behaviours and the interaction spaces appear to take a shape which provides the person with more privacy (…) This seems to be supported by the properties and affordance of the physical environment encouraging a certain type of behaviour." As mentioned, when looking at context, this also suggests a need for a recombinant approach to design. Here, place and what it "does"—its program—is not the result of a linear technological disruption and replacement but of a concerted or sometimes spontaneous convergence of its different built/physical, digital, and human aspects.

In the case of combining spaces—the concept of a place becoming many—we have looked at the tension between diverse appropriations and the lack of coherent meaning stemming from an overlapping of different contexts. We can also look at it through the lens of combining affordances multiplying a diversity of functions and uses within the same space. Facilitating appropriations can make a smart place design "democratic", yet the ability of digital technologies to override form and character and the cognitive coherence these offer to our experience of place needs to be taken into account. McCullough notes how "To the skilled tool-using mind, a set of external circumstances becomes 'about' something. A floor may invite dancing, just as a rake may invite gardening. As people learn from their settings, they come to associate them with particular states of intent (…) Intent shapes perception and, with it, discovery of affordances - possibilities for action afforded by objects or environments; conversely, intent is shaped by the presence of affordances" (McCullough, 2013; 72–73). Similarly, Rietweld and Kiverstein note how ecological the idea of affordance is and how it ideally requires a degree of coherence between possibilities and environment: "If the material environment did not offer the opportunities for action it does, our form of life would not include the practices it does" (Rietveld and Kiverstein, 2014; 339). All of this points again at the fundamental principle that we do not design in a void, and places and their physical artifacts hold memories, character, and cultural meanings that should not be ignored, whether we intend to comply with them or generate a contrast or alternative. There is a difference between overriding or ignoring place—which is what happens in technological add-on or replacement approaches where new things happen just because they can—and changing or reinterpreting it. The author has commented previously that "We can let people hunt for Pokémons anywhere, including for example in a religious building or a cemetery, but do we really want to, and why? What does it mean for that place, for how and why it was designed, and for its character, cultural and practical functioning, and its *raison d'etre*?" (Aurigi, 2017; 15).

An extreme way to look at this is through a perspective of affordance overload, which a proliferation of functions and possibilities, within the same place, can also lead to. While providing more reasons to be somewhere and use a space can be a positive way of revitalizing it, the importance of strategically and purposefully managing and limiting this through design cannot be understated. Exercising digital technologies' ability to boost in potentially unlimited ways what objects and spaces can do by code is an idea celebrated through the concept of "ubiquitous city" (Lee, 2009) and often uncritically practiced. This liberation of agency from form or location is however not without its pitfalls. In the cybernetics-inspired Superstudio's *Supersurface*, utopian vision "all edifices cease to exist" (Superstudio, 1972; 242–51) and a global grid provides everything needed wherever ones moves to. This comes at the price—desirable in such a utopia but debatable at least in urban design and placemaking practice—of the annihilation of cities and the establishment of a nomadic society where place design is fundamentally redundant. If the promise of "anything, anytime, anywhere" (Graham and Marvin, 1996; 88) is embraced uncritically, we risk challenging the very nature of designed place as a grounded—hence somehow specialized—one. Smart design therefore faces the challenge of programming place in a balanced way between diversity, appropriation, and the coherence proper to any established sense of place.

Conclusions: Ideas for a recombinant, holistic approach to smart place design

This chapter has challenged the view of the shaping of smart places as an exclusively technology-based add-on operation, where disruption is seen as so strong to dwarf any other factors playing a role in placemaking. The range of opportunities and threats introduced by the added fluidity and possibilities of the design of digitally enhanced spaces point at the need for clarity and direction, rather than any deterministic and simplistic assumption that high technologies are somehow an inevitable force for making better places. Designers can indeed rely on new tricks in articulating spaces and affordances, but how these are understood and used within a shaping process of our urban spaces is what makes the difference. Specifically, this chapter has argued that a good starting point to frame and enhance our smart place design processes can be the following:

Design smart places, not smart technologies

This might feel like a semantic distinction, but it implies a significant shift. A strategic and holistic approach, able to define appropriate challenges and aims, and reflect upon and coordinate how the various components of place recombine toward addressing these in a successful way, is key to the shaping of smart places. This above all means shifting the focus from the "adoption" of technology—as warned already 30 years ago by Guthrie and Dutton (1992)—and the configuration of it, to rather a deep look at how places, and "good" places, work, through what relationships. This in turn cannot be generalized. There is no such thing as the absolute good place, and any vision needs to be formulated for a specific context. This is something that most smart urbanism advocacy very often fails to grasp, focused as it is on "smart" as a series of products, and on too generic concepts based on "tech solutions," a general idea of "users," and looking at "the city" as a nondiverse entity. Having a vision that is properly contextualized will call for answering questions like "why does this place matter" or indeed "how could it matter more." This will work out answers through a strategy of actions that willingly and specifically address contextual opportunities and rethink and define affordances and relationships with a deep awareness of scenarios ahead. It will use all means, materials, and tools—not just hi-tech ones—available. This is different from "adopting" and asking ourselves "how can I introduce technology here?"

Know your context, and how it can be recombined

If designers start to form place, a good understanding of the existing situation, the spatial, temporal, and social relationships at play, and how these can be harnessed and modified, is essential. A deep analysis of place and its spatial, social, and cultural dimension, normal in good spatial design practice yet seldom relied upon in allegedly "disruptive" smart, is a good basis to build upon. Yet, Rietveld and Kiverstein note how "the determination of affordances directly requires the expertise of designers who have knowledge of the context in which the artefact or building will be used" (Rietveld and Kiverstein, 2014, 406). Armed with that awareness, proper and useful recombinations or places or elements of them can become part of a design strategy.

Program place, not devices

The recombination of place implies and relies upon extended affordances. These are not about what technology can do, but about what can be done and by whom in the place itself. New or extended affordances offer the possibility of expanding and complexifying programs, potentially in edgeless ways. But design intentions will embed specific meanings for place, and these can imply the opportunity of modulating, prioritizing, or even limiting what the place is about. This balancing act between freedom and constraint, open-endedness, and collective meanings is as delicate as important. And it is most definitely not happening on its own. Designing affordances, that is what a place and its elements facilitate, suggest or resist to, is political and strategic. But it is essential as far as placemaking is concerned, and results cannot be left to any naïve technological utopianism.

Designing smart space cannot therefore be seen as an add-on activity or any easier than, or divorced from, the rather wicked issue of tackling good architecture or urban design. This means that shaping smart places is more than simply a visual, material, or technological exercise but a recombined one that needs to rely on strategic clarity. We have the tech, but we are just beginning to understand how to make it part of our placemaking toolbox.

References

Aurigi, A., 2013. Reflections towards an agenda for urban-designing the digital city. Urban Des. Int. 18 (2), 131–144.

Aurigi, A., 2017. Foreword: Space is not a platform - foregrounding place in smart urban design. In: Zammit, A., Kenna, T. (Eds.), Enhancing Places Through Technologies. Edições Universitárias Lusófonas, Lisbon, pp. 7–17.

Awan, N., Schneider, T., Till, J., 2011. Spatial Agency: Other Ways of Doing Architecture. Routledge, Abingdon.

Benedikt, M., 1991. Cyberspace: First Steps. MIT Press, Cambridge, MA.

Bentley, I., 1985. Responsive Environments: A Manual for Designers. Architectural Press, London.

Betsky, A., 2015. The evolving landscape of architectural affordances. J. Am. Inst. Archit. 27 July 2015.

Bijker, W.E., Law, J. (Eds.), 1994. Shaping Technology/Building Society: Studies in Sociotechnical Change. MIT Press, Cambridge, MA.

Brand, S., 1997. How Buildings Learn: What Happens After They're Built. Phoenix, London.

Bullivant, L., 2006. Responsive Environments: Architecture, Art, and Design, V & A Contemporary. V & A Publications, London.

Bullivant, L. (Ed.), 2017. 4D Hyperlocal: A Cultural Toolkit for the Open-Source City. John Wiley & Sons, Oxford. Architectural Design Profile No. 245.

Caprotti, F., 2014. Eco-Cities and the Transition to Low Carbon Economies. Palgrave Pivot, London.

Cullen, G., 1961. Townscape. Architectural Press, London.

Dalsgaard, P., Halskov, K., 2010. Designing urban media façades – cases and challenges. In: Proceedings of CHI 2010, Atlanta, USA.

de Lange, M., de Waal, M., 2019. The Hackable City: Digital Media and Collaborative City Making in the Network Society. Springer, Singapore.

Design Council, 2020. What Is the Framework for Innovation? https://www.designcouncil.org.uk/news-opinion/what-framework-innovation-design-councils-evolved-double-diamond. (Accessed 20 May 2020).

Fatah gen Schieck, A., Penn, A., Kostakos, V., O'Neill, E., Kindberg, T., Stanton Fraser, D., Jones, T., 2006. Design tools for pervasive computing in urban environment. In: Proceedings of the 8th International Conference on Design & Decision Support Systems in Architecture and Urban Planning. Springer, Eindhoven, Netherlands.

Foth, M., Brynskov, M., Ojala, T., 2015. Citizen's Right to the Digital City: Urban Interfaces, Activism, and Placemaking. Springer, Singapore.

Fujimoto, K., 2006. The third stage paradigm: territory machines from the girl's pager revolution to mobile aesthetics'. In: Ito, M., Okabe, D. (Eds.), Personal, Portable, Pedestrian: Mobile Phones in Japanese Life. MIT Press, Cambridge MA.

Gehl, J., 2006. Life, spaces, buildings. And in said order, please. In: Moor, M., Rowland, J. (Eds.), Urban Design Futures. Routledge, Oxon.

Gehl, J., 2011. Life Between Buildings: Using Public Space. Island Press, Washington, DC.

Gibson, J.J., 1979. The theory of affordances. In: The Ecological Approach to Visual Perception. Houghton, Mifflin, Hopewell, NJ, USA.

Graham, S., Marvin, S., 1996. Telecommunications and the City: Electronic Spaces, Urban Places. Routledge, London.

Greenfield, A., 2006. Everyware: The Dawning Age of Ubiquitous Computing. New Riders, Berkeley, CA.

Guthrie, K., Dutton, W., 1992. The politics of citizen access technology: the development of public information Utilities in Four Cities. Policy Stud. J. 20 (4), 574–597.

Hespanhol, L., Häusler, H.M., Tomitsch, M., Tscherteu, G. (Eds.), 2017. Media Architecture Compendium: Digital Placemaking. Avedition, Stuttgart.

Hoggenmueller, M., Hespanhol, L., Tomitsch, M., 2020. Designing urban robots for hybrid placemaking experiences. Medium. https://medium.com/design-at-sydney/designing-urban-robots-for-hybrid-placemaking-experiences-b85574148f16. (Accessed 6 May 2020).

Hollands, R.G., 2008. Will the real smart city please stand up? City Anal. Urban Trends Cult. Theor. Pol. Action 12 (3), 303–320.

Hollands, R.G., 2015. Critical interventions into the corporate smart city. Camb. J. Reg. Econ. Soc. 8, 61–77.

Horan, T.A., 2000. Digital Places: Building Our City of Bits. ULI-the Urban Land Institute, Washington, DC.

Jacobs, J., 1992. The Death and Life of Great American Cities. Vintage Books, New York (Orig. 1961).

Kitchin, R., 2015. Making sense of smart cities: addressing present shortcomings. Camb. J. Reg. Econ. Soc. 8, 131–136.

Kitchin, R., Lauriault, T., McArdle, G., 2015. Knowing and governing cities through urban indicators, city benchmarking and real-time dashboards. Reg. Stud. Reg. Sci. 2, 1–28.

Lee, S.H., 2009. Ubiquitous City: Future of City, City of Future. Hanbat National University Press, Daejeon.

Lefebvre, H., 1991. The Production of Space. Blackwell, Oxford.

Luque Ayala, A., Marvin, S., 2020. Urban Operating Systems: Producing the Computational City. The MIT Press, Cambridge, MA.

Lynch, K., 1960. The Image of the City. MIT Press, Cambridge, MA.

MacKenzie, D.A., Wajcman, J. (Eds.), 1985. The Social Shaping of Technology: How the Refrigerator Got Its Hum. Open University Press, Milton Keynes.

Maier, J., Fadel, G., Battisto, D., 2009. An affordance-based approach to architectural theory, design, and practice. Des. Stud. 30 (4), 393–414.

Mattern, S., 2017. A city is not a computer. Places. February 2017.

McCullough, M., 2013. Ambient Commons: Attention in the Age of Embodied Information. The MIT Press, Cambridge, MA.

Melhish, C., Degen, M., Rose, G., 2017. The real modernity that is here: understanding the role of digital visualisations in the production of a new urban imaginary at Msheireb downtown, Qatar. City Soc. 28 (2), 222–245.

Mitchell, W.J., 1995. City of Bits: Space, Place, and the Infobahn. MIT Press, Cambridge, MA.

Mitchell, W.J., 1999. E-Topia: 'Urban Life, Jim–but Not as We Know It'. MIT Press, Cambridge, MA.

Mora, L., Bolici, R., Deakin, M., 2017. The first two decades of smart-city research: a bibliometric analysis. J. Urban Technol. 24 (1), 3–27.

Mora, L., Deakin, M., Reid, A., 2018. Strategic principles for smart city development: a multiple case study analysis of European best practices. Technol. Forecast. Soc. Chang. 142, 70–97.

Norberg-Schulz, C., 1971. Existence, Space and Architecture. Studio Vista, London.

Odendaal, N., 2018. Smart innovation at the margins: learning from Cape Town and Kibera. In: Karvonen, A., Cugurullo, F., Caprotti, F. (Eds.), Inside Smart Cities: Place, Politics and Urban Innovation. Routledge, London.

Odendaal, N., 2019. Appropriating 'big data': exploring the emancipatory potential of the data strategies of civil society organisations in Cape Town, South Africa. In: Cardullo, P., di Feliciantonio, C., Kitchin, R. (Eds.), The Right to the Smart City. Emerald, Bingley.

Oldenburg, R., 1999. The Great Good Place: Cafes, Coffee Shops, Bookstores, Bars, Hair Salons, and Other Hangouts at the Heart of a Community. Marlowe & Company, New York.

Page, S., Phillips, B., Siembab, W., 2003. The millennium city: making sprawl smart though network oriented development. J. Urban Technol. 10 (3), 63–84.

Pawley, M., 1998. Terminal Architecture. Reaktion books, London.

Picon, A., 2015. Smart Cities: A Spatialised Intelligence. John Wiley, AD Primers, Chichester.

Project for Public Spaces, 2018. Placemaking: what if we built our cities around places? Available online at https://assets-global.website-files.com/5810e16fbe876cec6bcbd86e/5a6a1c930a6e6500019faf5d_Oct-2016-placemaking-booklet.pdf. (Accessed August 2020).

Rekow, L., 2013. Including informality in the smart citizen conversation. In: Hemment, D., Townsend, A. (Eds.), Smart Citizens. FutureEverything, Manchester.

Relph, E., 1986. Place and Placelessness. Pion Limited, London.

RIBA, 2020. Plan of Work 2020 Overview. RIBA, London.

Rietveld, E., Kiverstein, J., 2014. A rich landscape of affordances. Ecol. Psychol. 26 (4), 325–352.

Rose, G., Willis, A., 2019. Seeing the smart city on twitter: colour and the affective territories of becoming smart. Environ. Plann. D: Soc. Space 37 (3), 411–427.

Shapiro, A.L., 1995. Street corners in cyberspace. Nation. 3-7-1995.

Shepard, M., 2011. Toward the sentient city. In: Shepard, M. (Ed.), Sentient City: Ubiquitous Computing, Architecture, and the Future of Urban Space. Architectural League of New York; MIT Press, Cambridge, MA.

Shepard, M., 2020. Bias in urban research: from tools to environments. In: Willis, K., Aurigi, A. (Eds.), The Routledge Companion to Smart Cities. Routledge, Abingdon, pp. 29–41.

Smith, P.F., 1977. The Syntax of Cities. Hutchinson, London.

Söderström, O., Paasche, T., Klauser, F., 2014. Smart cities as corporate storytelling. City Anal. Urban Trends Cult. Theor. Pol. Action 18 (3), 307–320.

Superstudio, 1972. Description of the microevent/microenvironment. In: Ambasz, E. (Ed.), Italy: The New Domestic Landscape. Achievements and Problems of Italian Design. The Museum of Modern Art, New York, pp. 242–251.

Tomitsch, M., 2018. Making Cities Smarter: Designing, Interactive, Urban, Applications. Jovis, Berlin.

Vande Moere, A., Wouters, N., 2012. The role of context in media architecture. In: International Symposium on Pervasive Displays (Paper No. 12). Presented at the PerDis'12, Porto, Portugal, 04 June 2012–05 June 2012.

Vanolo, A., 2014. Smartmentality: the smart city as disciplinary strategy. Urban Stud. 51 (5), 883–898.

Walsh, N.P., 2018. Carlo Ratti's prototype for sidewalk labs shows how the design of streets could change in real time. ArchDaily. 18/7/18 https://www.archdaily.com/898471/carlo-ratti-associatis-latest-prototype-shows-how-the-design-of-streets-could-change-in-real-time. (Accessed May 2019).

Willis, K., 2016. Netspaces: Space and Place in a Networked World. Ashgate, Farnham.

Willis, K., 2017. Digital neighbourhoods: hyperlocal village hubs in rural communities. In: Bullivant, L. (Ed.), 4D Hyperlocal: A Cultural Toolkit for the Open-Source City. John Wiley & Sons, Oxford. Architectural Design Profile No. 245.

Willis, K., Aurigi, A., 2017. Digital and Smart Cities. Routledge, London.

Yigitcanlar, T., Bulu, M., 2018. Urban Knowledge and Innovation Spaces: Insights, Inspirations and Inclinations from Global Practices. Routledge, London.

Zammit, A., Kenna, T. (Eds.), 2017. Enhancing Places Through Technologies. Edições Universitárias Lusófonas, Lisbon.

Author biography

Alessandro Aurigi is Professor of Urban Design at the University of Plymouth, having previously worked at the University of Newcastle and University College London (UCL). His research focuses on the relationships between our digital society and the ways we conceive, design, and manage urban space, to enhance and support place quality. He has published *Digital and Smart Cities* and *The Routledge Companion to Smart Cities* (Routledge, with Katharine Willis), the multidisciplinary book *Augmented Urban Spaces* (Ashgate, with Fiorella De Cindio), and *Making the Digital City* (Ashgate).

Chapter 3

Responsive public spaces: Five mechanisms for the design of public space in the era of networked urbanism

Martijn de Waal, Frank Suurenbroek, and Ivan Nio
Amsterdam University of Applied Sciences, Amsterdam, The Netherlands

Chapter outline

Introduction

In the last decade, various new interdisciplinary practices of urban design have emerged, using labels such as media architecture (Hespanhol et al., 2017) digital placemaking (Hespanhol, 2018) or urban interaction design (Brynskov et al., 2014) to describe themselves. What they have in common is that participants call for an approach to the design of urban space from an integrated or "hybrid" perspective (De Souza e Silva, 2006; Willis and Aurigi, 2011), combining the design of physical and digital infrastructures and experiences.

The Amsterdam-based research project Coreus (Cocreating Responsive Urban Spaces) aimed to contribute to this development by exploring how various disciplines can work together in the design of contemporary public spaces under conditions of what has been called "networked urbanism" (Blokland and Savage, 2008). Broader developments such as globalization, urbanization, shifting investment patterns, and the rise of digital technologies have led to the emergence of

Shaping Smart for Better Cities. https://doi.org/10.1016/B978-0-12-818636-7.00009-3

new urban typologies and practices, such as clusters of large-scale shopping, entertainment, and office complexes often located near multimodal access points, at the crossroads of highways and public transit. While these have been criticized as ill-fit for public space functions (e.g., Sorkin, 1992), we wanted to explore to what extent spatially designed responsive media could aid in bringing out public space qualities in the experiences of these places. What if urban design could integrate new digital instruments to reshape and activate public spaces? In the context of festivals and in controlled settings such as musea, interactive technologies have shown promise for the praxis of urban design. Yet their implementation in urban public space demands new strategies and above all new forms of cooperation between disciplines such as architecture, urban planning, and urban interaction design (Aurigi, 2013; Foth and Sanders, 2008).

In Coreus, we invited urban designers, interactive designers, landscape architects, representatives from the government of Amsterdam, and local stakeholders around the ArenA Boulevard—a mid-1990s development at the periphery of Amsterdam—to explore such an approach. We found that a first step needed was the creation of a common understanding of responsive technologies in relation to urban public space. In answer to that challenge, we have created a heuristic of five mechanisms that map possible applications of responsive technologies in relation to particular public space qualities and the affordances of networked urbanism (see also: Suurenbroek et al., 2019).

In the remainder of this chapter, we will discuss these five mechanisms in detail. To discuss the possible contributions of responsive technologies to the activation of urban public space, we will first set out the key qualities of public spaces that we find desirable as design objectives. This is followed by a brief analysis of networked urbanism as a particular contemporary condition. That will then have set the stage for a further in-depth exploration of our five mechanisms.

Qualities of urban public space

The difference between a city and a village, Jane Jacobs (1961) once famously wrote, is that in cities we always find ourselves surrounded by strangers. Strangers, as various urban sociologists have pointed out, with whom we somehow have to find a way to live together (Lofland, 1973, 1998; Hajer and Reijndorp, 2001). Public spaces have often been understood as the "mixing chambers" (Goldberger, 2003) that allow urbanites of different backgrounds to get to know each other at least categorically; exchange fashions, trends, ideas, and opinions; and create and experience shared understandings. In that line of thought, cities, Manuel Castells (2002) has argued, can be conceptualized as interfaces:

> *between individual and communal identities and shared social representations. It is their ability to organize this interface materially in forms, in rhythms, in collective experience and communicable perception that makes cities producers of sociability, and integrators of otherwise destructive creativity.*

In that light, many critics have theorized cities using the metaphors of the theater and the marketplace, with public space as the stage on which citizens perform their everyday lives, while at the same time forming the audiences for the performances of others (e.g., Mumford, 1937). Through these performances, citizens familiarize themselves with the rhythms of the city, the broad variety of cultural practices, and their fellow urbanites. Through these repeated interactions, for instance, strangers may become "familiar strangers," contributing to "public familiarity" and a broader sense of trust (Blokland, 2006; Van der Zwaard, 2010). As both Willis and Aurigi (2011) and Gumpert and Drucker (2001) have argued, building upon a large body of sociological research, public spaces on the one hand inspire and surprise us through their serendipitous encounters and playful interactions. On the other hand, they are places where social norms are produced and maintained, contributing to a sense of safety and control. Social relations and understandings thus come about through the numerous individual and collective activities and experiences that take place in public spaces.

It is through these repeated interactions that places themselves become meaningful as well and acquire a range of symbolic meanings that citizens can identify with. That is how spaces acquire a sense of place and bring out a feeling of being at home, meaning that citizens have the experience that they belong to that space while the space also belongs to them. Sennett (2018) theorizes this relation between the built environment and the experienced city as one between the *cité* (the infrastructures and built environment) and the *ville* (the layers of symbolic meanings, shared understandings, and imaginaries).

This relation can also be experienced more individually, as poets and theorists since Baudelaire and his infamous flaneur have repeatedly demonstrated. In this tradition, Walter Benjamin has extensively theorized the attitude of the flaneur as someone who wanders through the labyrinths of the city, exploring dialectical images that represent the sedimented complexities of urban life (Boomkens, 1998). More recently, Matos Wunderlich (2008) has described such an approach as a discursive way of walking, which she opposes to purposive walking.

The challenge then of urban design is to construct the conditions in the built environment that invite citizens to become inhabitants of the *ville*. In doing so, urban design has to deal with difficult trade-offs. For instance, public spaces are not only stages for social encounter but also an infrastructure for urban traffic. For public spaces to be successful as urban publics, urban design then has to focus on strategies that can (1) induce a sense of *belonging* through spaces that can be appropriated physically and symbolically by various publics and induce a sense of feeling at home. They must also—at least to a certain extent—feel safe and secure; (2) forge new *relations* by shaping spaces as stages for encounter, exchange, and confrontation; and/or (3) *provoking* inspiration, excitement, surprise, and a discursive and reflective orientation toward one's surroundings (Sennett, 1974; Giddens, 1984; Lofland, 1998; Boomkens, 1998; Hajer and Reijndorp, 2001).

Public space in the era of networked urbanism

Many of the theoretical points invoked above call for particular types of public space: those that function as meeting places for local communities. Typically, they offer a broad variety of functions that attract a broad variety of inhabitants to a central location. However, in the last decades, urban development and practices have taken a different direction. During these years, cities blurred into regional urban cityscapes, which scholars like Brenner and Theodore (2002) and Sassen (1991) addressed and analyzed as regional and postindustrial cities. Among others, Florida (2005) positioned these new cityscapes in a perspective of a global competition between cities, competing to attract new business and the creative class. As a result, in urban planning, cities started to develop new urban areas as nodes connecting local and global networks. Examples are the districts for professional service firms and developments with retail and leisure functions (de Hoog, 2012).

The rise of these sites has concurred with shifts in urban practices. Particularly, scholars with a focus on both the emerging blurring regional cityscapes and the influence of new technologies have argued for a new perspective on contemporary urban publics. Blokland and Savage (2008) described how "sociation" should no longer be understood as taking place in place-bound local communities centered around intensely used public spaces. Rather they see social relations in cities taking shape in a "decentralized diffuse and sprawling character which depends on multiple and myriad technological, informational, personal and organizational networks that link locations in complex ways" (Blokland and Savage, 2008). The authors use the label "networked urbanism" to describe this new pattern of urban usage. Their diagnosis is congruent with numerous analyses of the contemporary cityscape (Tordoir et al., 2015; Van Engelsdorp Gastelaars et al., 2006) that describe contemporary cities as pluralistic and polycentric, based on "more and more differentiated types of delocalized urban lifestyles" (Finka, 2006).

The ArenA Boulevard, developed in the mid-1990s at the periphery of Amsterdam, could be seen as an example of such developments. It contains a 50,000 seat stadium, a megaplex cinema, two concert halls, an array of big box stores, and a series of office towers and budget hotels. Adjacent to a train and subway station and in the vicinity of two major highways, it is easy to reach. Spatially the development is organized around a wide open pedestrianized public space that is designed to accommodate large crowds during events but can feel empty and deserted at other times. The ArenA Boulevard attracts different crowds of people—from soccer fans and concert attendees to office workers and tourists. In general the use of the space can be characterized as functional: visitors plan their visit purposely and don't linger much. Most of these groups have their own usage pattern, and there is little overlap between them. The Boulevard may be colonized by a flock of soccer fans on Sunday afternoon, while by Sunday evening, fans of a rock band have taken over, without much overlap between the two groups.

Digitalization plays an important role in the emergence of these new spatial practices. The rise of digital and mobile media started to personalize the use of the urban space in a subtle way. Wellman et al. (2003) has argued that networked media have enforced a broader sociological development of "networked individualism," in which citizens increasingly have the opportunity to self-select the social groups and networks they associate with (e.g., Tironi, 2010; de Waal, 2014). Each of these groups inscribes its own social geography on the urban space, where digital media function as the coordination centers for both physical meetups and online exchange. As Foth and Sanders have argued, these developments lead to a different understanding of urban publics and their usage of public space. Public space is much more than the central square in the geographical heart of a community, as "the contemporary interpretation of community is shifting from 'village' and 'neighbourhood' to 'social network' and 'urban tribe'" (Foth and Sanders, 2008).

More recently a number of authors have described how the rise of digital platforms such as Google Maps, Uber, or Airbnb has increasingly started to organize the interactions between citizens and their geographic activities. Critics have warned that such a "platform urbanism" as an interface to the city produces urban "filter bubbles" and practices of "software sorting" (Foth, 2017; Foth et al., 2018; Rabari and Storper, 2014; Widmer, 2016; Graham, 2005). Analysis of user data in social networks and mapping services could influence algorithms that recommend, reveal, or hide particular urban sites and networks, based on user profiles. These spatial, digital, and social developments could lead to a "networked parochialization" of urban space, with various urban publics shaping their own networked geographies of "parochial spaces" (Lofland, 1973), possibly decreasing the use of traditional public spaces as site where various publics temporarily overlap.

Such a personalized way of using urban space, in combination with the purpose-driven practices of visits to sites as the ArenA Boulevard, could undermine the vital social role of public spaces as the sites where citizens familiarize themselves with (and contribute to) the rhythms of the city and their fellow inhabitants. They may be experienced of what Auge (1995) has conceived of as "nonplaces," interchangeable sites without any local identity or grounding. Taken together, these developments could erode the necessary repetitive interactions that create a shared social fabric and the feeling of being at home among strangers, as well as the emergence of shared place-based symbolic meanings. Do these expressions of networked urbanism then break the link between *ville* and *cité* and undermine the qualities of public space, described earlier as belonging, relations, and provoking?

Perhaps, however ironically, these new frictionless urban spaces like the ArenA Boulevard may also hold the key to create new public domains in our spatially blurred and socially fragmented cities. As Willis (2016, p. 4) has argued, urbanites themselves do usually not perceive these sites as nonplaces, even though, as she summarizes, "places are increasingly becoming contingent on interactions occurring within technological meshwork." Rather, they use digital media to select relevant places to visit and constantly shift their attention between the here and now and the mediated content accessed through their mobile devices. However, that doesn't mean that their spatial experiences are not meaningful to them. The fact that they make use of urban spaces in a more fleeting and networked manner may also lead to a preference for spaces that are easy to reach. Referring to a study of the locative media app Foursquare, she concludes that it is sites like airports, railways stations, cafes, and restaurants that are highly valued. "[these] are the places that become most valued in the meshwork, and they are also characterized by their urban publics; they are places where people converge and then disperse; brought into being for the time in which the networked links connect" (Willis, 2016, p. 5). Well-connected and functional sites like the ArenA Boulevard are frequented by diverse groups of people, albeit not always at the same time, and for a diverse range of functions. Could those cultural-geographical patterns hold the key to revive it as a public space and reconnect the relation between *ville* and *cite*? In the next section, we explore how responsive media in urban design could play a role in that process.

Responsive technologies: Five mechanisms for public space

Our starting assumption was that interactive technologies, wireless networks, sensors, smartphones, and technologies such as the Internet of Things offer an entirely new, complementary set of instruments for urban designers and the way we approach public spaces. Of course, interactive technology has already been applied a great deal for individual objects, especially in the arts. In museum exhibitions, responsive installations add new layers to the story, experience, and immersion—while at the same time reshaping the relationship between the object and the visitor. Open-air artworks such as those of Studio Roosegaarde or the annual light festivals in Amsterdam or Sydney offer visitors spectacular experiences, reshaping public space temporarily. Academia is contributing to the development of this interdisciplinary approach through international institutes such as the Institute for Advanced Architecture of Catalonia (IAAC) in Barcelona and MIT's Senseable City Lab. Much progress has been made in the operation and properties of interactive installations in disciplines such as media architecture (Hespanhol et al., 2017) and urban interaction design (Brynskov et al., 2014) and through the debates around smart cities. In parallel a number of researchers and practitioners have also started to explore interactive technologies as a tool to activate urban public spaces as placemaking (de Waal, 2014; Pop et al., 2016; McQuire, 2008). Various terms are being used to describe these experiments. Already in 2006 Frenchman and Rojas (2006) used the term "responsive" to describe physical installations in public space that are able to adapt to different circumstances. Others have used terms such as "augmented urban space," highlighting the complex layering of digital and physical spaces (Aurigi and Cindio, 2008), and "active public space" (Markopoulou et al., 2017), referring to the notion that public space is not a given condition, but needs to be activated, for instance, by the use of digital technologies. Taken together the combination of interactive technology and urban design can be understood as an act aimed to invoke public domain qualities in new urban spaces—as well as an act of cocreation between multiple (design) disciplines. We have conceptualized these challenges as the *design of responsive public spaces* (Cantrell et al., 2015; Ratti and Claudel, 2016).

Such an integrated vision requires an interdisciplinary way of working, in which developers, urban designers, interaction designers, and local stakeholders collaborate toward the shared goal of establishing an active urban public domain. This means that these various parties need to have a shared understanding of the desired qualities of the public domain, as well as insights in the ways in which both spatial and interactive design can contribute to this.

Yet, as we found in our own 2-year action research on responsive spaces, these disciplines do not tend to have a tradition of pursuing this collaboration. Strikingly the need for such a shared frame of reference came to the light during a series of cocreation sessions in which various actors involved in the ArenA Boulevard were invited to start reflecting on the design of interactive installations. This resulted in a broad variety of perspectives and discourses on public space, different design practices, social interaction, and the use and meaning of technology, resulting in equally different design strategies and divergent expectations about the use of responsive technologies.

As a result, we worked with these various actors to construct a shared vocabulary. We constructed a heuristics meant to combine lessons learnt from existing interactive installations with the qualities of public space described earlier and the affordances of networked urbanism. How can we understand responsive media as an instrument in urban design to bring out these public space qualities in environments that attract various groups of citizens, albeit in different rhythms, and often in a purpose-driven mode?

Our heuristics was constructed in a number of cocreation sessions, where we discussed examples of existing responsive technologies. A number of studies were particular useful in introducing us to these projects, notably *What Urban Media Art Can Do* (Pop et al., 2016), *State of the Art and Best Practices Collection. Active Public Space* (Markopoulou et al., 2017); the proceedings from the Media Architecture Biennales held in Sydney and Aarhus (Dalsgaard and Fatah, 2014; Dalsgaard and Fatah gen Schieck, 2016); and the compendia and website archiving entries for the Media Architecture Awards (Hespanhol et al., 2017).[a] In addition, in other cocreation sessions, participants created paper prototypes of responsive media installations for the ArenA Boulevard, acting as conversation pieces for further discussion. Parallel, detailed spatial, and social analysis of the ArenA Boulevard was conducted, feeding the sessions with tangible problems and existing sociospatial properties.

---

a. Awards.mediaarchitecture.org.

This resulted in a set of five concepts that are meant to be used as point of departure in discussions between various stakeholders to discuss how public space qualities can be brought out in settings of networked urbanism with the aid of responsive technologies in urban design. We will now continue with the discussion of these mechanisms.

Sense of place

One of the functions of public space identified earlier is its ability to work as an interface between individual and communal identities or, to use Sennett's terms, to invite the users of the *cité* into the *ville*. Public spaces that fulfill these functions are sites where layers of symbolic meaning can emerge and where the rhythms of the city can be experienced. This can create a sense of belonging and provoke as a sense of excitement, curiosity, inspiration, and reflection. We identified "sense of place" as a mechanism that aims to activate these functions of public space. Responsive media in this category provide means to record, store, or attach meaning and associate these with a specific location and make these experienceable for others, who may or not be copresent at the same time and place. The sense of place mechanism attempts to capture and visualize a location's rhythms, to represent the collective identities and meanings of various publics that are connected with a location, and to make a location's stratified historical meaning accessible to individuals and collectives, both now and in the future. Conceptually, individual experiences are consolidated as collective experiences, and these collective stories, practices, and meanings are made experienceable again, so visitors to a location can learn about them, identify, or oppose them, even if the events they refer to are not occurring or immediately visible at that moment.

There are various ways to apply this sense of place mechanism. Rhythms and symbolic meanings can be traced through sensors in public space, the capture of data from other sources, such as social media, through crowdsourcing or by means of a curatorial practice of a designer, for instance, in collaboration with a local cultural institutions. In turn, these data, experiences, and stories can be brought back into public space in a variety of ways, ranging from the very prosaic (a measuring device that records exactly how many people passed by at specific points in time) to the very poetic (abstract images of rhythms that cannot be directly traced back to actual events). Some installations in this category show an "average"; others, by contrast, make a sequence of individual experiences visible that together add up to a unified whole. Sometimes the mechanism is invoked to make collective patterns visible or to reveal particular histories. It has also been applied to communicate experiences happening inside a particular building to passersby outside ("what happens inside") or to communicate experiences over time ("what happened yesterday").

The spatial positioning of these installations is also a mean to reshape urban space itself. The "sense of place" mechanism is a spatial stylistic device.

It shifts the built space from a neutral backdrop to a condition creating "enabler." Positioning the installation on a blind wall or inside a venue, for example, could instantly add cohesion and unity to a whole space. A central position in the middle of a square or two-thirds of the way up a space can create the same spatial effect as a fountain or statue, creating smaller subspaces and adding meaning along the way. The installation creates a shape and connects a space or differentiates it into subareas. The relationship with the user is also part of the spatial assignment. How does the object relate to the experience of arriving in the space: is it immediately visible or in fact waiting to be discovered "around the corner"?

There are by now many exemplary projects that have operationalized this mechanism, most of the time in the form of temporary art installations and at other times as more permanent fixtures in urban space. Many cities have featured "mood barometers" that in one way or another visualize various variables, from traffic movements to weather data in public space. In the field of culture, OKRA Landscape Architects used light projections in the city of Utrecht to highlight the outlines of a roman fort in the city center, relating its current urban form to its historic origins.[b] As an example of a contemporary and collaborative approach in her installation *Urban Alphabets*, Suse Miessner invited urbanites to use their mobile phones to make pictures of typefaces used in graffiti, shop signs, advertising and other texts found in the city. Taken together, these pictures create an alphabet that is projected on a screen in public space. As such, it encourages passersby to pay attention to typographic details in their surroundings as part of the development of a local identity.[c]

The sense of place mechanism can also be applied in a more critical or layered approach. For instance, the installation *Public Face* by Julius von Bismarck, Richard Wilhelmer, and Benjamin Maus measures the mood of a particular space by using video cameras with facial recognition software that captures the smile of passersby.[d] Their grins or grumpy faces are then assembled into a huge neon-lit smiley displaying the average mood of the day. Whereas this can be experienced as a collective rhythm or mood, it also brings its own mechanics to the attention. Is it acceptable that video cameras are used to measure people's moods in public space? Does it turn public space into an equivalent of Facebook where all our emotions are carefully datafied to be able to turn us into marketing targets? More directly critical is Nika Radić' *Office Cleaning*.[e] In this installation, Radić uses a "what happens inside" approach to project videos depicting office cleaners at work inside a building at the external façade. The installation makes the walls transparent, revealing a reality that is usually not visible in public spaces, featuring groups of people that are usually marginalized.

b. Domplein, Okra Landscape Architects. https://www.okra.nl/en/projects/domplein/.

c. Urban Alphabets, Suse Miessner. http://www.ualphabets.com/.

d. Public Face, Julius von Bismarck, Richard Wilhelmer and Benjamin Maus. http://juliusvonbismarck.com/bank/index.php/projects/public-face-ii/.

e. Office Cleaning, Nika Radić, 2008.

(Playful) interaction

Following the metaphors of "theater" and "marketplace," the (playful) interaction mechanism highlights the relational qualities of public spaces. Playful interaction turns public space into a stage or "magic circle" on/in which urbanites can interact with each other. Whereas the sense of place mechanism can be experienced in an "ambient" way while passing by, (playful) interaction actively invites visitors to take part in a particular dramaturgy. Yet passersby can also watch the playful interactions of others, assuming a more passive role as spectators. (Playful) interaction is thus about connecting people in public space, in various degrees of intensity. This could vary from brief encounters to more prolonged and in-depth interactions. Also the type of interaction could vary, from playful and affective experiences to democratic debates and the construction of issue publics.

As such, it draws upon different types of conceptualization of public space. In the first, public space is mostly seen as a site that builds familiarity and trust between urbanites. Urbanites get to know each other by observing each other from a certain distance, overhearing conversations, or having a chat with someone. The playful interaction mechanism contributes to this process. The exchanges facilitated by this mechanism do not necessarily lead to new friendships or a close homogeneous community but rather to a fundamental trust that perpetuates the idea of the city as a community of strangers. This might involve playing a game together that has specific goals, rules, and scoring, often with a competitive element and high scores. Other installations invoke the principle of free play. Like in a playground, there are no specific rules or established goals but rather an environment or "world" that reacts to players and provides a number of expressive tools. And just as the seesaw or duo-swing in a real playground invite players to coordinate their activities, these projects also invite players to work together or coordinate, resulting in a brief, shared choreography. In some cases such an approach departs from a critical appraisal, addressing the commodification of public space and the emergence of a "society of the spectacle." These examples take inspiration from the situationist movement that in the late 1950s and 1960s staged artistic interventions aiming to undermine a dominant culture of consumption. Instead, they proposed a new type of relationships between urbanites and with their surroundings.

In another tradition, public space is seen as a space for (rational) political debate and struggle, a site where issues can be brought in and discussed. In turn the staging of what Marshall Berman (1987) called "recognition scenes" can lead to the formation of a public around these issues. Here, different approaches are taken. In some cases interactive installations invite passersby to contribute to a public discussion. They can vote, contribute arguments, or voice their opinion that is then usually projected on a screen or projection in public space. The installation itself functions as an arena for debate. In a different approach the goal of the installation is not so much to mediate the discussion itself, but to provoke debate among bystanders by visualizing a particular topic or issue. In these examples the installation is not an arena for debate, but rather a conversation piece. A subgenre in this category concerns data visualization in public space. Data about, for example, electricity use or pollution are reproduced using an interface, not necessarily in real time, to raise awareness of an issue and help create "issue publics" (communities that concentrate on a defined issue such as sustainable energy, focusing on a particular issue; see, e.g., Claes, 2017). These responsive installations often revolve around the question of "citizen engagement": how can citizens become involved in social and local issues in new ways through responsive installations?

Spatially, these offer a different mechanism to shape the space. As their goals are much more orientated toward activating and lingering, the positioning of the installation helps to activate part of the spaces that had been overlooked or underserved, adding new nodes to the entire public space. Moreover, this kind of installation could be positioned as a "stepping stone" between the more remote areas and public transportation, enabling "safe passages," by clustering pedestrians and users.

Many responsive projects in this category have been employed as temporary installations in public space, often as part of festivals or events, sometimes also as a "guerilla" intervention. Others have been designed as an integral part of a building or public space. An example of this is *BruumRuum!*,[f] a large-scale interactive light installation at the Plaça de les Glòries Catalanes in Barcelona. Nearly 10,000 LEDs are integrated in the pavement of the square. They light up in reaction to sounds made at the square that are measured by sensors in large periscopes at the perimeter of the installation. This encourages passersby to experiment. By whispering, shouting, singing, or clapping, they can change the pattern of light on the square. It results in playful interactions, with passersby trying to control the pattern of light or enjoying the ability to influence it.

SMSlingshot is an example of an artistic intervention that addresses the commercialization of our everyday surroundings, leading to a domination of advertising messages instead of social or political interaction. In this installation, urbanites were given the opportunity to seize back their public space. Using a catapult, fitted with a mobile phone keyboard, they could type text messages and "sling" these onto the façade of an adjacent building.[g]

In some instances, the playground metaphor is used literally. The Canadian project *21 Balançoires*, for example, consists of a series of swings that make music when visitors move them back and forth. Coordinating their movements allows for particular compositions to emerge.[h] In other examples, variations on long-forgotten games and drama genres are used. Rafael Lozano-Hemmer's *Body Movies* installation, for example, uses the concept of 17th century shadow plays: passersby are invited to use their bodies to create silhouettes on a building's facade.

f. BruumRuum! David Torrents & artec3; LEDsCONTROL. https://summalab.com/BruumRuum.

g. SMSlingshot, VR/Urban - Christian Zöllner, Patrick Tobias Fischer, Sebastian Piatza en Thilo Hoffman. http://www.vrurban.org/smslingshot.html.

h. 21 Balançoires Daily Tous let Jours. https://www.dailytouslesjours.com/en/work/musical-swings.

While this free play provides an easy entry into the installation, there is a game mechanic hidden in the installation that is not immediately obvious. Players have to make an effort and really look or talk to others who already understand it and coordinate their moves to bring out a fresh background image.[i]

Personalization

The mechanism of "personalization" aims to contribute to a sense of belonging. It provides urbanites with ways to appropriate a space and claim it as theirs. It aids in navigating cities by bringing out relevant themes, locations, and connections while filtering out the noises, nuisances, and unfamiliar elements that can make public space inhospitable. Whereas sense of place and (playful) interaction as mechanisms are concerned with bringing out urban publics and relations between citizens, personalization addresses the reverse issue: how can urbanites—either as individuals or as collectives—stand their ground; how can they feel at home among strangers?

For a long time, critics and researchers have laid out how urbanites have developed a number of "defense mechanisms" to cope with the (over)stimulation of urban life in the modern metropolis. In the early 20th century, German sociologist Georg Simmel (1969) discerned a blasé attitude ("Simmels' Mask"). He noted that as a reaction to the information overload, citizens tend to withdraw into their private world in the middle of the public domain. For instance when they hide behind their newspapers in public transport, using the printed broadsheets to carve out a niche of personal space in a crowded environment. More recently the mobile telephone has been analyzed in a similar way. Ito et al. (2006) have shown how this device too can function as a "territory device" that allows users to create a private space in the public domain.

Other observers have pointed to more "offensive" tactics that urbanites have developed. Already in the 19th century, the French poet Baudelaire introduced *flânerie* as an attitude that allows urbanites to more or less intuitively select inspirational elements from the abundance of impressions in the city. Various philosophers have further contributed to the development of the now mythical figure of the *flâneur*. Walter Benjamin, for example, referred to the *flâneur*'s capacity for "absent-minded attention" (Boomkens, 1998). We usually perceive the multitude of impressions in the city unconsciously. At the same time, we have the ability to draw meaning from them and to select from all those impressions those that are relevant to us.

This is what the personalization mechanism aims to do: to help urbanites make a selection from a plethora of impressions and at the same time aid urbanites in carving out a place for themselves amidst the multitude of experiences. From a perspective of responsive media, this often takes the form of an app. The smartphone is a personal technology par excellence. It allows urbanites to obtain personalized information about a city and its inhabitants and can filter out

i. Body Movies Rafael Lozano-Hemmer. http://www.lozano-hemmer.com/body_movies.php.

those aspects that are of interest for the user. The other way around, visitors can use their mobile phone and its camera to capture an experience, make it their own, and share it online as an act of appropriation. It is a way to connect themselves with a space symbolically. There are also installations in public space that can operationalize the personalization mechanism. There are various tools or installations that allow users to mentally or physically isolate themselves temporarily in public space. Digital displays in public space could also give out personal information, and technologies such as eye-beacons can recognize and welcome users with personalized information. Although such an approach is not without its controversy, as it could also undermine the feeling of being at home. If responsive technology directly addresses an individual in public space, he or she is no longer anonymous. An individual might then feel observed or controlled by an invisible force.

There is a certain tension at play in this mechanism in relation to the experience of public spaces. Personalization allows urbanites to personalize their experience of the city and to select from the overwhelming choices on offer. While this makes the polyphony of the city habitable, it may also contribute to the further privatization of public space and the formation of parochial domains that are frequented by specific groups of people. Although this is a risk with regard to the successful functioning of public space, it does not necessarily need to have a negative effect. Tactics of personalization not only can be understood as a porous boundary that is raised temporarily but also could easily be lowered again. Looking up from the newspaper or mobile telephone for an instant is enough to burst the bubble. The very experience of such a temporary private space can provide people with the confidence and foothold they need to feel at ease in the public space.

Perhaps the most important design issue raised by this mechanism is how it could be operated in the spatial design. The interaction between users and their surroundings mostly occurs on the screen of a mobile telephone. How can the spatial design of the public space facilitate this dynamic? On the one hand, it is important for spaces to be organized in such a way that they can be appropriated by a variety of groups. Is it then possible to temporarily appropriate part of the space? On the other hand, specific cues in a space can make the presence of diverse publics possible or encourage users to appropriate the space online too. Think, in this regard, of the various selfie spots that are frequently photographed and shared via social media. The photograph might be considered an act of appropriation; the online sharing possibly contributes to a sense of place for a particular location or alerts future visitors to the attractiveness of that location.

This mechanism can also make it possible for a variety of groups to simultaneously use the same spaces. Groups of people with a similar identity or purpose do not necessarily need the symbolically loaded spatial cues that tell them that they are among each other and at home on their own turf. When the software of responsive media can forge a connection between people, they can meet up anywhere that is convenient for them. Following this logic, linked through an app or a website, various groups can arrange to meet at the same sites. A park becomes a meeting place for digitally organized and coordinated soccer teams, groups of friends having a barbecue, and parents with young children; a cafe becomes a venue where freelancers hold business meetings, school pupils do their homework, and two potential lovers encounter each other for the first time on a date arranged by a dating app. From within all these temporary bubbles, urbanites can in turn relate to each other. Through such a mechanism, also sites such as shopping centers or transit lounges that are often understood as nonplaces can be temporarily transformed into meaningful meeting places. Locations on or around transport hubs are particularly well suited for these sorts of digitally arranged meetings because they are easy to reach.

Routing and legibility

The routing and legibility mechanism addresses the ways in which responsive media can aid urbanites in their orientation on and navigation of public spaces. The term "legibility" refers to the seminal study *The Image of the City* by Kevin Lynch (1979), undertaken in 1960. In this study, Lynch looked at the ways in which city dwellers find their way through the city and make sense of their environments. Lynch found that urbanites construct mental maps of their cities, consisting of elements such as landmarks, edges, paths, and nodes. Legibility then is the extent to which these elements make the layout of and relation between places easy to grasp. This not only makes navigation easier but also opens up opportunities to make places more meaningful, as these elements become bestowed with symbolic meanings.

Ideally, Lynch argues that the design of such elements does not only make navigation easier, but also provides emotional security and a sense of (shared) identity.

In this, Lynch was not just interested in improving the ease of making routine trips. He also made the point that designers have a duty to encourage urbanites to explore new spaces outside their usual routines and familiar places. "The function of a good visual environment," wrote Lynch (1979, p. 109), "may not be simply to facilitate routine trips nor to support meanings and feelings already possessed. Quite as important may be its role as a guide and a stimulus for new exploration." In other words, making a location more legible should improve its quality as a public domain. A similar call has been made by Matos Wunderlich (2008), encouraging designers to invite urbanites to deviate from their purposive routes and engage them in a more discursively mode of exploring the city. As such the routing and legibility mechanism addresses all three public space qualities at the same time. It provokes inspiration and a reflective orientation and forges new relations by seducing urbanites to discover new territories, and it can induce a sense of belonging by the creation of landmarks and other visual elements that become meaningful over time.

Responsive media can activate the routing and legibility mechanism in a variety of ways. First, interactive installations can become meaningful landmarks themselves. *The Crown Fountain* in Chicago is an example of such an installation. The fountain, designed by Jaume Plensa, has become a central meeting point in the park at which people congregate. At the same time, it adds to the identity of the neighborhood, as the fountain displays brief, slow-motion video films of neighborhood residents.[j] Spatially the mechanism offers the possibility to link and familiarize a particular public space (i.e., square of street) to the quarters or cities larger network of public spaces, both visually and in terms of identification or orientation.

Other responsive media installations guide urbanites through the city, dynamically altering the routing information provided. This could vary from the prosaic traffic information systems to more poetic interventions in the spatial architecture highlighting a particular route through public space. Interactive kiosks and information pillars can offer suggestions and help visitors to orientate themselves in their surroundings. Experiments are taking place in Japan with robots that guide visitors around a shopping center. Studio Roosegaarde's *Van Gogh Path* is an example of a poetic approach. A cycle lane near the Dutch village of Nuenen (Van Gogh's residence during a part of the 1880s) is illuminated with thousands of reflective stones, highlighting the route itself and linking it to Van Gogh's painting Starry Night.[k]

Digital apps too play an increasingly important role as navigation tools. Apps such as Google Maps or TomTom navigation systems not only provide routing information but also can help urbanites to search for specific places like restaurants, museums, or shops. They often make use of live traffic data and public transport information to guide users in the most efficient way to these locations. However, this efficiency sometimes runs counter to the quality of the experience of public space. In our research, we found that tourists looking for a hotel at the ArenA Boulevard are guided through the unpleasant dark alleys at the rear side of the boulevard instead of along the spacious boulevard itself. This observation can be linked to a wider spread criticism of GPS, claiming that it could undermine our ability to get our bearings ourselves and build meaningful relationships with the spaces we traverse and make use of. Instead the argument went that people are slavishly following instructions, forgetting to look around.

Some apps and installations try to reverse such behavior. Many of these draw upon the rich legacy of the derive as a playful approach to navigating cities introduced by the situationists, a French avant-garde movement in the 1960s. Mark Shepard's *Serendipitor* for instance is a navigation app that provides users with interesting detours and playful interventions. Others use game mechanics to encourage people to explore public spaces. *Urban Code*, for instance, is a

j. Crown Fountain, Millennium Park Chicago, VS, Krueck + Sexton architects in collaboration with Jaume Plensa. 2004. https://jaumeplensa.com/works-and-projects/public-space/the-crown-fountain-2004.

k. Van Gogh Path, Studio Roosegaarde. https://www.studioroosegaarde.net/project/van-gogh-path.

game that encourages users to find specific art objects attached to walls in public spaces. Playing the game unfolds as taking a discursive walk through the city to complete one's collection.[l]

While the routing and legibility mechanism draws the attention to the affordances of responsive media to help people create meaningful relationships with their surroundings and can invite explorative and discursive modes of navigation, critics point out that these same affordances can also be used to undermine some of the qualities of public spaces.

To what extent are the algorithms guiding urbanites around controlled by commercial interests resulting in a focus on customized offers from these parties? And to what extent could this mechanism also contribute to "software sorting," a process in which people mainly link up with likeminded? These are important issues to address in the operationalization of this mechanism in the design of responsive media.

Control

Control is a rather controversial mechanism through which responsive media aim to control, regulate, or nudge social behavior in public spaces. This could contribute to better managed, safe, and more enjoyable places. At the same time, it could also severely limit the publicness of public space, by the introduction and enforcement of strict rules that may exclude particular behavior and groups and undermine values such as privacy. The use of this mechanism also raises serious questions. Who decides what constitutes desirable behavior? Is it not the essential quality of the public domain that it provides an enormous freedom for diverse lifestyles and political views to be represented? Doesn't such an approach contribute, above all, to even more privatization and commercialization of public spaces?

In a "light" version, this mechanism is operationalized in responsive media that stimulates particular behavior to keep public spaces clean and safe, often making use of "gamification" elements. For instance, *Street Pong*—also known as *ActiWait*—is a game that people can play while waiting at a red pedestrian traffic light. On both sides of the street, screens are installed at the traffic light, encouraging people to battle those at the opposite side of the street in a short game.[m] The goal of course is to encourage people to respect the red light. Similarly the project *Tetrabin* rewards people who throw away their garbage in the specifically designed trash cans. These are outfitted with screens on the outside, and when litter is thrown into the rubbish bin, Tetris-style blocks appear and animate.[n] As a last example in this category, *Northside beacons* was an installation deployed at the Northside festival in Denmark. Sixteen tall light

l. Serendipitor, Mark Shepard. http://serendipitor.net/site/?page_id=2
m. ActiWait, Indiegogo. https://www.indiegogo.com/projects/actiwait-a-smart-traffic-light-button#/
n. Tetrabin, Sencity Corporation. http://www.tetrabin.com/.

beacons spanning the entire length of the festival ground would flash when litter was deposed at the designated garbage cans, expressing a brief "thank you."[o]

A more extensive version of this mechanism can be found in the *Stratumseind* living lab. Cameras and sensors have been installed to map the behavior of nighttime visitors to the various restaurants and cafes in a bar street in the center of the Dutch city of Eindhoven. Various data are collected, ranging from the occupancy rate of the parking garages to the noise level on the street. Together, they provide an impression of the mood in the area. This is monitored from a central coordination point to take measures when necessary. For instance, research is currently under way whether these data can be used to adjust the color and intensity of the street lighting, with the aim of influencing the atmosphere in the area to reduce violence.[p]

Experiments using the control mechanism in China go much further. Cameras equipped with facial-recognition software record pedestrians who cross the street when the light is red. In Shenzhen, they are then publicly shamed: their photographs, together with their names and national identity numbers, appear on large screens in the public space. Next, offenders receive an automatic fine via SMS (Baynes, 2018). Similar experiments are being conducted in the Netherlands. Rotterdam's RET public transport operator was experimenting with facial-recognition software in trams and buses as early as 2011. If a camera recognizes a passenger who has been barred from using public transport due to previous misbehavior, RET staff receive an alert so that they can take action against the passenger (Van den Dool, 2011).

At the level of the city, this mechanism can be found in the many "urban dashboards" that have been introduced in various cities around the world (Kitchin et al., 2015). One of the best known examples in this category is the *Centro De Operações Prefeitura Do Rio*. This control room was set up by the city of Rio de Janeiro together with IBM in the run-up to the 2016 Olympic Games. Various information flows about the city—from the traffic situation on main roads to the weather forecast—are projected onto large screens in a hall. Officials from various municipal services monitor these data streams and can intervene if required.

Many of these measures are introduced to improve security in public spaces, but there is also considerable criticism of their use. They are part of a broader development that has seen public spaces increasingly dominated by control, with targeted groups often being excluded. Take, for example, "bum-proof" benches, designed to make lying on them uncomfortable. Critics argue that that these measures contribute to development in which public spaces are increasingly dominated by comfort and consumption; behaviors or publics that are thought to undermine these functions are discouraged from using the space. This could lead to public space losing its character as an inclusive meeting place, as a stage where all urbanites can be present, or as a place for political resistance. In short,

o. Kollision Northside Beacons. https://kollision.dk/en/northside-beacons.

p. Stratumseind Living Lab. https://www.tue.nl/universiteit/faculteiten/bouwkunde/onderzoek/smart-cities-program/collaboration/living-labs/stratumseind/.

this mechanism always involves a fine line between on the one hand enhancing the attractiveness of public spaces by making them safer and more comfortable and disciplining and even excluding visitors on the other.

Discussion

The perspective that we have taken carries a certain risk. We do not want to argue for an easy "solutionism" (Morozov, 2013) that through interactive technology we can restore the relation between *ville* and *cite*, magically transferring places that are referred to as "urban deserts" into vibrant public spaces, if only we had a responsive installation. Rather, we wanted to explore the qualities of responsive technologies as a new building block, unlocking and adding up to the traditional instruments of urban design. As such the point we want to make is that interactive installations or services should not be added as an afterthought, a quick patch to bring some liveliness to a lifeless place. Rather, we argue for urban design that starts from an integrated perspective and takes the notion of urban public space as its point of departure.

A prerequisite to the creation of responsive urban spaces that are able to animate public domain qualities in public spaces is the equal involvement of both spatial and interaction designers in the design process. Spatial design is necessary to embed the responsive technologies as a spatial element and spatial solution, instead of a singular object or artifact at a location. Similarly, interaction design is essential to design and materialize the mechanism of responsiveness. Moreover, as we found in our research, the involvement of local stakeholders and sociospatial analysis of the space in the design process are important as well, as the design takes place in existing urban spaces, building on top of memories, experiences, and realities. As a result the design process is one of cocreation. Yet, this blend of professions lacks a culture of collaboration.

Our five mechanisms are meant to help out here as a shared vocabulary, focused on public space qualities to which responsive media could contribute when integrally designed as part of the spatial design of the site. Again, they are not meant as easy solutions, but rather as a heuristic that can aid in the discussion on how public space qualities can be brought out in an integrated design approach. This is no easy process. As described earlier, many of the mechanisms carry a tension inside of them with regard to the qualities of public space. Control can make spaces more safe and pleasurable, but it could also exclude. Routing and legibility not only can help in the discovery of new territories but also can reify sociospatial stratifications. Playful interaction can bring out new social relations but also can play into an agenda that highlights public space as an exclusive leisure space for the creative class.

We hope that his heuristic can also contribute to future HCI research in relation to responsive public spaces and urban interaction design. It could complement a usability perspective with a framework that relates particular design patterns to their affordances in relation to public domain qualities.

Acknowledgments

This chapter is based on the findings of our research project Co-Creating Urban Spaces executed at the Amsterdam University of Applied Sciences, in partnership with offices for architecture and urban design, interaction design, developers, the city of Amsterdam, and stakeholders at the ArenA Boulevard. The outcomes have been described in more detail in the final report of this study by Suurenbroek et al. (2019) *Responsive Public Spaces. Exploring the use of interactive technology in the design of public spaces.* Amsterdam: Amsterdam University of Applied Sciences. Some of the insights from that report have been used here verbatim. The research project was funded through a grant from SIA-RAAK. See: www.responsiveurbanspaces.amsterdam.

References

Auge, M., 1995. Non-places: Introduction to an Anthropology of Supermodernity. Verso, London.

Aurigi, A., 2013. Reflections towards an agenda for urban-designing the digital city. Urban Des. Int. 18 (2), 131–144. https://doi.org/10.1057/udi.2012.32.

Aurigi, A., Cindio, D. (Eds.), 2008. Augmented Public Spaces. Articulating the Physical and Electronic City. Routledge, London.

Baynes, C., 2018. Chinese police to use facial recognition technology to send jaywalkers instant fines by text. Independent. March 29. https://www.independent.co.uk/news/world/asia/china-police-facial-recognition-technology-ai-jaywalkers-fines-textwechat-weibo-cctv-a8279531.html.

Berman, M., 1987. All That Is Solid Melts into Air: The Experience of Modernity. Verso, New York.

Blokland, T., 2006. Het sociaal weefsel van de stad: Cohesie, netwerken en korte contacten. Oratie Erasmus Universiteit Rotterdam.

Blokland, T., Savage, M., 2008. Social capital and networked urbanism. In: Blokland, T., Savage, M. (Eds.), Networked Urbanism Social Capital in the City. Ashgate, Burlington, pp. 1–22.

Boomkens, R., 1998. Een drempelwereld: Moderne ervaring en stedelijke openbaarheid. NAi Uitgevers, Rotterdam.

Brenner, N., Theodore, N., 2002. Cities and the geographies of "actually existing neoliberalism". Antipode 34 (3), 349–379.

Brynskov, M., Carvajal Bermúdez, J.C., Fernández, M., Korsgaard, H., Mulder, I., Piskorek, K., Rekow, L., Waal, M., D., 2014. Urban Interaction Design: Towards City Making. UrbanIxD/Booksprints. http://booksprints-for-ict-research.eu/wp-content/uploads/2014/06/Urban_Interaction_Design_Towards_City_Making.pdf.

Cantrell, B.E., Holzman, J., J., 2015. Responsive Landscapes: Strategies for Responsive Technologies in Landscape Architecture. Routledge, Londen/New York.

Castells, M., 2002. The culture of cities in the information age. In: Susser, I. (Ed.), The Castells Reader on Cities and Social Theory. Blackwell Publishers, Malden, MA, pp. 367–389.

Claes, S., 2017. Public Visualization: Engaging Citizens with Data through Situated Public Displays (PhD thesis). KU Leuven, Leuven.

Dalsgaard, P., Fatah, A. (Eds.), 2014. Proceedings of the 2nd Media Architecture Biennale Conference: World Cities. Vol. 1. ACM, Aarhus.

Dalsgaard, P., Fatah gen Schieck, A. (Eds.), 2016. Proceedings of the 3 Media Architecture Biennale Conference. Vol. 1. ACM, Sydney.

de Hoog, M., 2012. De Hollandse metropool. Ontwerpen aan de kwaliteit van interactiemilieus. Thoth, Bussum.

De Souza e Silva, A., 2006. From cyber to hybrid: mobile technologies as interfaces of hybrid spaces. Space Cult. 9 (3), 261–278.

de Waal, M., 2014. The City as Interface. Nai010 Publishers, Rotterdam.

Finka, M., 2006. In: Zschocke, F.E.M. (Ed.), Mediacity – Theoretical and Conceptual Contexts. Verlag der Bauhaus Universität, Media City Weimar.

Florida, R., 2005. Cities and the Creative Class. Routledge, New York.

Foth, M., 2017. The software-sorted city: big data & algorithms. In: Odendaal, N., Aurigi, A. (Eds.), Proceedings of Digital Cities 10: Towards a Localised Socio-Technical Understanding of the 'Real' Smart City. Troyes, France, 26 June 2017.

Foth, M., Sanders, P., 2008. Impacts of social computing on the architecture of urban spaces. In: Aurigi, A., Cindio, D. (Eds.), Augmented Public Spaces. Articulating the Physical and Electronic City. Routledge, London.

Foth, M., Mitchell, P., Estrada Grajales, C., 2018. Today's internet for tomorrow's cities: on algorithmic culture and urban imaginaries. In: Hunsinger, J., Klastrup, L., Allen, M. (Eds.), Second International Handbook of Internet Research. Springer.

Frenchman, D., Rojas, F., 2006. Zaragoza's digital mile: place-making in a new public realm. Places 18 (2), 16–25.

Giddens, A., 1984. The Constitution of Society: Outline of the Theory of Structuration. Polity Press, Cambridge.

Goldberger, P., 2003. Disconnected Urbanism. Metropolismag.com. November.

Graham, S., 2005. Software-sorted geographies. Prog. Hum. Geogr. 29 (5), 562–580.

Gumpert, G., Drucker, S., 2001. Privacy, predictability or serendipity and digital cities. In: Kyoto Workshop on Digital Cities. Springer, Berlin, Heidelberg, pp. 26–40.

Hajer, M., Reijndorp, A., 2001. Op zoek naar nieuw publiek domein. NAi Uitgevers, Rotterdam.

Hespanhol, L., 2018. Making meaningful spaces: strategies for designing enduring digital placemaking initiatives. In: The 1st International Conference on Design, Innovation, and Creativity, Bangkok, Thailand.

Hespanhol, L., Haeusler, H., Tomitsch, M., Tscherteu, G. (Eds.), 2017. Media Architecture Compendium: Digital Placemaking. Avedition, Stuttgart.

Ito, M., Okabe, D., Matsuda, M., 2006. Personal, Portable, Pedestrian: Mobile Phones in Japanese Life. MIT Press.

Jacobs, J., 1961. The Death and Life of Great American Cities. Random House, New York.

Kitchin, R., Lauriault, T.P., Mcardle, G., 2015. Knowing and governing cities through urban indicators, city benchmarking and real-time dashboards. Reg. Stud. Reg. Sci. 2 (1), 6–28. https://doi.org/10.1080/21681376.2014.983149.

Lofland, L.H., 1973. A World of Strangers: Order and Action in Urban Public Space. Prospect Heights. Waveland Press.

Lofland, L.H., 1998. The Public Realm: Exploring the City's Quintessential Social Territory. Aldine de Gruyter, New York.

Lynch, K., 1979. The Image of the City Cambridge. MIT Press.

Markopoulou, A., Farinea, C., Marengo, M. (Eds.), 2017. State of the Art and Best Practices Collection. Active Public Space. Institut d'Arquitectura Avançada de Catalunya, Barcelona.

Matos Wunderlich, F., 2008. Walking and rhythmicity: sensing urban space. J. Urban Des. 13 (1), 125–139.

McQuire, S., 2008. The Media City: Media Architecture and Urban Space. Sage, Thousand Oaks.

Morozov, E., 2013. To Save Everything, Click Here: The Folly of Technological Solutionism. PublicAffairs, New York.

Mumford, L., 1937. What is a city. Archit. Rec. 82 (5), 59–62.

Pop, S., Toft, T., Calvillo, N., Wright, M., 2016. What Urban Media Art Can Do: Why When Where & How. Avedition, Stuttgart.

Rabari, C., Storper, M., 2014. The digital skin of cities: urban theory and research in the age of the sensored and metered city, ubiquitous computing and big data. Camb. J. Reg. Econ. Soc. 8 (1), 27–42.

Ratti, C., Claudel, M., 2016. The City of Tomorrow. Sensors, Networks, Hackers, and the Future of Urban Life. Yale University Press, New Haven & London.

Sassen, S., 1991. The Gllobal City. Princeton Univeristy Press, Princeton.

Sennett, R., 1974. The Fall of Public Man. W.W. Norton & Company, New York.

Sennett, R., 2018. Building and Dwelling: Ethics for the City. Farrar, Straus and Giroux, New York.

Simmel, 1969. The metropolis and mental life. In: Sennett, R. (Ed.), Classic Essays on the Culture of Cities. Appleton-Century-Crofts, pp. 47–60.

Sorkin, M. (Ed.), 1992. Variations on a Theme Park: The New American City and the End of Public Space. Hill and Wang, New York.

Suurenbroek, F., Nio, I., de Waal, M., 2019. Responsive Public Spaces. Exploring the Use of Interactive Technology in the Design of Public Spaces. Amsterdam University of Applied Sciences: Urban Technology Series, Amsterdam.

Tironi, M., 2010. Gelleable spaces, eventful geographies. The case of Santiago's experimental music scene. In: Farias, I., Bender, T. (Eds.), Urban Assemblages. How Actor-Network Theory Changes Urban Studies. Routledge.

Tordoir, P., Poorthuis, A., Renooy, P., 2015. De veranderende geografie van Nederland. Ruimtelijk atelier Tordoir, Regioplan, Amsterdam.

Van den Dool, P., 2011. Proef met gezichtsherkenning in Rotterdams openbaar vervoer. September 10, NRC.nl. https://www.nrc.nl/nieuws/2011/09/10/proef-met-gezichtsherkenning-in-rotterdams-openbaar-vervoer-a1453041.

Van der Zwaard, J., 2010. Scènes in de copy corner: Van vluchtige ontmoetingen naar publieke vertrouwdheid. SUN Trancity, Amsterdam.

Van Engelsdorp Gastelaars, R.E., Hamers, D.A.F., Bletz, J.C.F., 2006. De nieuwe stad: stedelijke centra als brandpunten van interactie. NAi Uitgevers.

Wellman, B., et al., 2003. The social affordances of the internet for networked individualism. J. Comput. Mediat. Commun. 8 (3).

Widmer, S., 2016. Experiencing a personalised augmented reality: users of foursquare in urban space. In: Amoore, L., Piotukh, V. (Eds.), Algorithmic Life: Calculative Devices in the Age of Big Data. Routledge, London, pp. 57–71.

Willis, K., 2016. Netspaces Space and Place in a Networked World. Routledge, New York.

Willis, K.S., Aurigi, A., 2011. Hybrid spaces: presence, rhythms and performativity. In: 2011 Seventh International Conference on Intelligent Environments. IEEE, pp. 100–106.

Author biographies

Martijn de Waal is a Lector (professor) in the Play and Civic Media Research Group at the Amsterdam University of Applied Sciences. His work focuses on the experience and design of public spaces in a networked society. He is the author of The City as Interface (Nai010 Publishers, 2014), The Platform Society (Oxford University Press, 2018—with Jose van Dijck and Thomas Poell), and with Michiel de Lange he edited The Hackable City (Springer 2018). He collaborated with Frank Suurenbroek and Ivan Nio in the research project Co-creating Responsive Urban Spaces (2017–19). With Frank Suurenbroek, Michiel de Lange, and Nanna Verhoeff, he is part of the executive committee and general chair of the Media Architecture Biennale 2020.

Frank Suurenbroek is a Professor (lector) of Spatial Urban Transformation at the Amsterdam University of Applied Sciences, Faculty of Technology. His chair conducts multiple action-research projects on urban transformation, with a focus on the relation between the physical and the social fabric. Recent research takes a neuro-architectural approach (sensingstreetscapes.com). He collaborated with Martijn de Waal and Ivan Nio in the research project Co-creating Responsive Urban Spaces (responsiveurbanspaces.amsterdam). With Martijn de Waal, Michiel de Lange, and Nanna Verhoeff, he is part of the executive committee of the Media Architecture Biennale 2020/21.

Ivan Nio is a Senior Researcher at the Amsterdam University of Applied Sciences (Faculty of Applied Social Sciences and Law). He obtained his PhD in social sciences at the University of Amsterdam. In his research and publications, he has explored diverse themes on the interface of urban design and urban sociology. He collaborated with Frank Suurenbroek and Martijn de Waal in the research project Co-creating Responsive Urban Spaces.

Chapter 4

Smart plays

Ben van Berkel[a,b,c,d,e], **Filippo Lodi**[a,f,g,h], **and Wael Sami Batal**[a,i,j]
[a]*UNStudio, Amsterdam, Netherlands,* [b]*UNSense, Amsterdam, Netherlands,* [c]*Rietveld Academy, Amsterdam, Netherlands,* [d]*Architectural Association, London, England,* [e]*Harvard University Graduate School of Design, Cambridge, MA, United States,* [f]*MSc Civil Engineering and Architecture, Università di Bologna, Bologna, Italy,* [g]*Architecture, MA Advanced Architectural Design, HfBK Städelschule, Frankfurt am Main, Germany,* [h]*MBA, Universiteit van Amsterdam, Amsterdam, Netherlands,* [i]*University of California, Los Angeles, CA, United States,* [j]*University of Colorado, Boulder, CO, United States*

Chapter outline

During the digital revolution in architecture roughly a quarter century ago, the influence of digital technology extended deeply into the structures of society. Through mass production and automated fabrication enabled by computational design, it fueled years of intense inquiry in the paradox between the virtual and the physical.

As in David Cronenberg's 1999 film *eXistenZ*, where characters plug into a video game environment that is a virtual construction of reality, virtual and physical space were put in paradoxical tension to a point of near collapse as one environment was indistinguishable from the other in a deceptive cinematic experience of reality. Beyond the more mainstream sci-fi of the legendary *Matrix* (1997), *eXistenZ* placed viewers in a world where the natural and artificial were not separate conditions. While the film remains one of the most important reflections on the cultural anxiety associated with this collapse, it was a reflection of a mere possibility in some distant future, only science fiction.

Today, in the wake of the Internet age, innovations in sensor-based technologies, artificial intelligence and machine vision, have thrust the world into a paradigm where this possibility is now an actuality. The gap between the digital and the physical world as we know it is all but entirely blurred.

Shaping Smart for Better Cities. https://doi.org/10.1016/B978-0-12-818636-7.00002-0

With the embedding of sensory technologies, Internet of Things, data mining, and social media, the influence of digital technology on society today is spread deeply and literally into the most intricate fabrics of everyday life. What was once abstract data is now a material condition that can be owned, produced, and consumed.

Further, the speed with which this paradigm evolves is unparalleled by the previous industrial revolutions, making it increasingly challenging for society, public policy, economics, and essentially every industry to catch up and participate in shaping these technologies, rather than to be shaped by them. The effects of this challenge are arguably nowhere more visible than in current technocratic visions and rhetoric of "smart cities."

Many smart city initiatives vigorously promote a new model of infrastructure and urban planning that allow the city to operate with the highest degree of ***efficiency*** through the agglomeration of sensors, data analytics, machine vision, and deep learning. The focus of these initiatives is primarily the sustainable usage of resources, transportation, health care, and security, and therefore purports a structure within which societies can exist in the highest degree of harmony.

As urbanization rates continue to increase and the availability of natural resources continues to decrease due to climate change, the efficient monitoring and distribution of resources in densely populated cities are indeed among the most exigent issues of our time.

However, smart city initiatives are often broad-scaled gestures of good intentions and focus solely on the functionality of cities. In many ways, they are strikingly similar to the functionalist doctrines of the CIAM congress led by Le Corbusier, Walter Gropius, and many others between 1928 and 1956. With ideas extending from the Radiant City to the Athens Charter, such grand narratives for the *ideal* city of the future promised to secure quality of life for all through the ordered structuring of the city by splitting its functions into separate districts to ensure efficiency.

Many architects such as Aldo van Eyck—a prominent figure of the *Team 10* group that sought to subvert CIAM ideology—rejected this model of "top-down" planning in favor of in-between spaces that support "bottom-up" urban development.

In a reaction against the postwar functionalist city plan of Cornelis van Eesteren (longtime president of CIAM) for Amsterdam that aimed to partition the city into zones based on their function (separating recreation, living, and work), van Eyck designed nearly 700 playgrounds in the underutilized spaces of the city to transform them into ***zones of play***, socialization, growth, and creativity in children. In this sense, they took on a much larger project than just the boundaries of their sites; they were part of a coordinated urban and sociocultural intervention through which everyday life would be transformed. In the context of the highly ordered and rigid organization of van Eesteren's vision for the city, van Eyck's playgrounds were a strategy to splice informality, serendipity, learning, and growth as social ***activators within the network*** of the city.

If we consider this recent history of the CIAM congress' functionalist visions and its subsequent dismantling by projects such as van Eyck's playgrounds, "smart" cities are at risk of being complicit in a past ideology of *modernity* rather than a model for the future.

FIG. 1 Playgrounds designed by Aldo van Eyck for the city of Amsterdam. (A) Igloo and (B) climbing mountain. *(Images: Denisa Kollárová/Seventeen Playgrounds.)*

While "smart" city initiatives capitalize on immense amounts of extracted data to allow cities to function as optimally efficient operating systems, they may only offer the realization of a pragmatist dream. They have yet to consider the complexity of social interaction and the changing relations between people in society, and they often overlook the significance of infrastructures that support unpredictable human behavior such as ***creativity*** and ***experimental learning*** in a healthy society. As in Cedric Price's Fun Palace or Archigram's Plug-In City projects, technology can be instrumentalized to allow for publicly modifiable spaces and environments.

Rather than simply providing open spaces for passive leisure, green park space, or public spaces in a city, spaces of ***play***—as in van Eyck's playgrounds and orphanage in Amsterdam—offer a far more active way to engage with the inhabitants of a city (Fig. 1). While park space allows for people to connect with nature, play spaces allow for the leisurely development of cognition and imagination. ***Play*** is an essential function of society and for cultural production at large in that it provides the environment for free experimentation, abstraction, concept generation, and learning.

Therefore it is the imperative of this paper to assert the value of ***play spaces within the city*** and to address the activity of play more seriously as we explore future possibilities of healthy, resilient, and responsive cities that grow and evolve over time. As made visible by artworks throughout history and by the video game culture of today, play is intensely interwoven within the organizational networks of society. In parallel, play can affect society through infrastructures of experimental learning that encourage public participation and engagement in cultural change. In this sense, play is capable of both representing and affecting social conditions for cities in the future.

Play in culture and society

From 17th-century Dutch Golden Age paintings such as Dirk Hal's *Children Playing Cards* (1631) or Dirk van Santvoort's *Boy in White* (1661), the notion that play affects, interprets, and represents society can be seen in the way children are portrayed. Changes in the modern notion of *childhood* as a concept (in and of itself) were simultaneously explored through the allegory of domesticity, where complex ideas about social structures are visible in family life and in the household. The contest between ***liberty*** and ***obedience*** in the moral education of the child was of paramount importance to a broad understanding of society (Boult, n.d.). In Jean Siméon Chardin's *Boy with a Spinning Top* (1738), this contest is articulated with an elegant sleight of hand. Here the child is dressed as an adult while staring inquisitively and mischievously at the spinning top, wondering if his behavior is becoming of his formal attire.

The idea that play is deeply engrained within the structures of society can also be seen in Constant Nieuwenhuys' speculative New Babylon project from 1956 to 1974, which extends Johan Huizinga's philosophy of play as a cultural phenomenon. In his seminal text *Homo Ludens,* Huizinga claims that "...*play* is older than culture, for culture, however inadequately defined, always presupposes human society, and animals have not waited for man to teach them their playing...in play, there is something 'at play'..." (Huizinga, 1949). Huizinga focuses on contests, races, performances, tournaments, pageants, and masquerades, as forms of social play rather than pure playfulness, and offers some essential characteristics: "Play is *free*, is not ordinary or *real* life, it creates *order, is not connected to material interest or profit*," and "... it promotes the formation of social groupings which tend to surround themselves with secrecy and to stress their difference from the common world by disguise or other means" (Huizinga, 1949). To pay particular attention to this *secrecy* sheds light on current models of play and conditions of society today as this "difference from the common world" is no more apparent in virtual play, video games, augmented/mixed reality gaming, and a plethora of new models of play in the gaming industry at large. While much research indicates that virtual gaming is isolating due to this detachment from the real world, the *Develop* game developer conferences in Brighton, UK, show a keen interest in the opposite; the future of the virtual play industry is social, physically collaborative, and focused on "feel engineering" where the feelings of a player drive the overall design of the game dynamics (Stuart and Webber, 2015). Furthermore, online games such as Minecraft, Fortnite, Moshi Monsters, and Club Penguin operate just as much as a game as they operate as social media platforms, which makes the experience of digital gaming far less of an individual and solitary experience.

Scaffolds of experimental learning and play

According to Article 31 of the UN Convention on the Rights of the Child, ***play*** is a fundamental right of children, and for good reason, it is crucial for healthy brain development and the production of new knowledge.

Current research on the role of play in the neurological development of children shows that it not only increases problem-solving abilities and motor skills through physical training but that a major function of play is "to ***build pro-social*** brains…that know how to interact with others in positive ways" (Jaak Panksepp, neurological researcher at Washington State University, 2014) (Hamilton, 2014). Through the experience of learning to negotiate the rules of games among each other, children actively develop their social skills and abilities to collaborate. Social-pretend play also allows children to explore various social roles and act out characters—and characteristics—that they observe from adults around them.

While an essential element of play is that it is self-initiated and unrestricted *free play*, studies show that it is most beneficial to the development of a child and the growth of their knowledge when it is ***scaffolded*** by a "reiterative process… in which new skills are built on previous skills and are facilitated by a supportive social environment" (Yogman et al., 2018). One form of scaffolding is the guidance of an adult whose role is to support a child to do something that they otherwise could not do on their own. Rather than to instruct a child with a rigid structure, strict rules, and passive memorization, a child's creativity, growth, and development are better encouraged by an environment that actively engages their own desire for growth and their imagination through experimental learning (Yogman et al., 2018). This scaffold can take various forms and act in various stages of cognitive development. For example, it can be a social structure upon which a person experimentally tests ideas for resonance and reinforcement. More broadly, this scaffolding can also take the form of what philosopher Immanuel Kant called a ***schema***, which is a framework through which one structures learning and the way that they perceive the world. Sociological, political, and philosophical inquiry therefore—as in Dutch Golden Age paintings and Huizinga's research—can be understood at the scale of the child and the broader scale of society at large through the concept of play and learning.

In the educational and psychological development models developed by prominent psychologist Lev Vygotsky, whose work on the zone of proximal development is widely influential on contemporary theories of play-based experimental learning, this scaffolding takes the form of a "more knowledgeable other" (MKO). While the MKO is typically an older and more experienced teacher or parent, it can also be someone younger. Importantly, it can also be a machine, book, or other medium (Yogman et al., 2018).

The effects of playful environments for experimental learning throughout the city, therefore, cannot be overstated. It is an active binding element that facilitates social cohesion between children from a young age, which then extends to adolescence and further into adult life. In terms of physical health, it reduces obesity and related diseases in children and helps them manage stress, recover from trauma, and build essential learning skills by improvisational play (rather than instruction) and helps families to bond (Yogman et al., 2018). Noting play as essential to children's development, the International Play Association (IPA) Declaration of the Child's Right to Play in 1977, which was later adopted by the General Assembly of the United Nations in 1989, asserted that children have a right to "leisure, play, and participation in cultural and artistic activities" (International Play Association, n.d.).

We argue that it is the role of architecture and the city to facilitate, support, and "scaffold" the ***participatory engagement*** of people—of all ages—in producing new knowledge, schema, and perceptions about the world. In the following projects, this scaffolding is materialized into architecture that guides a participating audience into various play-based learning environments—some more physical, some more digital, and some in between.

Cable car projects: Gothenburg, SE; IJbaan Amsterdam, NL; Blagoveshchensk terminal, RU

Some of the essential elements of play in society are for people to freely perceive, engage with, and learn about their environment, both physical and digital. For the city of Gothenburg in Sweden, we designed an infrastructural network of six cable car towers and four stations to provide precisely this form of engagement.

As the first new mode of public transportation in Sweden since their subway system was built in the 1930s, the 3-km-long cable car line would offer aerial shortcuts across the river and a continuous stream of vantage points to experience the city from an aerial view.

Whereas observation decks on tall structures in cities often provide a panoramic view from a fixed point, the experience of the city from the cable car is far more cinematic as it provides a series of views along a journey that undulates in height with the curvature of the cable.

The pylon towers themselves were designed through a process of learning from the city. Drawing from the history of Gothenburg's skyline, the pylons playfully reference the city's shipyard cranes while articulating them in a new way with asymmetrically triangulated structures (Fig. 2).

In our home city of Amsterdam, the IJ is a body of water that is not only a source of symbolic pride for the region but also marks the geographic boundary between the fast-growing Amsterdam Noord and the city center south of the IJ. The IJbaan Cable Car project is a system of three slender pylon towers that span one and a half kilometers across the IJ, connecting Amsterdam West with Amsterdam Noord.

While the city is well known for its many forms of transportation—bicycles, street-level trams, underground subway, and regional trains—connectivity to the residential areas of northern Amsterdam is needed as the region continues to grow. The cable car cabins could carry 30 passengers on a 4.6-min trip across the IJ at 21.6 km per hour (13.4 miles per hour) (Figs. 3 and 4).

FIG. 2 Gothenburg cable car. (A) Pylon and (B) terminal building. *(Images: UNStudio/Plomp.)*

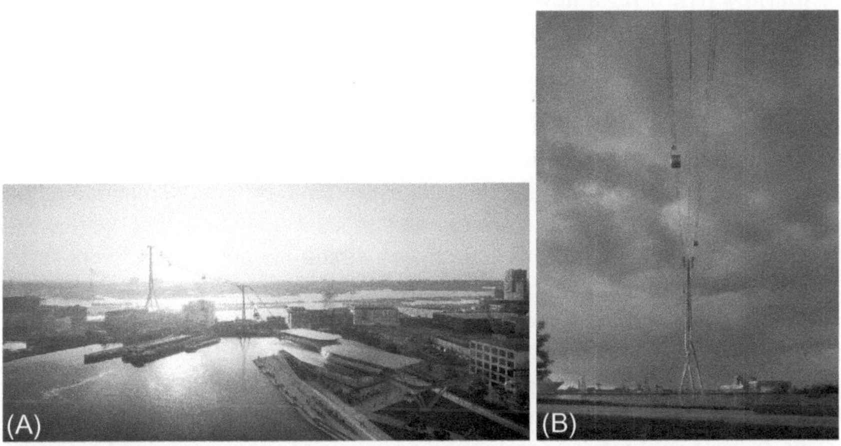

FIG. 3 IJbaan cable car, Amsterdam, NL. (A) Aerial view and (B) pedestrian view. *(Images: UNStudio/Plomp.)*

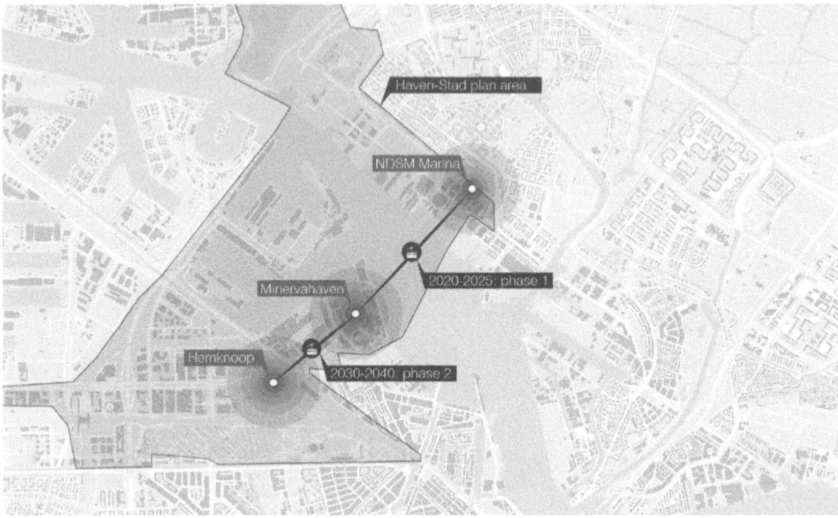

FIG. 4 IJbaan cable car site plan diagram, Amsterdam, NL. *(Image: UNStudio.)*

Between the Russian city of Blagoveshchensk and the neighboring Heihe in China lies the Amur River, which has defined the natural boundary between the two countries since the Treaty of Aigun in the mid-19th century. As one of the largest rivers in the world extending nearly 3000 km, it is a symbol of social connection in the countries' growing relationship. To further encourage social, cultural, and commercial exchange, a new cable car system will bridge across the 750-m-wide river, reducing the distance to a short 8-min journey. Similar to the Gothenburg and IJbaan projects, the journey will offer a new form of mobility that combines efficient connection with a playful and picturesque experience of Russian and Chinese landscapes.

For this project the terminal building reconstitutes the notion of "international" architecture and the notion of efficient infrastructure. The design fluidly combines highly efficient transit functions of passport control and customs, with meandering commercial and gastronomic programs, punctuated by spaces for pause, reflection, learning, and social connection. The visitor experience flows along a path that alternates directions and successively turns toward Russia and then toward China.

Weaving through three elongated volumes of arrivals, departures, and retail programs, travelers are drawn from end to end by an episodic sequence of curated views that frame particular moments of each city. The three volumes are stacked vertically and rotated toward the views, which creates a series of exterior roof terraces and interior pockets of space reserved for tourist

FIG. 5 Blagoveshchensk cable car terminal, RU. (A) Aerial view and (B) pedestrian view of main entrance. *(Image: UNStudio/Pyxid.)*

information points, tickets, galleries of cultural artifacts, and spaces for public events and lectures.

Through the playfully reiterative process that exposes travelers to framed views of various moments of each city, the terminal building is designed as an engine that scaffolds the development for prosocial relationships between two societies (Fig. 5).

In combination with a highly efficient form of urban mobility, the cable car systems for Gothenburg and Amsterdam and the international terminal building in Blagoveshchensk become sites of ***continuous learning*** and act as a ***scaffolding*** for individuals to constantly reevaluate and reconstruct their perceptions of the city. As such, they are ***social activators in the network*** of the city that engage the public through the concept of play.

Gow Nippon Moon: Japan

The Nippon Moon is a wheel that offers a place to observe and learn about the city on a dramatic circular journey.

With a focus on the senses, experience, and interactivity, the wheel supplements the physical experience of the capsule with a virtual world. Upon entering one of the capsules, passengers can download the Nippon moon app and become part of a social network connected with other capsules as they take part in a three stage journey: discovery, the ride, and the return. Via the app, passengers can communicate with each other about things that they see around the city and to share their own knowledge, stories, and histories.

Through this medium of communication, the dialogue that is produced is decidedly a collective experience that does not aim for a fixed or singular perspective. Rather, it celebrates and supports the sharing of various readings of the city, society, and connections with culture at large. In this project the physical and virtual are conflated to produce a hybrid experience of the city that is simultaneously real and simulated.

Suspended upon a hybrid tension wheel, thirty-two gyroscopically connected capsules are enclosed by fully transparent glass to provide 360-degree aerial views of the city on a 40-min journey. The openness makes the sky above easily visible and gives the sensation of floating in the atmosphere. The glass exterior is also the surface of an artificial interior with augmented reality projections and animations that are superimposed onto the city.

In this way the experience of one capsule is different than another, which encourages riders to keep coming back for a new experience and new ***afterimages*** of the city (Figs. 6 and 7).

FIG. 6 Gow Nippon Moon, Japan. *(Image: UNStudio/MIR.)*

FIG. 7 App interface with augmented reality games, social network, and trip data. Gow Nippon Moon, Japan. (A) AR phone application and (B) app interface. *(Images: UNStudio.)*

Brainport smart district (BSD)

In the Dutch municipality of Helmond, ten kilometers to the east of Eindhoven, 1500 new homes and 12 ha of business development will be built incrementally over the next 10 years in the Brainport Smart District.

The project initiates an evolution in "smart" urban development with a model that is not based on a fixed and predetermined masterplan that is built all at once. Rather, it is guided by the participation of those that live and work in the area and will be built gradually in an experimental form of urban planning.

FIG. 8 Brainport smart district. *(Image: UNStudio/Plomp.)*

FIG. 9 Brainport smart district site plan. *(Image: UNStudio.)*

A flexible grid—or scaffolding—system integrates energy generation, food production, water management, and transportation systems, which will be fused with the latest technologies and knowledge. While these advanced technological systems will provide efficient functionality for these infrastructures, the BSD moves beyond merely "efficient" models and aims to bring forward a profoundly social element of data application. For instance, in a "learn by doing" form of urban design, new models of ownership will be played out, explored, and developed in conjunction with underlying data platforms that facilitate shared resource schemes.

By activating the participatory engagement of the residents of the BSD in its ongoing development as a "Living Lab," ***play*** and ***experimental learning*** are essential factors for the integrated growth between local society, infrastructure, innovative technologies, and the district as a whole (Figs. 8 and 9).

Conclusions

Since smart city initiatives were first launched in the late 20th century, the conversations and discussions that fuel their continued research have evolved very little. Their primary ambitions remain fixed on an interrelated set of pragmatic goals: the optimization of infrastructural functionality, the efficient use and distribution of resources, and the pursuit of a perfect model for an ideal city enabled by state-of-the-art sensory technologies and artificial intelligence. If we consider that many of our daily interactions, behaviors, social relationships, and vast swaths of personal information can be measured by advanced sensory technologies to achieve these goals, smart cities often risk reducing an immense amount of data to fundamental, fixed, or absolute truths.

It would be disingenuous, therefore, to suggest that the technocratic and positivist rhetoric of *efficient* smart cities is in any way synonymous with discussions about the *future* of cities, as their underlying ideological structure is bereft of a rigorous understanding of the city. Beyond the infrastructures of efficiency, the societies that inhabit these cities require insights into many other social frameworks, including the infrastructures of *knowledge, creative speculation, and learning.* As cities around the world continue to densify at a rapid pace, these infrastructures are increasingly vital to facilitating social interaction and cultural evolution within large-scale communities.

According to recent World Bank research in "The World Development Report 2018 (WDR 2018): LEARNING to Realize Education's Promise," learning and education are in a state of crisis. "Worldwide, hundreds of millions of children reach young adulthood without even the most basic life skills…This learning crisis is a moral crisis." The reduced public spending on education and the misconception that schooling is the same as learning have resulted in lower educational outcomes. Further, limited access to learning in disadvantaged areas widens gaps between classes and social structures (World Bank, 2019).

More specifically, models of learning that develop ***creativity*** are making their way out of school systems in favor of models that increase standardized testing scores such as memorization, calculation, and strict instruction (Jeffreys, 2018; Ruiz, 2017; Kim, 2016). As technological innovations are rendering process-driven and routine jobs as increasingly obsolete, ***creative critical thought*** is paramount to educational models in the future. However, the notion of creativity itself is in need of a fresh look. In her research on "The New Creativity," architectural critic and historian Sylvia Lavin implicates creativity as a "diffuse" condition that has "produced a virtually uninterrupted continuum of creative space that houses everything from You've Got Talent to an infinite number of square feet of office space, the over-designed but equally ubiquitous counterpart of junk space." For Lavin the New Creativity is less interested in creativity as a generic attitude that one might find near artisanal bakeries or vintage clothing stores. Rather, the New Creativity is interested in the ability for a creative person to stimulate and provoke creativity in another (Lavin, 2015).

In this sense the development of architecture in the cities of the future cannot ignore the consideration of ***play*** as it helps to build prosocial relationships between people from a young age into adulthood, cultivate new knowledge, produce new perceptions about the world, and stimulate the creative imagination of society through ***experimental learning***. Through the hybrid combination of digital and physical mediums of communication, this engagement can occur in a manner that is unfixed to a particular time or a specific place, unfixed to the conventions of playgrounds in physical space, or the virtual world of video games or gaming platforms. Therefore spaces of play in the city—both digital and physical—are poised to become sites that enhance the self-motivated and self-guided ***participation*** of people of all ages and demographics in their own creative education in relation to society.

Acknowledgments

Gothenburg Cable Car Project Team:
UNStudio: Ben van Berkel with Arjan Dingsté and Juergen Heinzel, Ayax Abreu Garcia, Xinyu Wang
Local Architect: Kjellgren Kaminsky Architecture Ab
Advisors: Structure & Engineering: Knippers & Helbig
Lighting Design: Licht&Soehne
Visualizations (CGI): Plomp
IJbaan Cable Car Project Team:
UNStudio: Ben van Berkel with Arjan Dingsté, Jaap Willem Kleijwegt and Milou van Min, Juergen Heinzel, Dana Behrman, Bruno Peris, Pedro Silva Costa
Advisors: Arcadis
Visualizations (CGI): Plomp

Blagoveshchensk Cable Car Terminal Project Team:
UNStudio: Ben van Berkel with Arjan Dingsté, Juergen Heinzel, Wael Batal, Daniel Cely, Milou van Min, Bruno Peris, Evan Shaner, Dario Castellari
Advisors:Strelka KB: Competition Management, formulation economic and functional model
Visualizations (CGI): Pyxid
Gow Nippon Moon Project Team:
UNStudio: Ben van Berkel, Gerard Loozekoot with Frans van Vuure, Filippo Lodi and Harlen Miller, Jan Kokol, Wendy van der Knijff, Todd Ebeltoft, Tina Kortmann, Patrik Noome, Jeroen den Hertog, Mariusz Polski
Engineer: Arup Tokyo + Melbourne
Interactive design: Experientia, Italy
Animation: Submarine, Amsterdam
Visualizations (CGI): MIR
Brainport Smart District Project Team:
UNStudio: Ben van Berkel with Machteld Kors, Misja van Veen, Dana Behrman and Teun Bimbergen, Maria Zafeiriadou, Pedro Silva Costa, Chen Shijie
Advisors: Habidatum (data analysis and modeling), Felixx (landscape and ecology design), Metabolic (Circularity)

References

Boult, J. (n.d.) Portraits of Children in Dutch Golden Age Art. Academia.edu www.academia.edu/25357448/Portraits_of_Children_in_Dutch_Golden_Age_Art.

Hamilton, J., 2014. Scientists say child's play helps build a better brain. NPR. August 6 https://www.npr.org/sections/ed/2014/08/06/336361277/scientists-say-childs-play-helps-build-a-better-brain?t=1589289938332.

Huizinga, J., 1949. Homo Ludens: A Study of the Play-Element in Culture. Routledge & Kegan Paul, London.

International Play Association. (n.d.) The Child's Right to Play. www.ipaworld.org/childs-right-to-play/the-childs-right-to-play.

Jeffreys, B., 2018. Creative Subjects Being Squeezed, Schools Tell BBC. BBC News, BBC. January 30 www.bbc.co.uk/news/education-42862996.

Kim, K., 2016. The Creativity Challenge: How we Can Recapture American Innovation. Prometheus Books, p. 2016.

Lavin, S., 2015. California College of the Arts – CCA. "Lecture by Sylvia Lavin." (video file). Retrieved from www.youtube.com/watch?v=F62yegQKCWU.

Ruiz, M., 2017. How to Combat America's Creativity Crisis. The Greater Good Science Center at the University of California, Berkeley. January 27 https://greatergood.berkeley.edu/article/item/how_to_combat_americas_creativity_crisis.

Stuart, K., Webber, J., 2015. 16 trends that will define the future of video games. Guardian. July 23 https://www.theguardian.com/technology/2015/jul/23/16-trends-that-will-change-the-games-industry.

World Bank, 2019. The Education Crisis: Being in School Is Not the Same as Learning. www.worldbank.org/en/news/immersive-story/2019/01/22/pass-or-fail-how-can-the-world-do-its-homework.

Yogman, M., Garner, A., Hutchinson, J., Hirsh-Pasek, K., Micknick Golinkoff, R., 2018. The power of play: a pediatric role in enhancing development in young children. Am. Acad. Pediatr. 142 (3), e20182058.

Author biographies

Ben van Berkel studied architecture at the Rietveld Academy in Amsterdam and at the Architectural Association in London, receiving the AA Diploma with Honours in 1987. In 1988 he and Caroline Bos set up the architectural practice UNStudio, and in 2018 he founded UNSense, an Arch Tech company that designs and integrates human-centric tech solutions for the built environment. He has lectured and taught at many architectural schools around the world and from 2011 to 2018 held the Kenzo Tange Visiting Professor's Chair at Harvard University Graduate School of Design.

Filippo Lodi holds several master degrees in engineering, architecture, art, and business, making him an all-around thinker. Lodi, as Head of Innovation and Knowledge Management, leads UNSx, UNStudio's in-house innovation think-tank and experience lab, where he works, taking a human(ity)-centric approach, on the development of disruptive technologies for the built environment from a unique coating that cools down buildings, responsive lighting system—as well as providing consulting services to companies on their workflow and rebranding strategies.

Wael Sami Batal received a Master's degree in Architecture from UCLA and is an Architect at UNStudio where he has worked on numerous projects across scales and typologies. He believes that architecture is a cultural discipline that operates between theory and practice. As a Knowledge Platform Coordinator at UNS, he is active in the production of knowledge through writing and lecturing about the intersections between architecture, technology, media, and society.

Chapter 5

Snowfall on Piazza Castello: Stubborn dispositions and multiple publics in a (temporarily smart) Milanese square

Yulya Besplemmenova[a] and Andrea Pollio[b,*]
[a]*Oblo, Service Design and Research, Milan, Italy,* [b]*Institute for Culture and Society, Western Sydney University, Sydney, NSW, Australia*

Chapter outline

Introduction

Smart city projects that involve spatial transformations span from large-scale (re)developments, such as the Sidewalk Labs-governed Hudson Yards in New York (Mattern, 2016), to small interventions of tactical urbanism. Projects of this latter type have received just as much critical scrutiny (e.g., Foth et al., 2015; Cardullo et al., 2018; McFarlane and Söderström, 2017) as larger greenfield and brownfield initiatives. The examples of temporary, small-scale (Mould, 2014), pop-up (Bishop and Williams, 2014), DIY (Iveson, 2013), and urban designs that involve smart technologies are numerous, but their capacity to challenge dominant understandings, hierarchies, and powerful interests in the planning of cities is often questioned (Colomb, 2012; Mould, 2014; Brenner, 2015).

In this chapter, we interrogate the capacity of such experiments of smart urban design to mobilize the multiple histories of public space, to reveal its

* Current address: Future Urban Legacy Lab, Polytechnic of Turin, Turin, Italy.

Shaping Smart for Better Cities. https://doi.org/10.1016/B978-0-12-818636-7.00004-4

hidden power structures, and to multiply the possible meanings of its publicness (Iveson, 2011). To do so, we chart the genealogy and aftermath of a pedestrianization experiment that shares many of the characteristics mentioned earlier: a purportedly "smart" and bottom-up approach, a light and temporary nature, the use of ready-made smart products, and an explicit aim at challenging dominant uses of a specific public space. All these characteristics, we argue, depended on the multiple ways in which the physical space of the pedestrian experiment was conceived by the residents, the city planners, the designers, and through the smart technologies themselves. Specifically, we narrate how the experiment elicited different conflicting narratives and involved multiple temporalities of public space, revealing its most stubborn dispositions (Easterling, 2014)—that is, the hidden power structures that are materially enacted in space.

Our case study is the iconic Piazza Castello in Milan, Italy—a square that sits in one of the wealthiest postal codes in the country but has long been at the center of controversial debates about the meaning of urban publics in the city, ever since Bonaparte's grandiose plan for an enormous public forum. In 2014, as the city was preparing to host the World Expo 2015, the municipality decided to transform the square into a New York-style pedestrian space. Inspired by the transformation of Times Square led by pedestrianization guru Jan Gehl (whose consulting firm had delivered a purposefully "mobile" design solution—see McNeill, 2011), the city envisioned a large car-free area that was meant to be appropriated by various city actors, ranging from tourists to street vendors. The proposal found so little support that the residents eventually took the municipality to court to halt the process, restore car circulation, and forbid the new informal uses of public space that the pedestrianization had allowed.

In this contribution, we detail the design process, which followed the initial fiasco. Drawing on the failure of their top-down decision, municipal authorities envisioned a design competition that was later won by a team of architects, urban designers, and interaction designers who suggested an experiment of tactical urbanism: covering the square with an artificial snowfall, a layer of white pavement material that would erase all the previous traces of how the space had been used. The most important element of the proposal was not its architectural blueprint but the creation of an open, collective narrative, capable of reimagining the existing physical space of the square. As an example, the new pedestrian space also became an online and offline wiki collecting the contributions from hundreds of authors, and the temporary uses were managed through a collaborative online platform.

We begin the chapter with a short overview of current debates concerning the potential of tactical urbanism to challenge dominant understandings and uses of the city. As we chart the actual implementation of the snowfall experiment and its eventual downfall, we highlight how different narratives and temporalities produced actionable "active forms" for the *piazza*

(Easterling, 2014, p. 14), how different publics were enacted through these processes, and the conflicts that eventually anchored the future of the square to a specific idea of public space. In the concluding section, we offer some remarks on both the limits and the indeterminacy of smart city experiments of tactical urbanism.

Our chapter is based on an action research conducted by one of the authors, who was involved in a professional capacity between July 2014 and December 2015 as a public space designer, on archival research on the planning history of the square and on a longer engagement of both contributors with the making of smart city initiatives in Milan.

Tactical urbanism and the possibilities of space

For some years now, tactical urbanism has been a popular strategy for temporary interventions in many cities around the world. Among the most popular proponents of tactical urbanism as a set of tools for urban change, Lydon and Garcia (2015) explained that the adjective "tactical" refers to the strategic potential of small-scale experiments of change in the public space to serve a larger purpose of progressive urban transformation. Their argument is that these experiments, in spite of their ephemeral nature and small scale, may in fact address much larger urban challenges by foregrounding alternative visions and possibilities.

Precisely because of their light character, several tactical urban experiments have included smart digital devices—from sensors to mobile phones—which lent itself to be used as tools for temporarily expanding the possibilities of public space. In a sense, therefore, tactical urbanism not only has often worked as a test bed for smart city products, in particular for corporations seeking new markets for urban technologies (see Halpern et al., 2013), but also has also been, as Sassen (2012) has written, the site where digital technologies become "urbanized"—that is, they are reappropriated beyond the uses and the markets that they had been envisioned for.

Other commentators have argued that both tactical urbanism and smart city solutions, as in the case of this chapter, have emerged as a response to austerity policies in Northern cities (Tonkiss, 2013; Pollio, 2016), creating opportunities for change with low budgets, small investments, and unused urban economies in the neoliberal city. For some, therefore, tactical urbanism has also worked to legitimize the withdrawal of the state from investing and subsidizing the making of public space (Bela, 2015) and does not do much to challenge neoliberal austerity (Spataro, 2016). In some contexts, these temporary experiments may even contribute to future waves of gentrification, expanding the frontiers of real estate interests (Novy and Colomb, 2013). Mould (2014) argues that tactical urbanism has indeed become a brand of creative city-making, a postpolitical marketing tool that aligns with powerful urban interests and yet generates the enthusiasm of many urban actors.

As a response to the excitement for the potential of tactical urbanism, Neil Brenner has warned against faulting temporary change for long-term progress in the making of urban space:

> *even in [a] maximally optimistic framing of tactical urbanism, the big questions regarding how to (re)design the city of the future—its economy; its property and labor relations; its spaces of circulation, social reproduction, and everyday life; its modes of governance; its articulations to worldwide capital flows; its interfaces with environmental/biophysical processes; and so forth—remain completely unresolved (2015, n.p.).*

However, even more critical commentators agree on the fact that tactical urbanism does in fact function as a tool "for changing received attitudes towards spaces and behaviors" (Webb, 2018, p. 59). What is of interest for the scope of this chapter is that, in spite of its limits, tactical urbanism may work as an experiment for revealing what we call, after Keller Easterling (2014), "dispositions" of public space: the invisible political arrangements and ordering that are folded into the materiality of space. In particular, drawing on the possibility that also public space can be experimentalized (Marres, 2012), in the next sections we show how the #nevicata14 experiment not only revealed the long histories of power and public space-making in Milan, but it also incorporated different materialities in the process of opening the forms of democratic discussion about the city and its futures, and multiplying the possibilities of one of its most iconic public spaces.

Background

In this section, we chart some of the historical and contextual trajectories of Piazza Castello, because, as it became immediately clear to the designers that were called upon to rethink the public space, the histories and the contextual politics of the place were affordances that they, as designers, had to inevitably confront and attune to,[a] both to counter and expand the possible ways in which the square could be imagined.

After Piazza Duomo, built around Milan's symbolic seat of religious power—the cathedral—Piazza Castello has only recently become the second most important public square in the city. Once at the margins of the city, the castle overlooking the square was a military fortress that had hosted both Milanese ruling families (the Viscontis and the Sforzas) in the late Middle Ages and early modern times. During this period the building was enriched with artworks, including Leonardo da Vinci's frescoes in the living quarters. Visconti's early intention was to create a second civic center for the city around their castle, taking away some importance from the religious core in piazza Duomo, but the project was never supported enough by their court (Casati, 1876).

---

a. Here and after, we use the terms "affordance," "attunement," and "coordination" in the sense explained by Annemarie Mol (2010).

Once Milan lost its independence and briefly became French, then Spanish, and then Austrian, the castle was used as the seat of foreign power in town. The building became so deeply associated with foreign rule that when Napoleon Bonaparte entered Northern Italy in 1796, freeing it from Austrian domination, the leaders of the revolt signed a petition to demolish the whole building—like French revolutionaries had done with the *Bastille* in Paris. Napoleon's plan to replace the building with a large public forum—an enormous circular square surrounded by public buildings (a theater, public baths, a museum, schools, and so forth)—never took off, but a debate about its demolition reemerged when Milan became part of the kingdom of Italy, some 50 years later. Real estate interests aligned with the desire to remove the hated building, and one of the earliest Italian insurance and development companies, Società Fondiaria, which had acquired and developed the lots surrounding the castle, kept pressuring the administration to raze the castle and release the military land for residential uses (Fiorio, 2005). The decision was so contentious that Milan's first comprehensive urban plan, in 1884, featured both options as possibilities (see Morandi, 1992).

Eventually the battle between those who wanted the castle to go and those who thought it was a monument to be preserved was won by the latter. The arrival of a new technology of transport, the tramway, allowed real estate interests to spread across a much larger metropolitan area, and the need to urbanize the land on which the castle stood became less tenable. A large park was built on what had been the military parade ground. The surrounding area was developed as a residential suburb, with large apartment blocks and tree-lined boulevards, and became known as Foro Buonaparte (Iarossi and Conte, 2015).

During the 1950s, Piazza Castello was the largest empty space remaining in close proximity to the city center. As such, it became the main long-haul bus stop in the city, a place where both tourists and economic migrants would first encounter Milan. The construction of the city's metro in the 1960s further entrenched the Piazza's role as a multimodal transport node, connecting coach services and underground rail to what became a large car park and busy intersection. Through the years that followed, several projects and proposals followed one another. The dominion of cars over the public space, however, was never challenged—not even when the municipality decided to rebuild the 1936 fountain in the late 1990s.

It was only in the last decade that a radical rethink of Piazza Castello became again a matter of political significance for the city, as the politics of the public space enmeshed again with the broader politics of the city. This followed a long period during which Milan had boldly embraced its transition from a postindustrial to a global service metropolis (Foot, 2001), a transition that included, among other things, large-scale brownfield redevelopments and the bid to host the 2015 World Expo (see Magatti et al., 2005). As González (2009) argued, these initiatives were discursively justified by the need to compete on the global stage and aligned with the spatial politics of neoliberal urbanism, allowing property developers to monetize land in a city fragmented by multiple corporate initiatives. In such contexts of weakened planning (Bolocan Goldstein, 2002), Milan also witnessed a large number of community-based, self-organized initiatives, geared around the environmental and social challenges that the city was confronted by (Balducci, 2004; for a critique, see Rabbiosi, 2016). These emerging community projects, often based in the outskirts, filled the vacuum left by Milan's globally minded administration (Cognetti and Cottino, 2003) and, eventually, became crucial in the 2011 elections.

Challenging outgoing mayor billionaire Letizia Moratti, who had been instrumental in relaunching Milan's global ambitions and given her green light to a series of large controversial urban transformations, underdog Giuliano Pisapia could count of a coalition of community movements. Supported by what became known as the "Orange Movement," as inspired by Juščenko's Ukrainian campaigners, Pisapia surprisingly defeated the right-wing candidate seeking reelection. Among other things the Orange Movement demanded that the development plans envisioned by previous councils be rediscussed and reopened to include the myriad of community organizations that existed in the city (Pollio, 2016). It is in this context that the Pisapia's administration found the political mandate to rethink a series of public spaces in the city, focusing on the social and environmental matters that his supporters had campaigned for. As an example, a radical overhaul of the congestion charge system was launched in 2012, limiting car circulation in a much wider scheme.

A series of spaces of the city were also identified as potential sites of new urban politics. One of these sites was Piazza Castello, described by the then transport commissioner Maran as a "place with a weak identity" (interview with one of the authors, 31 July 2017). In the meanwhile a section of Piazza Castello (Largo Beltrami) had also been singled out as the location that would have housed Expo Gates, two glass pavilions that were meant to communicate the city's journey to Expo 2015, with exhibitions and community activities. Following the need to accommodate the increased pedestrian traffic expected with the Expo Gates, a decision was made to close access to the traffic in the whole square, for the period before (from April 2014) and during the Expo event (to November 2015). A consultancy from Gehl People on the development program had provided the framework for a city-wide strategy of pedestrian access and car reduction. In Piazza Castello, hollowed out of cars and with no specific plan for how the space would be used, a series of spontaneous activities started to take place, including a soon-very popular flea market.

These activities triggered a conflict with local residents over what kind of public space uses was suited for Piazza Castello. Property owners thus formed a committee "for the protection of Piazza Castello" (Longo, 2014) and took the council to court to restore car circulation and remove the market. Their claim was that the predestination had attracted activities and crowds that were not appropriate for a monumental, historical public space. The flea market became disparagingly known as the "Souq" and characterized as not suitable and "too ethnic" for the monumental identity of the square (e.g., Romano, 2014).

Ironically the long-hated castle was used to justify a hold over what values the public administration ought to prioritize in Piazza Castello. The same property interests that had wanted its demolition now insisted on a forged historical reading of the place, arguing that the square should be conceived, if anything, as a space of monumental contemplation. In other words the unplanned pedestrianization revealed some of the "dispositions" of Piazza Castello, some of its "covert, stubborn forms of power—political chemistries and temperaments of aggression, submission, or violence—hiding in the folds of infrastructure space" (Easterling, 2014, p. 73), in this case, the interests of wealthy residents wanting to limit the uses and the meanings of the public space.

The escalation of conflict led the municipality to announce a call for ideas for the square in the form of a public consultation—Atelier Castello, inviting 11 local architecture studios to take part in process. According to the brief, each studio had to perform a participatory design session in the square itself, inviting contributions from local residents, citizens, and experts. After this consultation made visible, each firm had to develop proposals to be presented to the public and eventually selected through a complex process that involved online voting open to citizens, local council discussions, and an expert jury selection, followed by the exhibition of all projects.

The pedestrianization experiment

Storytelling

After the selection process, #nevicata14[b] became the winning proposal, standing out from the list of competitors for the uniqueness of its narrative. In Italian, "#nevicata14" means "snowfall of 2014," and the hashtag underlined the strong online component of the project, which was framed by the narration of a peculiar meteorological event: a snowfall in the summer. This initial proposal suggested the possibility of covering the space with a layer of white pavement material, like snow covering all signs of previous uses: kerbs, traffics signs, and all the traces that cars leave on the city. This radical approach was inspired by the famous snowfall of 1985 (nevicata '85), which had forcibly pedestrianized the whole city with one meter of snow on the streets.

The need to *reset* the space felt especially appropriate to Piazza Castello, as the designers had observed that its previous pedestrian crossing signs, sidewalks, and other marks left from car circulation were making space users behave as if the road was still open to vehicular traffic. Importantly, #nevicata14 proposed a new white blank page for the imagination of citizens. Under this layer of artificial snow, a mesh of the technological

---

b. #nevicata14 was a project of architects Guidarini and Salvadeo with Interstellar Raccoons. It was financed by Comune di Milano through Triennale di Milano museum.

infrastructure was planned to permit flexible uses of the space, providing access to electricity, lighting, Wi-Fi, and all the other needs of formal and informal space uses.

The most important aspect of the proposal was not its visually striking design solution, but the will to challenge the politics of public space uses with storytelling. In this sense the creation of an open collective narrative of an unexpected meteorological event was used to open up new possibilities and visions of the square being transformed. From the very beginning, this participatory process of narrating #nevicata14 was articulated both offline (in the actual square) and online on social media. The multiple visions of nevicata14 were collected in a wiki that, at the end, featured the contributions from hundreds of authors. This approach was initiated during the public consultation process and continued throughout the following design stages and during the project's actual implementation in 2015, when the storytelling continued to gather momentum, assembling multiple authors and stories, which were made available to the public both online and in the physical pavilion.

Creative engagement

The first challenge of a participatory design session that had to take place inside one of the Expo Gates, for just one afternoon per studio, was the need to overcome the resistance of local residents. During several of the previous sessions, citizens had turned up to complain about the pedestrianization and had not engaged with the actual proposals. The storytelling of #nevicata14, however, created the possibility of a radical abstraction from the existing square and its current issues. The provocative narration functioned as a surprising device, which shifted the response of the participants from their grievances to imagining their proposals for the blank space that #nevicata14 presented. According to the designers, this was a way of challenging some of the "dispositions" of the square, manifesting as a resistance to change and diversity. Residents of the city center in particular, as documented in their submission against the municipality, felt that the pedestrianization had unleashed a series of uses that were not compatible with the purported elegance and monumentality of the square. To elicit different responses, participants were invited to literally draw their own idea on dozens of postcards featuring the square as an empty canvas (Fig. 1).

Participants to the workshop were also engaged in other ways, with missions in the square to explore it in unusual ways, theatrical performances articulating the possibilities of new pedestrian configurations, discussions, and more. All of these activities targeted the resistant skepticism of local residents with the aim of generating a more creative form of engagement with the possibilities of a temporary urban transformation. A few weeks earlier, the same process had been launched online, on social media, gathering contributions from other Milanese residents and city users who could not physically participate to Atelier Castello.

FIG. 1 A #nevicata14 postcard, asking workshop participants to imagine what could happen in the square.*(Credit: #nevicata14 team.)*

The combination of the online and offline activities in the square allowed #nevicata14 to stand out from the other proposals for the number of participants that they had been able to mobilize and the number of visions that were marshaled in the process of defining the final design blueprint. While the participatory approach was shared by all 11 studios, it was the scale of #nevicata14 engagement that, according to the final jury, contributed to its selection as winning proposal.

Urban design and technologies

Due to budget and time constraints, it was impossible to fully realize the initial concept of covering the entire square with white pavement material. However, #nevicata14's vision was never a purely architectural proposal: it was a device that "provoked" (see Lezaun et al., 2013) the imagination of a space supporting multiple uses, as opposed to a single, entrenched view of what the square ought to be and be used for. In translating the original vision to a feasible intervention of tactical urbanism, the designers prioritized precisely this aspect of the proposal: the possibility that multiple uses and understandings of publicness could be materialized in the square.

The actual project featured 12 circular islands of white pavement, which resembled spots of melting snow. Such strategy allowed to contain the cost of realizing the temporary experiment and, at the same time, created various zones with specific characteristics, accommodating equipment for various public uses: different types of seats, sun shelters, dispensers of fresh vapors, stage platforms for public activities or shows, small gardens, lighting facilities, signage poles, and so forth (Fig. 2).

FIG. 2 Piazza Castello before and during the #nevicata14 experiment.*(Credits: #nevicata14 team.)*

The 12 white islands purposefully rearticulated—in a material form—one of the main complaints of the local residents: the fact that the square had been pedestrianized without clear directions or rules for reconciling new public uses with their visions about the *piazza*. The physical presence of the islands, instead, suggested an order to the multiple activities that the square had begun to host. Their different designs (some islands featured long benches and canopies; others were conceived as smaller intimate groves of trees), including a calculated lighting strategy (which allowed space users to select the ambience via a mobile app), proved to be suitable for even more informal uses than those that had been already observed. From casual passersby who would stop for the public Wi-Fi or reading a book, to birthday parties and other large events, the white islands proved able to offer material support and a material coordination—that is, the ordering of the multiple logics of public space enabled by removing car circulation.

To the end of coordination, the whole engagement process that took place during the 6 months of Expo 2015 took place at different levels, beyond the visible white islands. One of the explicit aims of #nevicata14 was not only to allow multiple space uses but, importantly, also to document their coexistence, as a validation of alternative ways of conceiving the square's publicness.

Smart-city technologies, in this sense, became instrumental in layering additional modes of curating the experiment. The smart lighting system, for example, provided by multinational electronics giant Philips in partnership with the local utility provider A2A and combined with an online platform and the presence of suitable seating, augmented the visibility of Milan's public Wi-Fi. As observed by the community manager of the experiment, on user-generated tourists' websites, Piazza Castello became known as the "square with Wi-Fi," although free connectivity services were available in all the squares of the city.

The online platform was also crucial for introducing the "new" pedestrian space to the whole city. Through the webpage a shared calendar of events taking place in Piazza Castello reconfigured the accessibility of the square by ordering its formal and informal uses. On one occasion, for example, a giant dinner-in-white flash mob attracted thousands of visitors. A self-organized celebration of the National Day of Peru, on another day, showcased how local migrant communities could participate in the space publics without disputes over whether such activities were compatible with the monumental nature of the square. Similarly, events such as Street Tango competitions and Skateboarding festivals demonstrated the possibility of diverse communities inhabiting Piazza Castello together. Moreover the platform-mediated work of coordination allowed the piazza to gain a new a significant role in the public life of the city: occasions such as the exchange of gifts between the city of Milan and Guangzhou or the *Feeding 5000* event, which marked the signing of Milan Urban Food Policy Pact, brought together institutions, businesses, and policy activists in a space that had never had, in spite of Napoleon's intentions, an easy, organic, public role in Milan's long history.

The digital platform also constituted a gateway to other services that the city made available in the square—from the online resources of municipal libraries to the booking system for street artists and buskers—but it was not the only transmedial channel connecting the square to its users. In fact, through the hashtag, the online presence of Piazza Castello on social media became a way to chronicle its multiple uses as a pedestrian space. To curate and archive the user-generated content, the #nevicata14 team had developed special signage posts: trilingual signs detailing the modes and the rules for using the space, as well as the possibilities for engaging the experiment online. In practice, these signs, which were workshopped with local residents, functioned both as a wayfinding device and as a connecting tissue between the physical space of the square, the Hertzian space of the city's Wi-Fi network, and the social media platforms that extended the storytelling of the experiment much beyond the boundaries of Piazza Castello.

The signage was crucial in the process of "rewiring" the square's identity. As more and more stories of #nevicata14 appeared on social media and online news streams, more and more residents and tourists started gathering in the square. During the experiment, the team noticed that the online presence of the square shifted from photos of its monumental castle to snapshots of its myriad uses as a public space. So much attention was drawn to the piazza that it became attractive for advertisers and sponsors. The original budget that the municipality had pledged to support #nevicata14 and its engagement activities was almost doubled by private partners, who could use the square as a hybrid media platform or as a showcase of their products. In this sense the experiment also functioned as a prototype for a curated partnership between public and private actors. A strategic plan of collaboration between the city's communication team and the #nevicata14 team had been instrumental in coordinating the outreach of the various codesign activities and the communication of the actual experiment in the months of Expo 2015.

By the end of those 6 months, the #nevicata14 team had compiled a long, detailed report of all the multiple activities that had happened in the square, both formally and informally. With more than 10,000 photos, dozens of videos, hundreds of social media posts, Google Place reviews, and Streetview photos on Google maps, Piazza Castello had become the space with the most documented social life in the city of Milan. With the final report in their hands, the city government had now to take a decision of the pedestrian future of the square.

After the experiment

One of the most visible outcomes of #nevicata14, as documented in the final report to the city, was a radical shift in the perception of the possibility of pedestrianizing the square. A poll that followed Expo 2015 showed that the percentage of citizens in favor of precluding car circulation in the square had risen to 80%. As an experiment the project did achieve the goal that the administration had set for the design team: validating the possibility of pedestrianization and recording the modes of managing the new public space. The project had also showcased, with tracked metrics, the possibility of public–private partnerships in the delivery of placemaking strategies. However, the multiple meanings and uses of public space that #nevicata14 had successfully brought to the square were soon lost to a more traditional view of its role in Milan's public life.

Not long after the experiment, the physical structures of #nevicata14 were removed on schedule, without a clear plan for what would happen in the square afterwards. Pressures for eliminating the temporary structures became particularly strong as the festive season approached, because a Christmas market was traditionally held in the square. The Expo Gates structures, which had functioned as a community center for almost 2 years, were also dismantled soon after. As the city's head of the planning department and councilor explained to one of the authors in an interview, the city administration had chosen to remove all contemporary interventions in the area with the aim of preserving the square's historical identity. Rapid transformations and redevelopments in more peripheral areas of the city, the councilor explained, had made Milanese people skeptical about such kind of transformations taking place in the historical center.

Meanwhile, important changes happened in the political sphere of the city. As Italy and Europe were hit by the 2016 wave of right-wing populism, Milan elected a center-left coalition. Relations with the central government and the regional government became strained, and the city's new mayor abandoned some of the more radical pushes of the previous administration. For Piazza Castello an international architectural competition was launched, inviting proposals for the final layout of the square as an ongoing pedestrian space. The brief of the competition took little notice of the legacy of #nevicata14's processes and learnings, although these had been crucial in generating consensus around the pedestrianization itself. The team from #nevicata14 participated in the competition, proposing an intervention that aimed at replicating and expanding the temporary project on a long-term scale, but the final jury awarded the prize to another team.

Ironically the winning project featured a solution that, at first glance, had striking similarities to #nevicata14's original vision. In the artists' impressions a white pavement covered the whole area of the square. The project, however, could not be more different from the previous intervention, where the candid surface was meant to narratively provoke the many possible futures of the square. As in the first #nevicata14 proposal, the white paving envisioned by the landscape architects was *calcestre*—crushed limestone—a material that was very common in 19th century gardens and therefore compatible with the purported 19th century character of the square. The municipal council further explained that the transformation would bring "order and tradition"[c] back to Piazza Castello, prioritizing its historicity over other kinds of spatial readings.

Seemingly the tactical, experimental approach of #nevicata14 faded away from the final outcome and design blueprint. Construction works for the historicist project, which reproduces an ostensibly traditional public square, are yet to begin, but the decision has recently been confirmed as definitive (Mingoia, 2019). One of the challenges that the #nevicata14 designers had set for themselves—to confront one of the intrinsic dispositions of the square, its monumental and monolithic identity—also seemed to have failed.

However, as the new city administration's focus has shifted from the center to the peripheries—with special attention to the great abandoned areas of the former rail yards and public spaces in areas of low-income housing—it appears that tactical urbanism tools have become the norm to engage local citizens in the processes of transformation. With #nevicata14 being the first experiment of this kind, the city has launched a new phase of large urban renewals replicating the approach that was used in Atelier Castello: participatory workshops and temporary interventions that include tracking the residents' feedback and proposals. While it is still to be understood whether these large transformations will effectively tackle Milan's ecological and economic challenges (including a deep housing affordability crisis) or the long-standing "democratic deficit" (see John, 2001) in the planning of the city, tactical experiments have become ubiquitous in the city (Tacconi, 2019) and have brought a wide debate on the meanings of public space at the center of municipal politics.

Conclusion

As of the time of this writing, Piazza Castello has remained pedestrianized. Occasionally, local policemen and residents engage in fights with the informal street vendors who have taken back the space.[d] Secret stashes of selfie sticks and counterfeited Prada scarves are hidden in the decorative bushes, waiting to be

---

c. http://www.comune.milano.it/wps/portal/ist/it/news/primopiano/tutte_notizie/urbanistica_verde_agricoltura/piazza_castello_vincitore_concorso (Accessed 19 February 2019).

d. E.g. http://www.ansa.it/lombardia/notizie/2018/12/26/ambulante-abusivo-aggredisce-vigili_7f7d9461-94dd-4301-a50f-ac23b77bbc18.html (Accessed 19 February 2019).

sold to the tourists of the castle. The white spots of snow have become gray and now blend in with the asphalt color.

This chapter thus speaks to an important aspect of the literature on technology-enabled, tactical, and experimental forms of space design: the question of what counts as failure, which may be a contested terrain of evaluation but is often neglected to the advantage of a critique of how smart urbanism reproduces spatial inequalities (Barns and Pollio, 2018). Specifically, we suggested that while #nevicata14 was not able to permanently multiply the possibilities of piazza Castello as intended, but eventually only managed to create consensus around its pedestrianization, it did allow different visions of its publicness to coexist in the square, even though temporarily. According to Kurt Iveson (2011), this is what makes the city public: not just the physical making and occupation of space, but the process through which those makings and occupations become a matter of public interest and debate.

More specifically, our chapter has shown how a smart intervention of tactical urbanism, though temporary and partly immaterial, may reveal hidden forms of power that are embedded in the material histories of a place or show how these entrenched identities can be challenged to allow a multiplicity of alternatives. To do so, we have suggested that it is important to chart the modes through which a space is made public by urban design in its various forms: from participatory storytelling to the actual urban technologies that enable and measure its uses.

To achieve this goal the "augmented" composition of the design team—including architects, service designers, entrepreneurship, and marketing experts—was crucial in attuning to the affordances of the square and in challenging them. The multiple ways in which the project leveraged both the hardware and software of space expanded the initial brief of a purely architectural intervention. Such a tactical approach to smart city placemaking, which emerged organically as the project evolved, may seem at odds with the heavy infrastructural load of expected *smartness* or even with the monumental identity of a public space like Piazza Castello. However, as Easteling provocatively writes,

> *space is the underexploited opportunity or the low-hanging fruit. Not products and technologies circulating in space but space itself is the operating system to manipulate or overwrite. Spatial variables are the crucial active forms in an extensive shared platform—at once information, technology, product, and pawn. The space that has always been available for manipulation, when seen in this way, becomes a fresh territory for political action.*
>
> (Easterling, 2014, p. 232)

With its limits and its failures, we argued that an experiment such as #nevicata14 demonstrates how the possibilities of space itself can be hacked to become a radical site of material participation (see Marres, 2012), where urban design contributes to the everyday, mundane forms of democratic engagement that open space to multiple understandings of its publicness, often in spite of its stubborn dispositions.

References

Balducci, A., 2004. Milano dopo la metropolis: ipotesi per la costruzione di un'agenda pubblica. Territorio 29 (30), 9–17.

Barns, S., Pollio, A., 2018. Parramatta Smart City and the quest to build Australia's next great city. In: Karvonen, A., Cugurullo, F., Caprotti, F. (Eds.), Inside Smart Cities: Place, Politics and Urban Innovation. Routledge, London, pp. 197–2010.

Bela, J., 2015. User-generated urbanism and the right to the city. In: Hou, J., Spencer, B., Way, T., Yocom, K. (Eds.), Now Urbanism: The Future City Is Here. Routledge, London, pp. 149–164.

Bishop, P., Williams, L., 2014. The Temporary City. Routledge, London.

Bolocan Goldstein, M., 2002. Governo locale e operazioni urbanistiche a Milano tra gli anni '80 e '90. 119 Urbanistica, pp. 90–102.

Brenner, N., 2015. Is "tactical urbanism" An alternative to neoliberal urbanism? Post. New York. May 24 https://post.at.moma.org/content_items/587-is-tactical-urbanis.

Cardullo, P., Kitchin, R., Di Feliciantonio, C., 2018. Living labs and vacancy in the neoliberal city. Cities 73, 44–50.

Casati, C., 1876. Vicende edilizie del Castello di Milano ricercate dal Dott. Carlo Casati. Brigola, Milano.

Cognetti, F., Cottino, P., 2003. Developers of a different city new forms of community in laissez faire Milan. City 7 (2), 227–235.

Colomb, C., 2012. Pushing the urban frontier: temporary uses of space, city marketing, and the creative city discourse in 2000s Berlin. J. Urban Aff. 34 (2), 131–152.

Easterling, K., 2014. Extrastatecraft: The Power of Infrastructure Space. Verso, London.

Fiorio, M.T., 2005. Il Castello Sforzesco di Milano. Skira, Milan.

Foot, M., 2001. Milan Since the Miracle: City, Culture and Identity. Berg, Oxford.

Foth, M., Brynskov, M., Ojala, T. (Eds.), 2015. Citizen's Right to the Digital City: Urban Interfaces, Activism, and Placemaking. Springer, New York.

González, S., 2009. (Dis)connecting Milan(ese): deterritorialised urbanism and disempowering politics in globalising cities. Environ. Plan. A 41 (1), 31–47.

Halpern, O., LeCavalier, J., Calvillo, N., Pietsch, W., 2013. Test-bed urbanism. Publ. Cult. 25 (2), 272–306.

Iarossi, M.P., Conte, S., 2015. The forum of the desires. The area of Parco Sempione and Castello Sforzesco in Milan, between utopia and construction of the actual city. In: Drawing & City, Proceedings of the 7° Convegno Internazionale dei docenti della rappresentazione, Turin 17-18/09/2015. Gangemi, Rome, pp. 201–208.

Iveson, K., 2011. Publics and the City. Blackwell, Oxford and Malden.

Iveson, K., 2013. Cities within the city: do-it-yourself urbanism and the right to the city. Int. J. Urban Reg. Res. 37 (3), 941–956.

John, P., 2001. Local Governance in Western Europe. Sage, London.

Lezaun, J., Muniesa, F., Vikkelsø, S., 2013. Provocative containment and the drift of social-scientific realism. J. Cult. Econ. 6 (3), 278–293.

Longo, E., 2014. Milano. L'area del Castello Sforzesco chiusa al traffico. Ma ai residenti l'idea non piace. I Tempi, Milan. 1 May. Available at https://www.tempi.it/milano-area-castello-sforzesco-chiusa-al-traffico-residenti-non-piace/. (Accessed 28 January 2019).

Lydon, M., Garcia, A., 2015. Tactical Urbanism. Island Press, Washington, DC.
Magatti, M., Senn, L., Sapelli, G., Ranci, C., Manghi, B., Dente, B., Colombo, A., Ciborra, C., Ceruti, M., Balducci, A., Artioni, R., 2005. Milano, Nodo della rete globale: un itinerario di analisi e proposte. Bruno Mondadori, Milan.
Marres, N., 2012. Material Participation: Technology, the Environment and Everyday Publics. Palgrave Macmillan, Basingstoke.
Mattern, S., 2016. Instrumental city: the view from Hudson yards, circa 2019. Places J. April. Available: https://doi.org/10.22269/160426. (Accessed 15 January 2019).
McFarlane, C., Söderström, O., 2017. On alternative smart cities: from a technology-intensive to a knowledge-intensive smart urbanism. City 21 (3–4), 312–328.
McNeill, D., 2011. Fine grain, global city: Jan Gehl, public space and commercial culture in central Sydney. J. Urban Des. 16 (2), 161–178.
Mingoia, M., 2019. Piazza Castello, l'anno della svolta: verso la riqualificazione. Il Giorno, Milan. 3 January. Available at https://www.ilgiorno.it/milano/cronaca/piazza-castello-1.4373107. (Accessed 25 February 2019).
Mol, A., 2010. Actor-network theory: sensitive terms and enduring tensions. Kölner Zeitschrift für Soziologie und Sozialpsychologie. Sonderheft 50, 253–269.
Morandi, C., 1992. Italian planning in the late 19th century: the Milan example. Plan. Perspect. 7 (4), 361–390.
Mould, O., 2014. Tactical urbanism: the new vernacular of the creative city. Geogr. Compass 8 (8), 529–539.
Novy, J., Colomb, C., 2013. Struggling for the right to the (creative) City in Berlin and Hamburg: new urban social movements, new 'spaces of Hope'? Int. J. Urban Reg. Res. 37 (5), 1816–1838.
Pollio, A., 2016. Technologies of austerity urbanism: the "smart city" agenda in Italy (2011–2013). Urban Geogr. 37 (4), 514–534.
Rabbiosi, C., 2016. Urban regeneration 'from the bottom up' critique or co-optation? Notes from Milan, Italy. City 20 (6), 832–844.
Romano, L., 2014. Liberate piazza Castello da bancarelle e kebab. Il Giornale, Milano. 23 May. Available at http://www.ilgiornale.it/news/milano/liberate-piazza-castello-bancarelle-e-kebab-1021530.html. (Accessed 25 February 2019).
Sassen, S., 2012. Urbanising technologies. In: Burdett, R., Rode, P. (Eds.), Urban Age Electric Conference. London School of Economics, London, pp. 12–14.
Spataro, D., 2016. Against a de-politicized DIY urbanism: food not bombs and the struggle over public space. J. Urban. 9 (2), 185–201.
Tacconi, G., 2019. Milano e l'urbanistica tattica, la nuova piazza aperta a NoLo. Teknoring, Milan.
Tonkiss, F., 2013. Austerity urbanism and the makeshift city. City 17 (3), 312–324.
Webb, D., 2018. Tactical urbanism: delineating a critical praxis. Plan. Theory Pract. 19 (1), 58–73.

Author Biography

Yulya Besplemmenova is a service and system designer and researcher at Oblo, Milan. She holds a Master's degree in Product-Service System Design from the Polytechnic of Milan and is an alumni of the Terraforming Research Programme of the Strelka Institute of Media, Architecture and Design (Moscow).

Andrea Pollio holds a PhD from the Institute of Culture and Society (Western Sydney University) and is Marie Skłodowska-Curie Fellow at the Future Urban Legacy Lab (Polytechnic of Turin) and the African Centre for Cities (University of Cape Town).

Chapter 6

Designing for hyperlocal: The use of locative media to augment place narratives

Efstathia Kostopoulou and Ava Fatah gen Schieck
The Bartlett School of Architecture, University College London, London, United Kingdom

Chapter outline

Introduction

Digital media have become pervasive throughout our everyday lives. The last two decades, specifically, have witnessed a number of experimentations with various ways to digitally augment urban spaces. These are primarily led by artists and researchers working with cultural institutions and communities who collate and curate stories in public spaces to narrate local memory and share it with a wider public.

In this chapter, we explore the role of technological, spatiotemporal, and local parameters when designing for hyperlocal media to tell the stories of places. We focus on site-specific projects that draw on local stories through a

Shaping Smart for Better Cities. https://doi.org/10.1016/B978-0-12-818636-7.00016-0

variety of multimedia content (appearing in various digital forms such as images and text) that is conventionally stored in personal or institutional archives. Specifically, we ask how to contextualize local memory in the digitally augmented urban space? Here, we present two variations of site-specific projects, which use different technologies to share local stories in public space. Both can be experienced in situ within the same borough in East London. In the first one, spatial navigation and bodily movement is guided through archival material, in various media forms, viewed through Augmented Reality (AR) technologies. When moving between a selected number of points along a trail, this reveals local narratives that are spatially defined through the movement between these points of exploration, situated within proximity from each other. In the second project, behavior in urban space is guided through a fixed projection that annotates a physical memorial with digital moving images of archival material for an event of commemoration. We suggest that there is real potential in the mediation of physical space through digital interfaces that preserve local stories. We emphasize the role of the local scale where these tactful interventions can take shape. They trigger agency by contextualizing digital multimedia content to be experienced at the actual place that it refers to. We further argue that it is in local places rich in physical infrastructure of material interfaces where the immaterial digital annotations can create the possibilities for augmented places, allowing in part local stories and knowledge to be shared and experienced in a site-specific way. In the following sections, we outline concepts of digital and situated mediation, and spatial configuration, before we illustrate the two site-specific projects in more detail.

Background

Cities are complex constructs with urban spaces that are inherently social (Gehl, 2010; Whyte, 1980). In this respect, locative media offer new possibilities, through their support of site-specific digital information, to tell the stories of places and people (Julier et al., 2016; Martin and Lane, 2007) and facilitate meaningful encounters (Memarovic et al., 2016; Wouters et al., 2015). This has so far manifested in various city-scale initiatives within wider smart city agendas and city participatory management (Crooks et al., 2017; van Leeuwen et al., 2018), to the creative industries agenda (Tech and Interactive, 2019), to playable cities (Ackermann et al., 2016; Afonso and Fatah gen Schieck, 2019; Colley et al., 2017; de Lange et al., 2019; Nijholt, 2017), and to heritage places (Avram et al., 2019; de Waal, 2016; Ritchie, 2014).

Focusing on the latter, we observe a greater interest in sharing memories of places in site-specific ways through the use of digital experimentations with locative media. Heritage, in this instance, can be better understood as a process related to human agency and action (Harvey, 2001). Its complexity allows the inclusion of associated stories of people who interact with the places over time. These stories can be intertwined in the material aspects of the specific site, mediated by tools that allow not only rich interpretation but also engagement, both with the stories and the physical site. In this context, engaging with our physical environment, and its past stories and memories, enables our social and cultural navigation through these places that further shapes agency.

Locative memories and digital mediation

Van Dijck observes that media and memories are not separate entities but rather the first enhances, corrupts, and mediates the latter (van Dijck, 2007). Our lived memories seem to be increasingly shaped by digital technologies (Huyssen, 2003), and how we create and preserve local memories has been closely related to the use of media tools. For instance, photographs and videos are all shaped by the specific technologies used for their production. Archives exist in various forms, physical or digital, from traditional archive buildings, to shoe boxes, to the web.

Yet, digital encounters with the physical site can be considered as a different category, bringing an added value and contribution to our ability to engage with places. Here, the interaction of content and form encourages us to move beyond the opposition of physical and digital encounter with the past, looking into the precise location of value in the process of digital engagement (King et al., 2016).

Digital technologies can be used as tools to communicate and amplify social and spatial storytelling, providing spatial modes for engagement with site-specific stories, in new ways. In particular, locative technologies such as AR and large digital projection can combine computer-generated digital content with real spaces (Feiner et al., 1997; Vlahakis et al., 2002; Vande Moere and Wouters, 2012). They do so, by overlaying the digital immaterial content on the material physical space. As such, they spatialize digital media content. In this respect, the value of engagement with the context arises from the local situation, in defined configurations of space and time coupled with technological mediation (McCullough, 2007). Such mediation through digital technologies, mobile or fixed, enhances our perception and experience of place, allowing us to form new connections with the context. We suggest that this is even more relevant when it comes to the localities we inhabit and the preservation of their local memory (Julier et al., 2016).

Hyperlocal content and spatial storytelling

Given our focus on local augmentation of memories, it is important to introduce the concept of "hyperlocal content" as the place-based information that is mediated through digital technologies, static or mobile. This, it has been argued, by being highly relevant to its spatial context, allows local inhabitants to better relate to their locality and enrich their experience of place (Wouters et al., 2015). When mobile technology meets physical structures in the urban fabric, the experiential qualities of such situated interactions expand to the various physical structures in space (McCullough, 2007). Physical objects in place, such as monuments and buildings, are no longer fixed realities, in a sense that there is no singular assigned meaning or at least everlasting meaning to them but it is something that can evolve over time (Huyssen, 2003). The latest rise of digitally annotated locations has given a wealth of projects that tell stories of places by contextualizing the multiple meanings of the urban physical structures.

An example of a movement-based interaction is Street Museum, a mobile application commissioned by the Museum of London. It brought the experience of viewing historic photographs of London to the actual locations, where they were initially captured, by using GPS and AR technologies that enabled the overlay of physical spaces with digital image. The VisAge project is another mobile example based on movement between selected points of interest, a collaboration between the Regency Town House (RTH) charity with researchers from UCL (Julier et al., 2016; Kostopoulou et al., 2018). The team developed a prototype that told the story of people from the past. This was achieved by using some of the RTH charity's rich archival material (including images, text, maps, and plans) that were superimposed on existing physical structures and places in the city of Brighton (United Kingdom). The archival material was experienced in situ through the deployment of AR technology along a defined route in the city center. Another good example of telling stories from the past, albeit through large projection on a building, is the projection on the Irvine Maritime Museum. A large collection of archival images, related to the maritime history of the place, were projected on the façade of the Irvine Maritime Museum for St. Patrick's Day during the Harbour Festival of Light (2017). In these cases, the digital storytelling was part of a local event adding to the narrative and celebration of these places, acknowledging their local social history by drawing on relevant visual archival material to engage the public with it.

Spatial configuration and situated engagement

What all the previous examples have in common is that they require situated engagement at the actual site where the narrative belongs. These situated digital interactions can transform the notion of place, supporting it as a social and cultural construct, through the hyperlocal content they mediate (McCullough, 2007). Most of the digital encounters are short in duration (de Waal, 2014; Giaccardi, 2011). They are supported by ephemeral layers that mediate our experiences with the physical world during brief encounters with the artifact that they annotate while moving in space. However, it is the specificity of the context, augmented or amplified through the digital mediation, that allows relational associations with the space and renewed excitement about the physical engagement (King et al., 2016). Drawing on this, we suggest that a holistic approach to the design of "mediated" urban experiences should further take into consideration the spatial and social affordances of place through its materiality and agency (Kostopoulou et al., 2018).

In the following section, we outline two cases and explain how they make use of the two distinct types of digital mediation, through the movement-based AR and the fixed projection with the intention to situate local stories, taken from archival repositories, in the actual spaces, and offer new perspectives for the narrative of place. We then discuss the spatiotemporal and social dimensions of the digital mediation and their role in generating encounters with local memory in public spaces. We conclude with examining the importance of contextualizing relevant local content to be experienced in the locations.

Contextual urban experience: Two approaches of mediated spatial storytelling

In this section, we outline two projects that apply a holistic approach to design to mediate hyperlocal stories, taking into consideration the spatial and social context coupled with the affordances of two different types of technologies.

Mobile augmented reality

"Augmented Reality in Whitechapel" was developed by researchers from University College London where encounters with narratives from the past were designed as a trail-based urban experience that revealed, through the use of AR technology, hidden stories about the site of augmentation (AR in Whitechapel, 2016). The aim of the project was to interact with various layers of the past at various points of interest in Whitechapel, an area that is undergoing rapid transformation. AR technology was deployed to superimpose local narratives in situ using some of the abundant archival material of text and images collected by the Survey of London as part of the Whitechapel area study (Histories of Whitechapel, 2019).

The design process was iterative with multiple layers of engagement with both the archive and the Survey of London and the research team. Specifically, the team intended to explore how AR can enable experiencing and contextualizing the archival material within the urban space. The project further sought to understand how best to present archival content in situ at the actual places it represents, supporting spatial storytelling that augments four selected stops, unfolding along a continuous urban route in Whitechapel Road in East London.

The selection of the different physical sites for digital annotation was intended to introduce variety within an urban route, with locations in proximity to each other, allowing the whole experience to unfold within a timeframe of 30 min. The digital content on the screen consisted of both textual and visual material, where the text was formed in short sentences distributed into two to three speech bubbles. This was based on a previous exploration by the research team, with multimedia storytelling through AR, which was evaluated within a university context (Javornik et al., 2019). The augmented experience in Whitechapel brings a dense context within a short route, guided by the researcher. This reveals the richness and multilayered character of this side of East

London with variations in function that include an urban park (Altab Ali Park) and industrial and religious buildings (the Bell Foundry shop and workshop and a disused synagogue).

In creating the trail, along the four points of interest, the route was structured around the main high street, Whitechapel Road, and centered on locations that are spatially highly integrated within the urban scene (Hillier, 1996) and adding the depth of one urban block when navigating to the points of interest. Yet the moment of pause and engagement with the hyperlocal content, that is, when the position and orientation of the body in space is fundamental to experience of this content, was amplified through the affordances of the AR technology in use, as outlined in detail in the following text.

The first destination of the trail is the Altab Ali Park. The spatial encounter here (in both physical and digital modes) is staged over two steps, making use of physical signage that already exists in the park. By pointing the mobile device at the sign at the entrance of the park, the participants could reveal a map of the area superimposed on the ground showing all four stops. Unlike a conventional 2D map, where one might feel lost if they have difficulties orienting, the mediation through the AR supports natural navigation and orientation in space as the digital notation is combined with the actual location rather than replacing it. The second sign is related to the remnants of the old chapel. The digital information reveals an interface of textual material accompanied by images, where one learns about how it was known as the medieval white chapel, first built in 1282 and rebuilt in several occasions and eventually ruined by fire bombs in 1940 (Fig. 1).

The second stop annotates the façade of an existing building at the main Whitechapel Road, situated right after the park, superimposed with an image of the interior of the Bell Foundry shop that was operating here until 2017. The link to the activity was amplified through an audio loop of the bells triggering

FIG. 1 Altab Ali Park, site of augmentation. *(Credit: Author's original.)*

FIG. 2 Bell Foundry, site of augmentation. *(Credit: Author's original.)*

an ambient sound (Fig. 2). At the time of the project, this building had been continuously operating in this location since the 1570s. The Bell Foundry produced a large number of bells sold worldwide. The building itself is generally unknown to the public, albeit there is a sign on top and on the ground floor the glass windows allow visibility. The mediation through AR reveals significant, albeit hidden, aspects of this rather unassuming building.

The third stop is a few steps behind the Bell Foundry shop (second stop) by the side street. This is the Bell Foundry workshop, a large industrial warehouse typology building that extends behind the shop. The digital content consists of a number of images of the foundry activity to illustrate the textual narrative (Fig. 3). The fourth and last stop brings the participants behind the main high

FIG. 3 Bell Foundry workshop, site of augmentation. *(Credit: Author's original.)*

FIG. 4 Disused Synagogue, site of augmentation. *(Credit: Author's original.)*

street in front of a disused synagogue, revealing an image of the interior of what was one of the last active synagogues in Whitechapel (Fig. 4).

Situated large projection

Unlike "AR in Whitechapel" the "Bethnal Green Memorial" projection was superimposed on a static physical structure. The memorial is located close to the entrance of Bethnal Green Park, next to the underground station, replacing a plaque (from 2017) that commemorates the loss of 173 people that were trapped and suffocated in Bethnal Green underground station in a domino fall during an air ride call in 1943 (Butler, 2015). The new memorial consists of a concrete trunk and a wooden canopy, the latter representing the actual underground staircase where the incident had occurred (Figs. 5–7).

A double projection, one on each side of the canopy, was scheduled to run during the evening hours between the 3rd and 7th of March 2018. Each projection consisted of the same loop of images (Kostopoulou, 2018). The projection content consisted of archival images as photographs, depicting the victims in their ordinary lives, primarily as individual or group portraits. One projection was outside onto the long side of the wooden canopy, while the second one was inside within the canopy structure. The purpose of the positioning of the two projections outside and inside was to initially attract attention and further bring the audience under the canopy (staircase miniature) within the representational space where the event took place. Overall,

FIG. 5 Bethnal Green Memorial 75th commemoration with double projection, 2018. To celebrate the new physical memorial, a crowdsourcing initiative was launched to fund a projection for the 75th commemoration of the event. *(Credit: Copyright of Arboreal & Stairway to Heaven Memorial Trust.)*

the main drivers of this initiative had to overcome a number of challenges involved in realizing the projection in public space, such as the placement of the projectors, which, unlike AR mediation, were fixed in space throughout the duration of the project.

As for the memorial location, it is closely placed to one of the four exits of the London Underground station where the original incident occurred. Proximity to this underground station was important for the creation of the situated experience and knowledge. Further, working with the local scale, the new installation was intended to be significantly visible across the street. Due to the nature of the experience, the light projection was scheduled from sunset until late night hours, capturing attention from peak movement flow hours to the late slower and lingering times in this urban space. During the evening hours of the projection, the highest flows were expected to happen closer to rush hours as people commute to and from the tube station and local bus stops. However, the scheduling time of the projection until the late hours allowed for the creation of a rich atmosphere in this part of London and attracted attention to this visual activity until late at night (Fig. 6).

FIG. 6 From the exploratory site observations, we identified three vantage positions to experience both the physical installation and the digital content. (A) A position directly under the memorial installation, giving visibility only to the enclosed surface of the wooden structure. (B) A position in the middle pedestrian island of Cambridge Heath Road when exiting the west gate of Bethnal Green station and moving at the street level toward the entrance of Bethnal Green Park. That allows visibility only to the external projection of the triangular side of the wooden canopy.

(Continued)

FIG. 6, CONT'D (C) Steps behind the park exit of Bethnal Green station (traffic light position) on Roman Road park side, allowing movement all along and positions of simultaneous visibility of both projection sides as per Fig. 5. *(Credit: Author's original.)*

FIG. 7 The Bethnal Green Memorial during day time. *(Credit: Author's original.)*

(Continued)

FIG. 7, CONT'D

Context, mediation, and the facilitation of social encounters with the past

Drawing on the two cases outlined earlier, we discuss the main factors that affect the design for hyperlocal in urban space, addressing both physical and digital aspects of mediation.

Mediation of space and narratives

It has been argued that introducing narratives in public space supported via technology enhances our experience of actual places (McCullough, 2007). It augments our engagement with the physical site via digital layers of information that, as active agents, we seek to reveal and experience. However, it has also been observed that mediating space can cause a "displacement" effect (Aurigi and de Cindio, 2008), a sense of losing the actual relationship with our physical surroundings. Here, we argue that the use of site specific content can minimize a sense of displacement from physical space by anchoring relevant digital content to the actual site. It does so through approaches that support new ways to create a relationship between the content, the content viewer, and the physical surroundings (Kostopoulou et al., 2018).

In both cases presented here, the digital content consists of significant archival material that has been resourced and collated by experts who have been researching the specific sites, including local experts who have invested interest in the localities, the Survey of London for the project in Whitechapel, and "Stairways to Heaven" for the project in Bethnal Green. Other studies focusing on using digital mediation to engage with cultural heritage on sites (Avram et al., 2019; Javornik et al., 2019; Kostopoulou et al., 2018) have further addressed similar challenges, albeit from a curatorial perspective in relation to how the digital content makes sense of the specific site or artifact it annotates. In the case of Avram et al.'s research, they further stress the importance not only of site identity and character but also of people's own knowledge, attitudes, and relevant past experiences of similar sites that might leave a gap between expectation and actual experience. Similarly, Javornik et al. address how participants to their AR experience trail have interpreted some imagery and cues in a different way of what the purpose of the design selection was intended.

In all these cases, this specificity of the context seems to allow relational associations with space and renewed excitement about the physical engagement with the site (King et al., 2016). Moreover, due to its short duration of encounters with the digital content, the experience only annotates and complements rather than overshadow the physical site itself.

Rhythm and temporality in the construction of encounters in space

Given our emphasis on the context and its social and spatial affordances in framing the mediation of "hyperlocal content" and in allowing people to better relate to a locality, we argue that urban structures, like the Bethnal Green Memorial, provide multiple levels and modes of engagement with the site and its stories. Here, we outline two different modes of engagement: the first one is a more fixed one happening through the physical structure, and the second one is a dynamic, hence temporal, and defined through the moving images projected onto the physical structure. This, however, also depends on the spatial configuration of the surrounding streets network, which play a key role in defining the interactions and encounters around this emerging type of urban media interfaces.

The Bethnal Green Memorial project highlights how its physical structure played a role in reinforcing the engagement with the site and its stories, including its construction and deconstruction through the use of the various parts to mediate these stories. It also illustrates the importance of our positioning, as viewers, through physical and digital annotations and interactions in the actual location.

When it came to annotating the Bethnal Green Memorial through digital projection, the sculptural three-dimensionality of the physical structure, with the names of the victims engraved onto the external side of the wooden canopy, was creatively used to introduce rhythm and visual narrative by deconstructing the physical structure into multiple projection surfaces where archival material was conveyed in a digital form. The projected images with their transparency effect offered another layer in the existing narrative, where textual content started blending with visual. This form of mediation was set in a spatiotemporal realm that engaged the viewer, where the analogue fused with the digital and their relationships were in constant reconfiguration via the move and pause of the viewer. Introducing rhythm in the presentation of the projected images created a flow that worked well specifically in some positions like in Fig. 5 with simultaneous visibility of the projection surfaces.

When addressing rhythm, the effect of repetition of images as loops, in whole or fragmented modes, defined a temporal frame for visual interaction. That, despite its visual effects, could still not overcome the powerful impact of the flowers, the physical elements that worked more successfully in making people stop. Here, supportive practices of both material and digital nature seemed to work better together, as the multiplicity of the object was more powerful, drawing on the familiarity with the performative gesture of placing flowers, an embodied practice and carrier of knowledge that was related to our remembrance rituals.

Temporality was thus constructed through the various rhythms of urban flows and curated digital content. It was in the instances of pause and interaction with the digital content, the flowers, and the memorial on site for the chance encounter of the passerby. But for the already engaged audience, there was a gradient of engagement modes on offer, prolonging the duration of interactions and engagement practices.

From a spatial perspective, it was the public space and its configuration that afforded chance encounters to happen and unfold over time (Hillier, 1996). We then further expanded the role of spatial configuration, in the instances of pause and mutual presence within the physical structure, through the provision of digital information. By doing so, we contextualized space at the local scale through new ways of experiencing local memory. We suggest that moving from a linear and sequential fruition of the stories to a configurational approach of the narrative through the street network of the city, within which this particular type of urban infrastructure was located, was what enabled social and cultural encounters and interactions with all its assigned meanings.

Material and immaterial interface

It has been argued that in spatial storytelling, the materiality of the fixed urban structures is used in linear and progressive ways to tell stories that can be understood based on a curated spatiotemporal experience (Ritchie, 2014). However, when considering our urban environment and its complexity in configurations of artifacts and assigned meanings, the possibility to connect with a specific place became stronger through mediation that engaged its existing particular context. In the case of the Bethnal Green Memorial, for instance, the material artifact has not only a representational but also a symbolic value, as the carrier of a very specific meaning of remembrance, located in the place where the actual event took place. Here, there were two different types of technological intervention: the material, manifested in the construction of the physical monument itself, and the digital, through the projection. Its enactment via the rituals of remembrance and the daily encounter involved both individual and shared experiences. It is in this interplay of both the situated (material, spatial) and relational (symbolic meaning) ways of encountering the memorial that the technological mediation via the projection utilizes for the narrative to unfold. All of this is dependent on those shared moments of presence on site.

Drawing on pause and interaction, the "Whitechapel Augmented Reality" trail provided an alternative to a guided exploration. Moving beyond the museum and the gallery space to the open public space, immaterial digital narratives told through multiple layers of information started to overlay and annotate our material world. This exploration required a mobile device with a viewing system as simple as a camera and a defined material interface to track a specified location so as to reveal digital content. This did not, however, substitute the more traditional and fixed material interfaces of the signs and plaques that became an anchor for the digital annotation. It rather enriched the experience providing more detailed information through the digital content that was spatially linked to the material interface of these signs. Notwithstanding current limitations of AR technology such as lighting conditions and marker suitability (Javornik et al., 2019), this enabled, however, new possibilities for the city's material interfaces to participate in the curation of stories that could be experienced at the actual location and without losing connection to the surrounding environment. Through this situated practice of annotation that linked directly to the actual places where the events took place, the possibilities to facilitate new encounters in space were created and potentially intensified. Especially when it comes to historic content or shared stories and memories about the place and the people who inhabit it, the structure of the narrative can prolong or create an engagement with the content and ultimately with the site itself.

The augmented experience of public space

The two cases outlined in this chapter represent two very different types of digital annotations of public space. In the first case, experimenting with AR technology, and despite the imposed technical limitations around the way the narrative could be structured and framed, seemed to support a configurational approach to design and hyperlocal experience in the urban space.

The Whitechapel Augmented Reality trail was based on a configuration of buildings and sites along a continuous urban route with nodes of exploration and relevant content that could be explored sequentially, one after the other, or in a more fragmented way, as independent stops. It offered a hidden layer in the narrative of the city that called for conscious action to move across points to reveal content. This was triggered through movement and the right physical positioning in space. It further depended on the use of a mobile device that could trigger the hidden digital narrative via positioning itself and its camera to activate the physical sites. Although groups could interact, the main interaction still stayed within the intimate sphere of the individual when we consider the mobile device as the main medium. When we think about the overall spatial configuration, the interaction space expanded to include all the chance encounters at the actual moment of pause and reveal (Kostopoulou et al., 2018). Therefore the actual experience in space expanded beyond the single human agent and the medium to include the physical-material space and all social interactions that were activated during these moments of copresence.

In the case of the Bethnal Green Memorial projection, on the other hand, and although the projection was situated, the experience was structured in a way to trigger movement and eventually draw the viewers to the memorial itself. The external projection in the triangular side of the canopy (Fig. 5) acted as a main attractor from distance to bring the passerby closer to the structure before moving under the canopy into a more intimate space to follow the same narrative, uninterrupted and in full frame.

Hyperlocality and private-public thresholds

In the growing world of digital annotation and especially when it comes to the use of technologies such as AR or projection, materiality is therefore still important. But what is also important is the notion of private-public thresholding addressed through (a) issues of management and accessibility to the actual interfaces and (b) issues of agency and actual interaction.

Considering the urban space, seeing the private-public relationship as a binary concept can be challenged. An associated challenge is also encountered around institutional or traditionally top-down management of space (de Waal, 2016). In the case of Bethnal Green Memorial, what is traditionally a remembrance practice took a completely new shift via the use of various forms of situated media to support new framing for it. This calls to rethink again the way we traditionally manage a monument, a private building, or even a façade when it becomes an anchor for digital annotation. Previous studies (Julier et al., 2016; Kostopoulou et al., 2018) indicate that various practicalities that emerge in the actual steps of implementation need also to be taken into account as part of the whole process of design. Here, we suggest that at this hyperlocal scale of digital implementation, more tactful interventions are needed that take into account the ethical requirements of consent and agency.

In the two projects the interactions required the positioning of the participant in exposed public spaces with high visibility. However, there were differences in how private-public thresholds unfolded. In the case of Whitechapel, while the participant could be highly exposed in public space, the digital content with which they engage via the AR stayed hidden. The interaction with the content happened at a private level, and it required activation by the participant who sought to reveal this content. The projection onto the Bethnal Green Memorial, on the other hand, was designed around visibility and vantage. The content of the projection itself was predefined by the community, the charity that produced it, in a sense that there was no further interactivity in adding, commenting, or complementing the projection. As such the installation invited for a private reflection and awareness in a highly exposed public place, something that the physical memorial itself already signified through its permanence and publicness.

Conclusion

Working within the specific sociocultural context and the locale requires technological implementations to adapt to the place. The ubiquity of digital network coverage allows for plug-ins and local experimentations with either low-tech solutions or more advanced novel technologies. In both aforementioned projects the stories and the imagery are anchored to the sites of memory, calling us to experience the actual locations. This site specificity of content and experience draws the attention to the locale through supportive practices of both a material and an immaterial nature, the latter being digital that work better together to renew our excitement for our physical engagement and connection with the site.

These projects in their very nature can thrive on local partnerships and the actions of human agents. It is their agency and materiality that further shape the affordances and relevance of the medium in use. They act as two different representations of digital annotation of urban space with a focus on local memory. They augment space with digital information, whether hidden or scaled up and visible from vantage points in urban space. They both require an active-passive participant who can only experience them in situ. The two projects outlined in this chapter, one mobile, the other fixed, they both activate content by body movement and positioning in space. Both situations call for experience in the actual locations where the event they narrate took place and assign relevant digital content, whether as image, sound, or text, where digital and physical elements work together in a reconfigured augmented space.

We suggest that by moving from a linear to a configurational approach to the spatial curation and digital annotation of local stories we share in public space, we trigger agency and enhance sociocultural encounters on site. We stretch the role of spatial configurations in the instances of pause and mutual presence with the physical urban and digital flows. By doing so, we contextualize space at the local scale producing new ways of experiencing local memory. The focus here is on the design for the hyperlocal spatiotemporal conditions to trigger agency to actively seek to reveal content (in the mobile experience) or offer a gradient of passive-active engagement (in situated projections). Thus local memory is digitally reconfigured through sociocultural and material parameters that draw the attention to the locale in urban space.

Acknowledgments

We would like to thank the members of Survey of London at UCL for supporting this study with providing access to archival material and content for the AR trail in Whitechapel. The Whitechapel study that was presented in this paper was partially supported by the Bartlett Enterprise Innovation Fund (BEDF) and Beacon Public Engagement, UCL 2016.We would further like to thank the Stairway to Heaven Memorial Trust, the Bethnal Green Memorial Project, and Arboreal for their contributions and support.

References

Ackermann, J., Rauscher, A., Stein, D. (Eds.), 2016. PLAYIN' THE CITY: Artistic and Scientific Approaches to Playful Urban Arts., https://doi.org/10.25969/mediarep/1643. Navigationen: Playin' the City, Jg. 16, Nr. 1.

Afonso, A., Fatah gen Schieck, A., 2019. Play in the smart city context: exploring interactional, bodily, social and spatial aspects of situated media interfaces. Behav. Inform. Technol. https://doi.org/10.1080/0144929X.2019.1693630.

AR in Whitechapel, 2016. https://www.youtube.com/watch?v=zi0vYMbQcE0. (Accessed 25 June 2020).

Aurigi, A., de Cindio, F. (Eds.), 2008. Augmented Urban Spaces: Articulating the Physical and Electronic City. Aldershot, Ashgate.

Avram, G., Ciolfi, L., Maye, L., 2019. Creating tangible interactions with cultural heritage: lessons learned from a large scale, long term co-design project. CoDesign. https://doi.org/10.1080/15710882.2019.1596288.

Butler, T., 2015. The 1943 Bethnal Green Tube Disaster: An Oral History. University of East London, London.

Colley, A., Thebault-Spieker, J., Lin, A.Y., Degraen, D., Fischman, B., Häkkilä, J., Kuehl, K., Nisi, V., Nunes, N.J., Wenig, N., Wenig, D., 2017. The geography of Pokémon GO: beneficial and problematic effects on places and movement. In: CHI, 17 Proceedings of the 2017 CHI Conference on Human Factors in Computing Systems, pp. 1179–1192.

Crooks, A., Schechtner, K., Dey, A.K., Hudson-Smith, A., 2017. Creating smart buildings and cities. IEEE Pervas. Comput. 16 (2), 23–25. https://doi.org/10.1109/MPRV.2017.23.

de Lange, M., Synnes, K., Leindecker, G., 2019. Smart citizens in the Hackable City: on the datafication, playfulness, and making of urban public spaces through digital art. In: Smaniotto Costa,

C., et al. (Eds.), CyberParks – The Interface Between People, Places and Technology. Lecture Notes in Computer Science. Vol. 11380. Springer, Cham.

de Waal, M., 2014. The City as Interface: How Digital Media Are Changing the City. Nai010, Rotterdam. Print. Reflect (Rotterdam, Netherlands); #10.

de Waal, M., 2016–2017. Heritage as platform. In: Street Values (Straatwaarden). Reinwardt Academie, Amsterdamse Homeschool voor de Kunsten, pp. 116–132.

Feiner, S., MacIntyre, B., Höllerer, T., Webster, A., 1997. A touring machine: prototyping 3D mobile augmented reality systems for exploring the urban environment. Pers. Technol. 1 (4), 208–217.

Gehl, J., 2010. Cities for People. Island Press. Washington, Covelo, London.

Giaccardi, E., 2011. On pause and duration, or: the design of heritage experience. In: Proceedings of the 25th BCS Conference on Human-Computer Interaction. British Computer Society, Swinton, UK.

Harvey, D.C., 2001. Heritage pasts and heritage presents: temporality, meaning and the scope of heritage studies. Int. J. Herit. Stud. 7 (4), 319–338. https://doi.org/10.1080/13581650120105534.

Hillier, B., 1996. Space Is the Machine: A Configurational Theory of Architecture. Space Syntax, London. Electronic Edition. Chapter 7.

Histories of Whitechapel, 2019. https://surveyoflondon.org/. (Accessed 25 June 2020).

Huyssen, A., 2003. Present Pasts: Urban Palimpsests and the Politics of Memory. Stanford University Press, CA, Stanford.

Javornik, A., Kostopoulou, E., Rogers, Y., Fatah gen Schieck, A., Koutsolampros, P., Moutinho, A.M., Simon, J., 2019. An experimental study on the role of augmented reality content type in an outdoor site exploration. Behav. Inform. Technol. 38 (1), 9–27. https://doi.org/10.1080/0144929X.2018.1505950.

Julier, S., Fatah gen Schieck, A., Blume, P., Moutinho, A., Koutsolampros, P., Javornik, A., Kostopoulou, E., 2016. VisAge: augmented reality for heritage. In: Proceedings of the 5th ACM International Symposium on Pervasive Displays, Oulu, Finland, pp. 257–258.

King, L., Stark, J.F., Cooke, P., 2016. Experiencing the digital world: the cultural value of digital engagement with heritage. Herit. Soc. 9 (1), 76–101. https://doi.org/10.1080/2159032X.2016.1246156.

Kostopoulou, E., 2018. Rethinking the urban memorial. In: Proceedings of the 25th EVA Berlin Conference, Berlin, Germany.

Kostopoulou, E., Javornik, A., Koutsolampros, P., Julier, S., Fatahgen Schieck, A., 2018. Mediated spatial narratives: experiencing archival material and shared memories in urban space. In: In Proceedings of Media Architecture Biennale 2018, Beijing, China., https://doi.org/10.1145/3284389.3284395.

Martin, K., Lane, G., 2007. Making glue: public authoring in urban tapestries. In: Position Paper for Workshop 'Shared Encounters', ACM Conf. Human Factors in Computing Systems CHI 2007.

McCullough, M., 2007. New media urbanism: grounding ambient information technology. Environ. Plan. B Plan. Design 34 (3), 383–395. https://doi.org/10.1068/b32038.

Memarovic, N., Fatah gen Schieck, A., Schnädelbach, H., Kostopoulou, H., North, S., Ye, L., 2016. Longitudinal, cross-site and in the Wild: a study of public displays user communities' situated snapshots. In: Proceedings of the 3rd Conference on Media Architecture Biennale (MAB). ACM, New York, NY, USA, https://doi.org/10.1145/2946803.2946804. Article 1, 10 pp.

Nijholt, A., 2017. Towards playful and playable cities. In: Nijholt, A. (Ed.), Playable Cities. Gaming Media and Social Effects. Springer, Singapore.

Ritchie, J., 2014. The affordances and constraints of mobile locative narratives. In: Farman, J. (Ed.), The Mobile Story: Narrative Practices With Locative Technologies. Routledge, London.

Tech and Interactive, 2019. https://creativeconomy.britishcouncil.org/themes/technology/. (Accessed 25 June 2020).

van Dijck, J., 2007. Mediated Memories in the Digital Age. Stanford University Press, CA, Stanford.
van Leeuwen, J.P., Hermans, K., Jylhä, A., Quanjer, A.J., Nijman, H., 2018. Effectiveness of virtual reality in participatory urban planning: a case study. In: Proceedings of the 4th Media Architecture Biennale Conference (MAB18). ACM, New York, NY, USA, pp. 128–136, https://doi.org/10.1145/3284389.3284491.
Vande Moere, A., Wouters, N., 2012. The role of context in media architecture. In: Proceedings of the 2012 International Symposium on Pervasive Displays (PerDis'12). Association for Computing Machinery, New York, NY, USA, pp. 1–6, https://doi.org/10.1145/2307798.2307810. Article 12.
Vlahakis, V., Ioannidis, M., Karigiannis, J., Tsotros, M., Gounaris, M., Stricker, D., Gleue, T., Daehne, P., Almeida, L., 2002. Archeoguide: an augmented reality guide for archaeological sites. IEEE Comput. Graph. Appl. 22 (5), 52–60.
Whyte, W., 1980. The Social Life of Small Urban Spaces. Conservation Foundation, Washington, DC.
Wouters, N., Claes, S., VandeMoere, A., 2015. Investigating the role of situated public displays and hyperlocal content on place-making. IxD&A 25, 60–72.

Author Biography

Efstathia Kostopoulou is an architect whose work spans across urban design and heritage to media architecture, infrastructure, and the publics. She is currently a doctoral researcher at the Bartlett School of Architecture, University College London. Her research looks into affective digital and physical experiences that relate public spaces to local memory and culture. Efstathia further works in place management, commissioning and facilitating a number of placemaking projects and activities in East London.

Ava Fatah gen Schieck is an educator, architect, and researcher. She is associate professor in Media Architecture and Urban Digital Interaction and leads the studios: Body as Interface and City as Interface on the MSc Architectural Computation at the Bartlett School of Architecture, UCL. Ava is the director of "Architectural Space and Computation" PhD Program and the departmental tutor for the Architecture and Digital Theory PhD Programme, UCL. She is the principal investigator of "Screens in the Wild: Exploring the Potential of Networked Urban Screens for Communities and Culture," resulting in a unique "living lab" environment of four interactive networked screens (connected through video feed) in London and Nottingham (United Kingdom).

Chapter 7

Place-based design as method of accessing memories and meanings: Historical augmentation in the harbor promenade of Lahti

Aale Luusua[a], Henrika Pihlajaniemi[b], Mika Hakkarainen[c], Petri Honkamaa[c], Eveliina Juntunen[c], and Sami Huuskonen[d]

[a]*INTERACT Research Group and Oulu School of Architecture, University of Oulu, Oulu, Finland,* [b]*Oulu School of Architecture, University of Oulu, Oulu, Finland,* [c]*VTT Technical Research Centre of Finland, Otaniemi, Espoo,* [d]*Tehomet Oy, Kangasniemi, Finland*

Chapter outline

Introduction

In this chapter, we present an analysis of participant experiences of an augmented reality (AR) application that was tested in the harbor promenade area (Satamaraitti) in the city of Lahti, Finland. The application was designed to digitally augment the historically interesting promenade and its surrounding area, which is a popular site for outdoor enjoyment.

Shaping Smart for Better Cities. https://doi.org/10.1016/B978-0-12-818636-7.00006-8

Our case study was a part of the SenCity project, which built intelligent LED lighting and digital service platforms in six cities in Finland and tested smart outdoor lighting in different kinds of urban environments (SenCity, 2020). This concept was built on the insight that the electric supply and communication technologies that are available in smart lighting fixtures make them an attractive support for other technologies, such as data collection or base station installations, as these can be integrated with the lighting infrastructure. The SenCity project focused on three aspects: (1) to study users' needs and experiences of intelligent lighting; (2) to develop and test new technologies such as sensing, data analysis, and communication needed for user-centric designs and services; and (3) to generate business opportunities for smart data–based services. In the project's case studies, the aim was to employ lighting infrastructure as a service platform—an IoT backbone—in intelligent cities. Together, separate pilot studies in six cities around Finland created a living lab ecosystem for developing and testing innovative solutions. The SenCity project group consisted of collaborating research institutions, cities, and companies in lighting and ICT fields.

To design and implement these IoT infrastructures in a sensitive manner in our case study, we chose to utilize a place-based, participatory, and user-centered approach.

In our earlier works, we have explored place in the context of urban technologies, such as urban lighting (Pihlajaniemi et al., 2012; Pihlajaniemi, 2016; Luusua et al., 2016) and urban screens (Luusua et al., 2017) and highlighted the importance of this as an approach (Luusua, 2016). These studies have shown the intertwined relationship between place and technology; one must necessarily affect the other, as both the place and the user experience are mutually dependent. Similarly, understanding users' needs and experiences is a central research theme in the developing and understanding of intelligent technologies in our research and the SenCity project of which this study was part. Thus a user-centered design and development process was carried out involving city representatives, business partners, researchers, and users of the area. Through the participatory design approach, we aimed specifically to gain insights from local people in regard to the present state of the area and their ideas for its development.

In the Lahti case study, we applied this place-based, experience-driven, and participatory approach to augmented reality (AR), an application area that is rapidly gaining popularity. AR presents us with interesting opportunities, as it is very obviously tightly entangled with its context, augmenting users' perception of place with digital content. Next, we will briefly discuss AR as a technology to situate our study within the context of both more widely related research and our specific approach toward it.

Understanding AR through the concepts of embodiment and emplacement

Since 2007 the advent of smartphones equipped with cameras and various sensors has enabled the widespread application of personal and mobile AR technologies (Haugstvedt and Krogstie, 2012). AR has been defined in several ways; for our purposes here, we consider it to be a real-time visual and sensory overlay to a physical environment through any mobile digital device. AR applications can be seen as being part of a larger family of mixed reality (MR) technologies; however, we do not consider pure virtual reality (VR) as being a part of these technologies, unlike Milgram and Kishino (1994). While MR and AR largely retain and augment users' sense of the surrounding physical space, VR replaces this with a completely alternative spatial experience that does away with the users' sense of their current physical locale as much as is possible. For our purposes here, we are interested in the phenomenon of augmenting place. Similarly, we do not consider it necessary for the digital visual addition to be three-dimensional; rather, what is important to us here is the overlaying of multisensory information on top of the normally experienced lifeworld in a portable, digital manner. AR as a media is technically able to utilize as many dimensions as it wishes; as is the case with all media, designers or artists are free to decide which dimensions (including time as the fourth dimension) are necessary for what they are trying to convey. We wished to incorporate two-dimensional information, as this made sense for our project of using historical photographs. The visual component in general, however, can be deemed a crucial aspect of AR applications, since it is obvious that mere background music or sound effects do not constitute a full AR experience; thus AR experiences are usually built around the visual experience, and other stimuli may complement it. Nevertheless, it is up to AR designers and artists to explore the range of possibilities within this medium.

AR has an extremely wide array of useful real-world applications, such as education (Balog and Pribeanu, 2010; Bower et al., 2014; Koutromanos et al., 2015), manufacturing (Ong et al., 2008), and medical assistance purposes (Sielhorst et al., 2006), to name only a few. What interests us here, specifically, is the overlaying of experiential information, sensory or otherwise, on urban environments. There is an existing literature of research on the use and development of AR technologies in urban places. Indeed the use of AR in the context of tourism and urban heritage has grown into closely connected subfields of their own. The purpose of these applications is to offer tourists enjoyable experiences and the ability to get to know the area and its intangible aspects quickly and easily (Jung and Han, 2014), as information can be drawn from various sources to complement the experience (Kounavis et al., 2012). It has also been argued that AR can offer a way for tourists to enjoy sensitive sites without excessively damaging the place (Jung and Han, 2014). Also the introduction of 3D material can make the site more readily understandable by nonprofessionals; furthermore, it can aid in the creation of novel tourism opportunities (Garau, 2014).

Much of the literature on AR in tourism and urban heritage focuses on technical, user interface, and technology acceptance factors (e.g., tom Dieck and Jung, 2018) with some authors focusing on the heritage aspects (Garau, 2014; Westin et al., 2018). Our starting point in this chapter is to examine the use of an AR application in an urban site, which undoubtedly has tourism potential; however, instead of focusing on tourism, we wish to adopt a more generally place-based, even place-centered approach, examining AR as a part of an everyday urban environment for all potential users of the place. We see heritage and history as an important part of the Lahti AR experience, but we assume a holistic viewpoint, where temporal, material, spatial, sociocultural, and personal viewpoints are equally of interest. For these reasons, our approach to understand these experiences makes use of the concepts of emplacement and embodiment. These concepts attempt to account for the mind-body-environment complex in a holistic manner, i.e., they maintain that the mind and body are not separate (unlike in much of Western thought) and similarly that this embodied mind is not separate from the environment that it is situated in. Furthermore, it also highlights the experience of temporality as the present experienced reality, as future expectations, and as subjective and intersubjective histories. The mind, the body, and the place all exist in time, and this fundamentally informs our experience of them.

The concept of emplacement was first developed within the social sciences to highlight the importance of place in embodied experience. This can be seen as a development within the larger, well-established theory of embodiment (Merleau-Ponty, 2012 [orig. 1945]), which, again, is located in the phenomenological (Heidegger, 1977) tradition in philosophy. The phenomenological tradition in general is much used in architectural theory and research (e.g., Norberg-Schultz, 1988); within the field the focus has indeed been on place and our experience of it, often from a very subjective, internal point of view. However, in the social sciences, much of the work on embodiment has focused quite literally on the body, with little emphasis on place (Pink, 2011). Yet, we would argue that the social sciences have been able to usefully theorize and discuss embodiment also from an intersubjective point of view (Luusua, 2016) to employ it as an analytical tool in empirical studies.

Our work aims to combine the aforementioned viewpoints. Thus, in our case study, the place, the participants, and the technology are considered as being situated and intertwined in a complex way with rich sensory and historical connections. Each also have their own agency in the situation, affecting the experience in expected and unexpected ways. Importantly the concept of emplacement highlights the significance of place. For our purposes, this is very useful, since our design and research has emerged from a place-based perspective.

Materials and methods

Our case study was situated within a larger design and development process of the harbor promenade area, which was inherently informed by participatory design. We have collaborated with the city of Lahti in a lengthy, still ongoing participatory process where citizens have been engaged in the development of the area. Our involvement in the development focused especially on the design of lighting environment, the applications of intelligent lighting, and user-oriented digital services. The user-centric design process had five phases: (1) preliminary analysis of the design area and design scenarios; (2) preliminary interviews with local people on-site; (3) cultural probe tasks; (4) codesign workshop; and (5) design implementations based on users' input, including our case study here, the history AR application. One outcome of phases 1–4 was the identification of the need to make the rich history of the area more visible through the means of digital services. This notion was the impetus for the AR work in phase 5. The process is continuing with the deployment of a web-based questionnaire concerning needs and ideas for using the intelligent lighting system for digital services (Fig. 1).

To continue with our participatory design process, we utilized walking in situ interviews to explore participant experiences of the AR application. These had to be executed by our team within a single day due to logistical reasons. Altogether, 14 individuals participated in our study, and these interviews resulted in approximately 300 min of audio/video data gained from the mini-HD video cameras that hung in small pouches around the interviewers' necks. The interviews enabled us to explore participants' opinions, ideas, and memories that the digital augmentation brought up in them, in situ. The video enabled us to see our participants' reactions, both verbal and nonverbal, in their context, including their positioning, surrounding environment, and weather conditions. These audio/video materials were also subsequently transcribed, enabling us to examine them more carefully. While these interviews recorded quickly gauged first reactions, a surprisingly rich set of materials was nonetheless achieved. Some of the interviewees had already been participating in the earlier phases of the participation process. As such, they had a personal history of involvement in the development of the area as codesigners.

FIG. 1 Evaluating participant experiences of the AR application. Still images from a publically available (with participant consent) video recorded on-site. *(Credit: Jaakko Hietala and Henrika Pihlajaniemi.)*

To understand these materials, we first explored them through video and then read the transcriptions through from top to bottom, as individual accounts, allowing themes to emerge without any need to apply concepts to them. Then, we took these themes, as well as our theoretical concepts, and examined the interviews "horizontally" across them to compare and contrast the emergent themes. Finally, we utilized the concept of emplacement to discuss and connect these findings to our earlier works and to the larger literature on embodiment and emplacement.

Our approach naturally has its limitations as well. Among these, we must first point out what while we consider the context-specific nature of our study an important feature of our work, it also means that the specific reactions of our participants are also grounded in a certain time and place. They cannot be immediately removed from that context; rather, they must be understood in context, and any conclusions must be carefully considered when attempting to apply the insights elsewhere. Similarly, we opted for the fairly in-depth approach of walking interviews. While this approach avoids some of the pitfalls of doing, for example, a survey with a larger number of participants, it also means that the results here are not to be handled in a statistical manner. Our purpose here, then, is to rather examine participant views more in depth rather than in breadth.

Research setting and AR application

The site of our research was a pedestrian route, 1.5 km in length, spanning from the Sibelius Hall in the main harbor area to the Lahti Sports and Fair Centre at the other end. The area sees many visitors especially during exhibitions, various winter sports events, and ice hockey matches. During the summer the activity in the harbor and numerous events enliven the area. Currently the area is being developed into a lively place of leisure activities for citizens and tourists alike (Fig. 2).

Our AR application was designed and researched within this ongoing developmental process; within it the entire lighting of the harbor promenade will be renewed, and intelligent lighting solutions will be introduced through the addition of sensor and fiber optic networks. In the custom-made and designed lighting poles, there are adequate installation space and connections for further assembly of different kinds of sensors fulfilling the needs of additional stakeholders. In addition to basic street lighting, the design includes DMX-controlled RGBW lighting for enabling an additional level of experience through color and dynamics. Furthermore an open Wi-Fi network and web cameras will be installed (Fig. 3).

The context for the history augmentation application contained several target locations where users were able to see historical images from the Lahti City Museum's archives superimposed over the real-world camera view. The desired target location and the associated historical image were selected from an interactive map layout. To successfully align the photo to the environment, the

FIG. 2 Aerial view/map of the case study area. *(Credit: City of Lahti/Lê Anh Huy and Henrika Pihlajaniemi.)*

FIG. 3 AR application. Still images from a publically available (with participants' consent) video recorded on-site during testing. *(Credit: Jaakko Hietala and Henrika Pihlajaniemi.)*

viewing location needs to be near the place where the original photo was taken. For this purpose the user was guided by showing both the map location and the historical image. When the matching scene was recognized by the application, the photograph automatically snapped to the correct location and orientation in the camera view. In the application the user was also able to change the photo transparency to easily compare the scene from the photo to the current scenery. Within the AR application the target scene was recognized and tracked using precreated point clouds and the VTT Technical Research Centre of Finland's Augmented Reality SDK (ALVAR SDK). The point clouds were created by taking several reference photographs and reconstructing the current scene using the Agisoft PhotoScan application. Historical photographs were aligned to the point cloud so that distinctive visual features would align as well as possible. Using this method the photo alignment was able to yield a good viewing experience even when the user's viewing location was not entirely correct.

Analysis

Our participants ranged from middle-aged to senior citizens with various professional backgrounds; these included an entrepreneur, two architects, a daycare teacher, a licensed practice nurse, and an engineer. Notably, two of our interviewees also worked as Lahti tour guides, and one was an architect who had been involved in the planning of the area. The participants were free to choose to come alone or accompanied by another person. Most people came alone (P1, P4, P6, P7, and P9), with some also opting to come with their significant other (P2a and P2b; P3a and P3b); also, one group consisted of three friends (P8a, P8b, and P8c).

Personal histories gave depth to participant experiences

One of the immediately emerging themes across our interviews was the significance of personal experiences with the place in question. An outlier here was the group of three people (P8a, b, and c) who had professional experience in project development. Understandably, they focused quite singularly on the application itself at the expense of personal reflections. Most participants, however, were eager to offer up their personal memories and knowledge of the places in question. In the following exchange, one of our team members is demonstrating the application to an elderly couple. After a lengthy explanation and a visual demonstration, the participant simply replies with his own experience, ignoring the technological aspect of the demonstration completely:

> *Interviewer: This is a map, all these things here are (…) old photos and their locations and you can see our location as a blue [symbol] and by touching on that it will load it, for example, we're looking at the Vesijärvi railway platform, so if we load that the picture will appear. (…) You can see here that this was a railway platform. Then there's this feature—you can put that on hold for now—you can go like this and it will become transparent so you can look through it. Now it's [both] the way it is at this moment and what it's been before.*

> *P3b: Those train cars, yeah. I've taken the train from here to Heinola over there. It must be 20 years ago. [Actually, the train service stopped in the 1980s, so almost 40 years ago.]*

In our mind, this is a positive outcome, since the technology is not supposed to be at the forefront. Rather the purpose of the app was to facilitate place-based experiences. This seemed to work especially well with senior participants, who had more experience with the place and less experience with the technology. The following represents a fairly typical comment:

P1: I looked at many of the images, and I told myself, many things are still were they used to be. [Yet] I'm surprised that there are these train tracks [in this image], that the train really went through there, the tracks were right next to us. (…) I'm beginning to wonder about a lot of things. Like, for example, that warehouse over there, was it really a wastepaper collection site (…) when I came to Lahti.

This comment highlights what was essential to many of our participants' answers. Overall, there was a recognition of some key elements of place from the past, particularly buildings. However, participants might not have been able to recollect other important facts about the place, like the existence of train tracks, which also were a central element of the place previously. In some participants' responses the professional and the personal became intertwined:

Interviewer: Why don't you hold this for a while now, I'll put my mittens on (…)

P2a: It actually looks exactly the same now when you really look at it!

Interviewer: Yeah, surprisingly similar.

P2a: That building, they've done almost nothing to it.

P2b: There's been an effort to maintain it and restore it, there was an architect called [name] who really delved into the history of the building (…) [Only] that door over there is new.

P2a: Are we on the train tracks now then?

P2b: Probably … no, actually, we're just about … see here there's this old base of a construction crane. There was a crane here when I was little, in the 50's, there was a crane there.

Even when the interviewer made a slight mistake and stopped P9 in the middle of his sentence, P9 ignored this and persisted with his personal memories of the place:

P9: Yeah, there was no …

Interviewer: Shall we move on to the next [picture]?

P9: Yeah, I didn't remember this thing at all, that this warehouse platform had been here like this (…)

Almost all participants were able to tell stories about what they did in the places, and while perusing the AR app, they actively made connections with their personal histories and the history of the place. This is very interesting in itself and potentially useful for designers. Overall, then, the AR application was able to connect the various personal histories of these individuals with the history of the place, acting as a catalyst for our conversations with them. As a result, nearly all participants regarded the historical AR application in a very positive light.

Commonalities in perceived place identity across individuals' accounts

Most participants identified the harbor promenade as the "city's living room"—an epithet that participants used on several occasions voluntarily.

P4: I usually describe this place as the living room of Lahti people [laughs], like this is the place that's happening and now that this has been opened after all the industrial estates [have left] now there's all these events for the people of Lahti and in the summer you're just here all the time. I mean, there's no one at the city center of Lahti. Well, they're here in the harbor.

According to this participant the harbor area was even a more significant place to spend time at than the city center itself. Many reported that they frequented the place fairly often, opting to come there for either outdoors activities, such as walking and skiing, or to meet friends and acquaintances at the restaurants and cafés that surround the area.

Interviewer: How would you describe this place?

P3b: Well this is ... listen, I used to work here, there was a lumber yard.

P3a: Rauma-Repola.

P3b: Yeah, the Rauma-Repola lumber yard, I did some work for them, metal stuff, this was a terrible place back then.

P3a: Now this is a living room in the summer for the people of Lahti. I think this is a place where people meet each other.

Similarly, participants were keenly aware of the closed-off nature of the place as a harbor area. One elderly person reminisced that his mother had forbade him to go there as a boy, as the still functioning harbor was a dangerous place—yet he had not been able to resist the urge to jump on the logs that had been piled there for storage. Here, P9 immediately offers his view on the history of the place:

Interviewer [tells about the project and the study at some length]

P9: "This was kind of a closed off area when I was a little boy (laughs)"

For this participant, the experience of the harbor's prior state apparently overrode everything else; without any prompt, he volunteered this memory. Participant P7 gave a strikingly similar response at the beginning of her interview:

Interviewer: Okay, well that sounds good. Shall we go over there and see more? So, this is a new kind of an experiment where you can [look at] these images in a way, you kind of go where the picture has been taken from [to see them].

P7: It's a good idea.

Interviewer: There's one, if I remember right.

P7: This place was off limits before the area was built.

Once again, without asking, the participant volunteered this information through an abrupt nonsequitur, hinting that this was key information about the place to our participant. The shift in the use patterns and identity from an industrial area to a gentrified outdoor living area had apparently been very significant. The identity of the place, then, was heavily historical, and this manifested in the participants' interests as well; there was a clear self-selection bias toward individuals who had an interest in historical topics. For our purposes, this was not problematic, since the service would be geared toward these individuals anyway.

Perspectives on usability were affected by participants' backgrounds and the outdoor context

The conceptual design of the application seemed easy for our participants to understand, and the images were seen as being very relevant and interesting. The tour guide offered also a more nuanced and critical view yet confirmed that the applications might work for tour guide purposes.

P4: I think it's a good idea. But I'm thinking all the time professionally as well, how could I use it, so [from that perspective] it might not be suitable for large groups unless the app is publicly available (…)

K: Everyone would view it through their own …

P4: Yeah, and then, wow, it could work. But I like it this way too, on my own.

The importance of history was highlighted, especially by more senior participants, and bringing local history to life was deemed a worthy subject. One participant also made the connection with the earlier tradition of newspapers publishing pairs of images where contemporary images of locales were contrasted; she had collected these images before as a hobby.

P3a: At one time, I collected those "Lahti now and then" pairs of pictures that were in the [local newspaper] (…) I thought it was interesting to see what everything used to be like and how everything's changed.

This connection with earlier and existing practices relating to images of the built environment helped make the application's concept easy to understand.

Obviously, as we have pointed out in our earlier works (Ylipulli et al., 2014), climate is a central issue in usability and urban computing, as highlighted in the following, slightly humorous exchange:

Interviewer: Let's go back. Ah, it went away. Did I press stop with my thumb ...?

P4: Be careful not to drop it [laughs]. I'm terrible with these devices (...)

Interviewer: Yeah, and [this device] is slippery too. That's why I wore leather gloves.

P4: Okay, it's the right spot now.

Interviewer: It's probably slow because of the cold.

P4: It must deplete the battery (...)

Almost all participants mentioned the cold and the significance of gloves. This is a feature that is inescapable in a northern climate (we have explored this before in (Ylipulli et al. (2014)), as people usually need to wear gloves for about 4–6 months of the year. Navigating the new application in the cold without gloves can create situations where interaction with the technology necessarily comes to the forefront of the experience. However, users did not have significant difficulty using the device in our study setup. For some of our participants, previous examples of AR such as Pokémon GO was already familiar and made it easy to interact with the application. For some, AR was not familiar at all. Indeed, some participants were hesitant to hold the device, opting instead for the researcher or another participant to use the device as much as possible.

P2b: Would you like to hold it; your hands are steadier?

P2a: Alright, I'll hold it.

[Later in the interview]

Interviewer: Why don't you hold it for a change.

However, our priority was not usability testing; rather, we wanted to explore the possibilities of AR from an emplaced point of view. Thus the walk-along (and use-along, in this case) method served us well, since it enabled participants to better focus on the content. For a finalized application, more typical user testing should be added to the next stage of the design process.

Participants offered various ideas for development of urban AR

Participants also came up with some possible additional uses for locational AR applications. The idea of expanding the photographs somehow into a three-dimensional, more immersive view was suggested by some. These would involve narration and soundscape. Also the idea of deliberately using pictures from the various seasons emerged. Suggestions for other uses also emerged, for instance, people looking to build their own houses might be interested in checking whether a specific lot is for sale while walking or driving by. Further historical images and stories, focusing on the Second World War, were also suggested. One also suggested that the natural geological formations of the area be visualized. This idea would create a novel kind of a natural history AR application.

AR might also be used to tell the stories of new residential areas and important buildings, such as the Sibelius House in Lahti, as suggested in the following:

> *P1: The new housing area there, there's a lot of stories there. There's a lot of people who've come from Helsinki there now and they have their stories. There's a lot of stories relating to building it and visions for its development and also thoughts on why it looks like this now. The Sibelius building in its entirety. Why has it been built like that and why do people say it's the finest wooden building when it's a box of glass from the outside (…) And there really is a lot of that wood in there, and those details could be explained. I mean, as a building it must be very significant to the architects. So, I think there might be several places like this here. A water tower (…) might have a lot to say about why it's in that place.*

This freewheeling commentary on the built environment was insightful in many ways. Firstly, P1 brings out the fact that also people who have recently arrived in the area "have stories" too, not just people who've had long established relationships to the place. This is a notion that is often overlooked in urban development and urban travel. Second, P1 openly wonders about the quality of the built environment. Why are some buildings considered better than others? What ideas and realities lie behind contemporary architectural creations? It would be immeasurably useful for citizens and visitors alike to be able to pull up a video or audio of the building's architect explaining their thinking to passersby. Finally, while P1 began with the idea of telling peoples' stories, she ends up with the poetic idea of buildings also having their stories. Even a utilitarian building like a water tower "might have a lot to say" about itself. The stories of people, projects, and buildings were some of the themes that AR could highlight, then.

Travel and tourism emerged on several occasions, unsurprisingly, but there were some useful nuances to some of the tourism related ideas, as in the following:

> *H8a: Sure, and in general all those tourist type sights. Like ski jumping, like there could be live footage in the summer so you can see people ski here in the winter [laughs]*
>
> *H8c: And in the winter you could see like, hah, people have been swimming here, there's a pool.*
>
> *H8a: Or art tours. Or why not just shopping (…) Nature in general is interesting, and maybe to foreigners, how the place looks in the winter or summer.*

The suggestions ranged from visualizing the past, to various aspects of the present, to the future. Importantly, while the suggestions were not surprising as general design ideas, they represented localized versions of them; many of these aspects of the place had been unknown to us as out-of-towners.

Overall, our participants wanted to be better connected to place. They wanted to know what had happened in the place, who was in the place, who had arrived, and what their stories were. They wanted to know more of what they could do in the place and why and how the environment had taken that shape. As a central way of relating to the place, they engaged their own memories, reminiscing at length about the places we visited and the historical images they saw. Despite obvious challenges with weather and climate, AR proved a useful tool, providing participants with accurate visual portrayals of the places to spark and complement subjective memories.

Conclusions

Our results contribute to the experiential, intersubjective understanding of digitally augmented public urban places, a phenomenon that is still poorly understood. The empirical findings we gained correspond with earlier theorizations concerning human relationship to technology. While in use, the technology remained mostly in the background (Heidegger, 1977; Weiser, 1991), while experiences in place and about place (Pink, 2011) were the prominent aspect in participant interviews. This interplay of digital technology and place further strengthened the idea that the design of smart city technologies should indeed be approached with an emplaced, participatory, and human-centered design approach (Aurigi, 2017; Willis and Aurigi, 2017; Luusua et al., 2016).

Overall, our results highlight the histories of people and places and how these can play a significant part in the design of urban technologies. While it is impossible to design for each individual user of a system, designers can of course utilize empathy-driven expert methods to attempt to cater to the needs of different individuals, constructing various kinds of theoretical user groups and needs. A more laborious, empirically oriented method is to conduct participatory design and studies, if resources allow for it, to gain an idea of real individuals' hopes, needs, and ideas. However, the place-based approach offers a third method that is still vastly underutilized in urban technology design. We argue that designers and developers would do wisely to consider this approach at the outset of their projects.

The reason for this is that our results show that place was the touchstone, the common denominator between our different participants, and, importantly, it enabled them to connect their personal memories and meanings to the technology at hand. This had the power to render the experience of digital augmentation much more meaningful to each participant, even though the designers knew nothing about these future users in advance. In other words, *meaningful connections were created between participants' individual histories, the history and identity of the place, and the affordances of the technology.* For our purposes, this connection also produced participant reflections that enabled the designer-researchers to deeply understand the place-people-technology relationship in the participatory process.

Furthermore the AR also acted as a medium that highlighted temporal aspects of place that would have otherwise been invisible. This was an important motivator for the technology. Cultural heritage, unsurprisingly, is an important focus of AR research, though this is often geared at benefiting visitors. Our study contributes to this by focusing on local people instead of tourists, demonstrating that residents are also a potential interested user group for these types of AR applications.

Even more importantly, though, this study strived to make a contribution to the study of urban technologies that would depart from the usability and technology-oriented focus of existing research. Our starting point was that since AR is a heavily space and place-based technology, it would be pertinent to also approach AR from an emplaced point of view. Our study, consisting of the design and the participatory evaluation of the application, was thoroughly emplaced in its goals and methods. We deem this approach successful and argue for its thorough integration in the field of AR technology research and design. Our analysis similarly was driven by an overall emphasis on emplacement, and this was a productive lens in understanding our materials.

From the point of view of urban places, AR offers a way to deepen urban inhabitants' relationship to their environment through digital augmentation. Our participants used the application as an opportunity to talk about what was important to them and what their relationship to the place was. Within a wider range of possible urban projects, the development and implementation of a place-based AR may be an intriguing, respectful, and resource-efficient way for cities to render visible what makes their city, its places, and people unique.

References

Aurigi, A., 2017. Space is not a platform – foregrounding place in smart urban design. In: Zammit, A., Kenna, T. (Eds.), Enhancing Places Through Technology. Lusófona University Press, pp. 7–17.

Balog, A., Pribeanu, C., 2010. The role of perceived enjoyment in the students' acceptance of an augmented reality teaching platform: a structural equation modelling approach. Stud. Inform. Control 19 (3), 319–330.

Bower, M., Howe, C., McCredie, N., Robinson, A., Grover, D., 2014. Augmented reality in education – cases, places and potentials. Educ. Media Int. 51 (1), 1–15.

Garau, C., 2014. Augmented Reality to Support of Cultural Heritage. Utopia or Plausible Reality? A Planner's View. LAP Lambert Academic Publishing, pp. 1–92.

Haugstvedt, A.C., Krogstie, J., November 2012. Mobile augmented reality for cultural heritage: a technology acceptance study. In: 2012 IEEE International Symposium on Mixed and Augmented Reality (ISMAR). IEEE, pp. 247–255.

Heidegger, M., 1977. (transl. Lovitt, W.). The Question Concerning Technology, and Other Essays. Harper & Row, New York.

Jung, T.H., Han, D.I., 2014. Augmented reality (AR) in urban heritage tourism. e-Review of Tourism Research 5.

Kounavis, C.D., Kasimati, A.E., Zamani, E.D., 2012. Enhancing the tourism experience through mobile augmented reality: challenges and prospects. Int. J. Eng. Bus. Manage. 4 (10), 1–6.

Koutromanos, G., Sofos, A., Avraamidou, L., 2015. The use of augmented reality games in education: a review of the literature. Educ. Media Int. 52 (4), 253–271.

Luusua, A., 2016. Experiencing and Evaluating Digital Augmentation of Public Urban Places. Doctoral dissertation, University of Oulu Graduate School.

Luusua, A., Pihlajaniemi, H., Ylipulli, J., 2016. Northern urban lights: emplaced experiences of urban lighting as digital augmentation. In: Architecture and Interaction. Springer International Publishing, pp. 275–297.

Luusua, A., Ylipulli, J., Kukka, H., Ojala, T., 2017. Experiencing the Hybrid City: the role of digital technologies in public urban places. In: The SAGE Handbook of New Urban Studies. SAGE, London, pp. 535–549.

Merleau-Ponty, M., 2012. Phenomenology of Perception. Routledge, London (orig. 1945).

Milgram, P., Kishino, F., 1994. A taxonomy of mixed reality visual displays. IEICE Trans. Inform. Syst. 77 (12), 1321–1329.

Norberg-Schultz, C., 1988. Architecture: Meaning and Place. Electa/Rizzol, New York.

Ong, S.K., Yuan, M.L., Nee, A.Y.C., 2008. Augmented reality applications in manufacturing: a survey. Int. J. Prod. Res. 46 (10), 2707–2742.

Pihlajaniemi, H., 2016. Designing and Experiencing Adaptive Lighting. Case Studies With Adaptation, Interaction and Participation. Doctoral dissertation, University of Oulu Graduate School; University of Oulu, Oulu School of Architecture, Acta Universitatis Ouluensis H 3.

Pihlajaniemi, H., Luusua, A., Teirilä, M., Österlund, T., Tanska, T., 2012. Experiencing participatory and communicative urban lighting through LightStories. In: Proceedings of the 4th Media Architecture Biennale Conference: Participation, pp. 65–74.

Pink, S., 2011. From embodiment to emplacement: re-thinking competing bodies, senses and spatialities. Sport Educ. Soc. 16 (3), 343–355.

SenCity, 2020. SenCity Project Website. 201Retrieved 15 January 2019 http://www.sencity.fi.

Sielhorst, T., Bichlmeier, C., Heining, S.M., Navab, N., October 2006. Depth perception – a major issue in medical AR: evaluation study by twenty surgeons. In: International Conference on Medical Image Computing and Computer-Assisted Intervention. Springer, Berlin, Heidelberg, pp. 364–372.

tom Dieck, M.C., Jung, T., 2018. A theoretical model of mobile augmented reality acceptance in urban heritage tourism. Curr. Issue Tour. 21 (2), 154–174.

Weiser, M., 1991. The computer for the 21 st century. Sci. Am. 265 (3), 94–105.

Westin, J., Foka, A., Chapman, A., 2018. Humanising places: exposing histories of the disenfranchised through augmented reality. Int. J. Herit. Stud. 24 (3), 283–286.

Willis, K.S., Aurigi, A., 2017. Digital and Smart Cities. Routledge.

Ylipulli, J., Luusua, A., Kukka, H., Ojala, T., 2014. Winter is coming: introducing climate sensitive urban computing. In: Proceedings of the 2014 Conference on Designing Interactive Systems, pp. 647–656.

Chapter 8

Designing smart to revitalize a multicultural shopping street

Ummu Sakiinah[a], Ingrid Mulder[a], Annemiek van Boeijen[a], and Rudi Darson[b]

[a]*Faculty of Industrial Design Engineering, Delft University of Technology, Delft, The Netherlands,* [b]*Shop owners' association of Middelland, Rotterdam, The Netherlands*

Chapter outline

Introduction

This chapter presents a local contextually grounded design project, illustrating the barriers and enablers of a culture-sensitive design approach in the context of a smart city. The focus in this project is on how a placemaking perspective can push the current envelope of the smart city debate. Within this, we refer to a broader concept of a *sociable smart city* (Mulder, 2014, 2015) that enables transforming society into a more participative domain where innovation takes place. More precisely a sociable smart city combines the best of two worlds: on the one hand, a social city that is people centered, values active citizenship, and embraces community-driven innovation, and, on the other, a smart city that utilizes and fosters technology-driven innovations. The current project takes liveability as a starting point but embraces digital innovation as one of the many

Shaping Smart for Better Cities. https://doi.org/10.1016/B978-0-12-818636-7.00014-7

factors that play a role in urban revitalization process. The resulting holistic approach includes the current functions of the shopping street, such as shopping, living, working, and leisure, and differentiates between the following entities: the hardware, software, and mindware.

The chapter's focus on placemaking is specifically related to the economic revitalization of the Middellandstraat, a multicultural shopping street in Rotterdam, The Netherlands. Through a growing influx of immigrants, the Middellandstraat has evolved into a unique multicultural neighborhood with a similar mix of shops. However, this area went into decay due to economic challenges and inadequate maintenance, resulting in a decline of the number of visitors and customers. This was due to competition from Internet suppliers; outdated business models of the shops; lack of knowledge how to use cheap marketing instruments, such as Facebook and Instagram; diminishing buying power of existing customers; and, as a consequence of all, diminishing income for shop owners. Therefore liveability was at stake, and only few shops could sustain themselves. The local government then tasked several civic servants with assessing the area, and their findings led to consider closing down the shops in the street. However, the government decided to turn the problematic situation into an opportunity for change and instead challenged and invited the various urban stakeholders (residents, shop owners, and property owners) in the Middelland district to come up with innovative ideas to revitalize the area. One of those selected ideas was to entrust a designer (the first author of this chapter) outside the group of stakeholders with the task to come up with a solution that would help with the revitalization of the shopping street while respecting the rich context of the area. The next section introduces that rich design context. The third and fourth sections of the chapter discuss the design approach and the resulting concept, respectively. The chapter concludes by reflecting on the lessons learned and outlines possible design principles for smart urban design and implementation.

Context of study

As part of the transition toward becoming a smarter, more sociable, and a more resilient city, Rotterdam has embraced cocreation and experimentation as a new way of participatory city making. The central government has selected the Middelland district as one of the pilots for a large-scale experiment of democratic innovation, embracing the value of cocreation between citizens, local entrepreneurs, and the municipality. This experiment has been financed with nine million euros including 2.5 million for the shopping street and ran for three years, from 2016 to 2018.

Mooi, Mooier, Middelland

Mooi, Mooier, Middelland is the title of the resulting joint initiative of residents, entrepreneurs, and officials of the city to improve the sense of ownership and liveability in Middelland, a neighborhood in Rotterdam West.

The goal of the program is to improve quality of life through physical improvements of public places and spaces, as well as building social structures in the neighborhood. The program operates with specific working groups on different topics, including both citizens and entrepreneurs in the area, as well as civil servants. This initial top-down decision process (on a local budget) was hijacked by a select group of citizens and then redirected by free agents within the municipal government to a process endorsed by many people and organizations throughout the neighborhood. Another key role was performed by the "District Advisory Committee" (in Dutch: Gebiedscommissie), a body consisting of elected representatives, entrusted with the task of facilitating the participation of all Middelland residents. It was also endorsed by Mayor Ahmed Aboutaleb, who created the conditions for an alternative approach by framing it as "an experiment." The title "Mooi, Mooier, Middelland" was coined by a local poet, Derk Otten, a detail that exemplifies the appreciation for creativity in the program. Translated into English, it refers to "a continued betterment of Middelland." Mooi, Mooier, Middelland can be seen as a cocreation process or as a cluster of bottom-up experiments that continued beyond the initially financed program. In other words the cocreation of public policy can be seen as an exemplary attempt of democratic innovation in an urban context.

Within the program the government also challenged local urban stakeholders in the Middelland district to come up with bottom-up and contextualized proposals to improve the area economically. The call for improvement did not promote the closing of existing shops, as was suggested by the initial advisory report. The government did not simply delegate the challenge to local stakeholders, but intended to experiment with new ways of participatory city making (Puerari et al., 2018; Concilio et al., 2019).

Middellandstraat

The Middellandstraat shopping street is a culturally diverse and an ethnically mixed area in Rotterdam. More than 12,000 inhabitants live in the Middelland district, which Middellandstraat is part of. When walking along the street, one can hear various languages in casual conversations. Rotterdam is generally highly multicultural, and half of the population has non-Dutch parents (Municipality of Rotterdam, 2017). This applies also to the shop owners in Middellandstraat, who have many different cultural backgrounds, including Moroccan, Turkish, Surinamese, Pakistani, Chinese, Indonesian, and more. There are over a hundred, mainly privately owned shops in Middellandstraat, such as butchers, bakeries, clothing shops, and supermarkets. Within the range of small shops, the food sector is particularly well represented. This is somehow not surprising, as food experiences are a clear expression of cultural diversity. Besides, it is easier to start a shop in the food sector than in most other sectors, as there is a high demand for it and the immigrant shop owner knows the market and can manage the logistics of sourcing the foods.

The Middellandstraat is located after West-Kruiskade, also known as Chinatown, where a large variety of mainly Asian ethnic shops are located (see Fig. 1). From Rotterdam station walking to the right, visitors will first arrive at West-Kruiskade, but only fewer visitors will continue their trip to the Middellandstraat. One reason might be that they are already satisfied by West-Kruiskade. Another might be that they do not continue crossing the main thoroughfare, dividing the Middelland (Delfshaven) district from the Oude Westen (city center) one. As explained in the next section, the thoroughfare is a border between two separate former municipalities.

Fig. 2 shows the street of Middellandstraat, a double lane street with various shops on each side with a well-known community center in the central part. While the open space enables people to explore, having the traffic in the middle street does not encourage people to visit those shops at the other side of the road (see Fig. 3).

Revitalization as a cocreation challenge

Over the years, many urban stakeholders have been involved in the revitalization of Middellandstraat, with the local government being a key stakeholder. However, until March 19, 2014, Rotterdam's fourteen districts had the formal status of local municipalities under the Dutch Municipalities Act. The main rationale for decentralization was to bring government closer to the people: the district's residents directly elected their own local municipal council as being responsible for many activities such as the revitalization of streets. Until 2014, only citywide affairs, such as major infrastructural projects, remained within the jurisdiction of the central municipal council. In 2014 local municipalities within one large city were abolished by Dutch law. However, the city of Rotterdam decided to maintain its districts in the form of directly elected "area committees." These area committees no longer have autonomous powers however and act primarily as advisory and participatory bodies for the central municipal council.

Understanding this history explains on the one hand the motivation for democratic experimentation to encourage citizen participation and on the other hand the differences between Middellandstraat and West-Kruiskade, as they have a long history as belonging to two different local municipal governments, namely, Delfshaven and Rotterdam City Centre. Differences between these two streets are also apparent in property ownership. Whereas at the West-Kruiskade, one housing corporation is generally the owner, this is not the case for Middellandstraat, where citizens own most of the properties. Consequently, it was easier to make changes aiming at the revitalization of the West-Kruiskade area, as the local government could make and implement any decision easily with the housing corporation, which in turn could invest a lot of money in its properties. The Middellandstraat buildings, though, are mostly owned by private property owners. This fragmented stakeholder situation makes communication and getting agreement and permissions for change in favor of the common good more complex in comparison with the situation in West-Kruiskade.

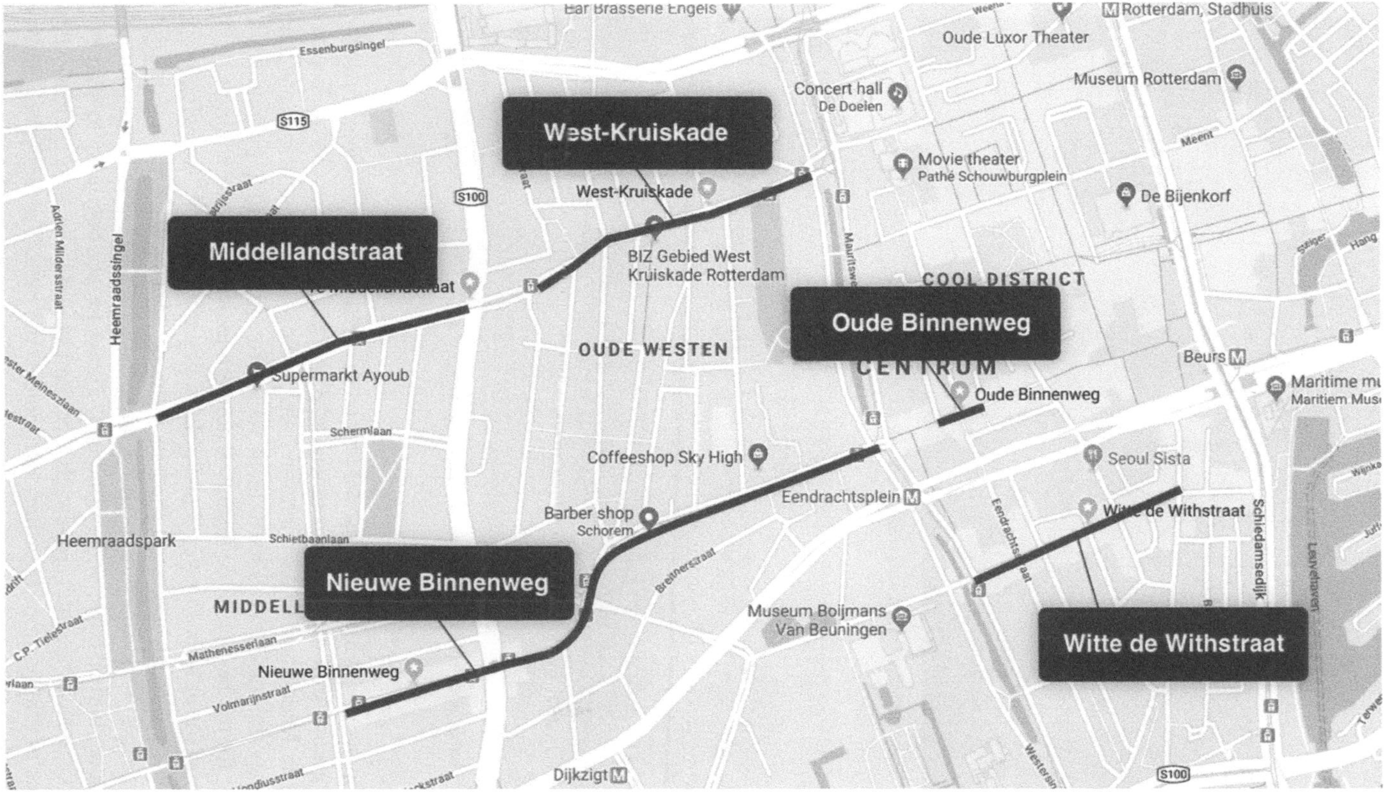

FIG. 1 Map of Rotterdam (district Middelland and Oude Westen) with the main shopping streets highlighted in *red* (dark gray in print version) (Google Maps, 2017).

FIG. 2 The Middellandstraat area. *(Source: Authors.)*

FIG. 3 Usable open space in the area. *(Source: Authors.)*

Shop owners in Middellandstraat rely on the area to make a living. However, the poor condition of the environment affects their businesses. Most property owners in Middellandstraat, therefore, are willing to renovate the shops, and when they do, they tend to leave them vacant for relatively long periods while looking for the best possible tenants. These are tenants who are willing to sign a contract for at least five years and able to pay a higher rent, allowing property owners a quick return on their investment. Mostly, they are looking for entrepreneurs who want to start shops other than the usual café/bar-restaurant, hair salons, barbershops, and call houses. These are not only easy to start but also easy to terminate and are already overrepresented in the street. Clearly the presence of empty shops does not contribute to the overall revitalization ambitions for the street, reinforcing instead a negative image of Middellandstraat.

Therefore the local shop owners' association coordinated a collaborative search for ideas to revitalize the multicultural shopping street while avoiding a negative form of gentrification that would exclude the current—and mainly economically vulnerable—shop owners and scare off another important stakeholder: the local residents. This was not a straightforward task. In many top-down pushed strategies, including similarly top-down smart city visions, local residents and private property owners are not necessarily in the driving seat (Foth et al., 2015).

The resulting cocreated concept of revitalization was through innovation using smart design and redesign of the *hardware* (the *architecture* of the shops, the street, and street furniture), the *software* (activities to enhance shopping, living, working, and leisure; developing new business models; entrepreneurship; strategy; branding), and the *mindware* (creating positive and valuable experiences, new perceptions, and a positive sense making) of the shopping street (Darson and Veen, 2017). In other words, at the heart of the revitalization in Middelland was the design challenge to use all the components making up the street environment, alone or in combination, to achieve the economic goals—healthy shop and district economy—to make the area safer and more liveable.

Although the program of Mooi, Mooier, Middelland enabled the general process of cocreation, it did not guarantee an easy implementation of cocreated ideas. Clearly, making proposals among people with different interests and agendas is not straightforward, and agreeing on which direction to take was not easy with so many different interests and voices. Consequently the shop owners' association took the initiative to facilitate a new perspective to frame the problem by bringing in a designer who had not been involved in the ongoing cocreation process of Mooi, Mooier, Middelland for the shopping street. The assumption in the current study was that this role of being an outsider would be an advantage, realizing that it could also be a disadvantage (Banks, 1998; Hao, 2019) as, for example, the designer might not be able to understand the deeper meaning of what people said and did.

Design challenge

The design challenge raised by the shop owners' association was to use design as a means to improve and revitalize the street (Darson and Veen, 2017). Aside from this assignment a variety of other projects and consultations was taking place on how to design or redesign hardware and software to influence the functions of living, shopping, working, and leisure toward economic revitalization. Fig. 4 shows a matrix highlighting the design challenges in relation to the various functions of the street.

As mentioned before, it was decided that a product/service designer would be asked to find new solutions for the shopping street, taking into account the design components detailed in Fig. 4. This approach is described in the following section. For the briefing the shop owners' association decided through its representative to focus on the food sector. This sector's various Rotterdam-based

		Functions of an urban shopping street			
		Shopping	Living	Working	Leisure
Design components	Hardware	Shop facades Architecture Interior design Street furniture	Parking lots Bicycle storage Street furniture	Working spaces in shops Street furniture	Benches Squares Playing ground Street furniture
	Software	Business model Services Information Virtual presence Internet Shopping activities and cultural aspects	Street activities Safety	Internet connectivity Pleasant working spaces	Street activities Information
	Mind ware	Attentive Innovative Ambience	Sociable Cozy Ambience	Mobility Innovative Ambience	Child and visitors' friendliness Ambience
	Results	Attractiveness Word of mouth marketing Increase in visitors and sales	Attractiveness Liveability In demand	Attractiveness Word of mouth Increase in visitors and sales	Attractiveness More visitors More families with children More sales

FIG. 4 Matrix highlighting the design components versus shopping street functions.

immigrant shop owners are representing well the culturally diversity of the district. This seemed to be a promising starting point also because food and culture have an intimate relation and food is a potent symbol and foundation for a collective identity (Bessière, 1998). Consequently, after consulting with its members, the shop owners' association decided to take "food experiences" as a strategic theme to attract more customers to the street. A noted positive precedent was the Witte de Withstraat (see Fig. 1), a shopping street in Rotterdam, which for its revitalization solely focused on food and turned itself into one of the most vibrant and economically successful streets in Rotterdam. Moreover, food can have a very strong relation with religious beliefs, and celebrating these is important for a multicultural area located in the Netherlands, a country traditionally adhering to different values. For many people, food has become an important and integral part of experiencing culture, and in accordance with that, this was a promising theme to design for. The question posed by the shop owners' association and partly leading this research was: how can food shops, such as a halal butchery, a Turkish bakery, and an ethnic African supermarket, be made interesting and unique for new visitors? What can design do? The shops were not attractive and informed by old business models—and the whole shopping street was perceived as unsafe (Mooi Mooier Middelland, 2016). Also the shops were not visitor friendly enough, at least not for those who needed to be attracted, but were not familiar with the habits and culture of the shop owners. The shop owners were not hospitable enough and had the habit of ignoring customers if they did not know them already. The physical look of the shops, the "hardware," was also not attractive because of the lack of maintenance and of attractive design. Besides, potential visitors who were not local to the neighborhood would not know about the variety of culture-specific food that was sold in Middellandstraat and that could make the street unique. The products and services offered in these stores could not be found anywhere else in Rotterdam, yet such possibilities were relatively invisible or not attractive enough.

A culture-sensitive design approach

The design approach was informed by the double diamond model (Design Council, 2005) as exemplified in Fig. 5. This process consists of four phases: discover, define, develop, and deliver. In each phase, there are activities for "diverging" (opening up the design "space" by increasing the possibilities for finding solutions) and "converging" (narrowing down the design "space" by making design choices and coming to conclusion) and multiple iterations according to the basic cycle of design (van Boeijen et al., 2020). At the heart of the assignment was the focus on the improvement of the shopping function (see Fig. 4), with a particular attention to the software and mindware. The shop owners and others were mainly interested in the design of the hardware instead.

In the discover phase the cultural backgrounds of stakeholders were studied into detail, not only to attune final solutions to culture but also to appropriate design methods and tools used in cocreation sessions. Cultural differences may

	Discover	Define	Develop	Deliver
Goal	Gathering insights from the shop owners, visitors, potential visitors Gathering related information in the context	Identifying underlying problems in the current situation of the shopping street Brainstorms ideas and get feedback from stakeholders	Creating few concepts and select the concept Evaluating and iteratively improving the solution for its implementation	Describing the final design Defining the implementation plan for the idea
Method and tools	Literature research about the area Contextual interviews with food shop owners Interviews with local people Observation in the street and in the shops Context mapping with visitor and potential visitor Creative trend research Literature research about the multicultural food adaptation history in The Netherlands	SWOT analysis Search areas Problem formulation Persona Brainstorming Project presentation with stakeholders	Concept review and feedback with shop owners and local community Idea discussion with fellow designers Creative session Expert interview Concept evaluation with target user List of requirements and wish list Weighted Matrix select ideas LIterature review Business model canvas	Implementation Roadmap Interview with related stakeholders

FIG. 5 The design process for the project.

not only cause barriers but can also create opportunities. A culture-sensitive approach was needed, where special attention would be paid to the influence of culture on the design process (*designing*) and on the outcomes (*design*). This meant that people were not only seen as individuals, with their specific needs and dreams, but also as members of groups, sharing values and practices. The designer needed to look through a cultural lens in a comprehensive mode, not only at people but also at the sociocultural meaning of things (as a material culture researcher), while taking into account the influence of his or her own cultural background (*designer*) (van Boeijen, 2015, 2020; Razzaghi et al., 2009). Culture sensitivity—or intercultural sensitivity when emphasizing the interaction between two or more cultural groups—is the genuine interest in the cultures of others and having an interest in interpreting signals from others (Bennett & Bennett, 2004). In keeping with Bates and Plog (1976), we refer to culture as a system of shared beliefs, values, customs, behaviors, and artifacts that the members of a society use to cope with their world and with one another and that are transmitted from generation to generation through learning (not inherited). Culture sensitivity is needed to take a stance in what design can do to affirm a specific culture or change it, to bridge two or more cultural groups, or just to tune with the existing culture to avoid mismatches between the design and its intended user group(s) (van Boeijen, 2015). To gain an empathic understanding of a cultural context, designers need to study it, partly through literature, but also through fieldwork: observations, interviews, and participatory activities (Hao, 2019).

In the *design* process, cultural sensitivity is needed when matching the designer's attitude with the design research methods (van Boeijen, 2015). First, *designers* with their cultural backgrounds influence the design; second, *designing*, the process and activities such as conducting the research, conceptualizing, and testing, needs to be tuned to the cultural context; and third, the *design* itself, that is, the products and services or other manifestations of design, might influence the cultural context.

Design process and resulting design

Discover: Understanding locality

The aim of the first phase was to discover and understand the design scope of the assignment given by the shop owners' association. Sources for insights were the shopkeepers themselves, local residents, visitors, and potential visitors (people who knew Middelland but would not visit it), with the information gathered through interviews, observation, and participatory sessions. A deeper insight into the preexisting process of cocreation in Middellandstraat was gained through analyzing internal documentation on the Mooi, Mooier, Middelland project. Furthermore a competitor analysis, creative trend research, and a historical study on multicultural food in the Netherlands were conducted. Besides, we analyzed the website that represented the shops in the street. These insights elicited the underlying problems in the street and possibilities for new product design. The focus and the *design goal* of this project were defined based on these insights on the underlying problems.

As part of the culture-sensitive design approach, we had to consider the language differences and different communication types between the (sub)cultures in Middellandstraat. At first, interviewing the people from the area was difficult due to these barriers. Therefore a participatory design exercise was chosen as a tool to bridge communication between the various groups. For this, physical artifacts were used as tools to project the participants' thoughts (Sanders and Stappers, 2012). According to Sanders and Stappers (2012), the technique enables people to let out their thoughts and feelings. It is also essential to make it not overwhelming to the participant; thus a simple map of the area was used as a tool to stimulate conversation (Fig. 6). The participants were asked to indicate their favorite store on the map and share their stories. The tool helped to bring out personal stories, and as the participants were also asked to draw and write on the map, this made those stories tangible and sharable.

The municipality of Rotterdam, especially through the area networker, was also involved in the process to get to know its concerns, target, and the project scope. Internal studies by the municipality were also made available and used (e.g., Mooi Mooier Middelland, 2016). The visitors, the target market for this effort, were also involved in the project and included native Dutch people and residents of the nearby area. They shared their concerns, needs, and hopes for the area to host restaurants with food from diverse cultures using "context mapping." Context mapping involves research techniques geared to gain insights and a rich understanding of the context for the usage of products by people in their everyday lives (Sleeswijk Visser et al., 2005). Fig. 7 shows the context mapping sessions and tools used.

The outcome of this phase was to gain a better and richer understanding of the area and the shops through an overview of the input from all stakeholders involved. This phase also focused on understanding the visitors' view on the street and their behavior when doing food shopping. Lastly, it also helped to identify global food trends reflected locally, as well as outlining the main problems and opportunities for the area as a center of food diversity.

Findings

Shopkeepers' perspective

Findings from the field research highlighted three significant attitudes and approaches of the shop owners. Firstly, they generally focused on short-term goals to generate direct revenues. If they could not foresee the immediate effects of an idea, they would avoid implementing those initiatives or take part in them. Some activities, such as taking part in street bazaars that would not necessarily increase their sales immediately, were considered a waste of time. Secondly, they were scared of change and saw it as something that it could turn current customers away. This was particularly felt for radical changes, such as changing

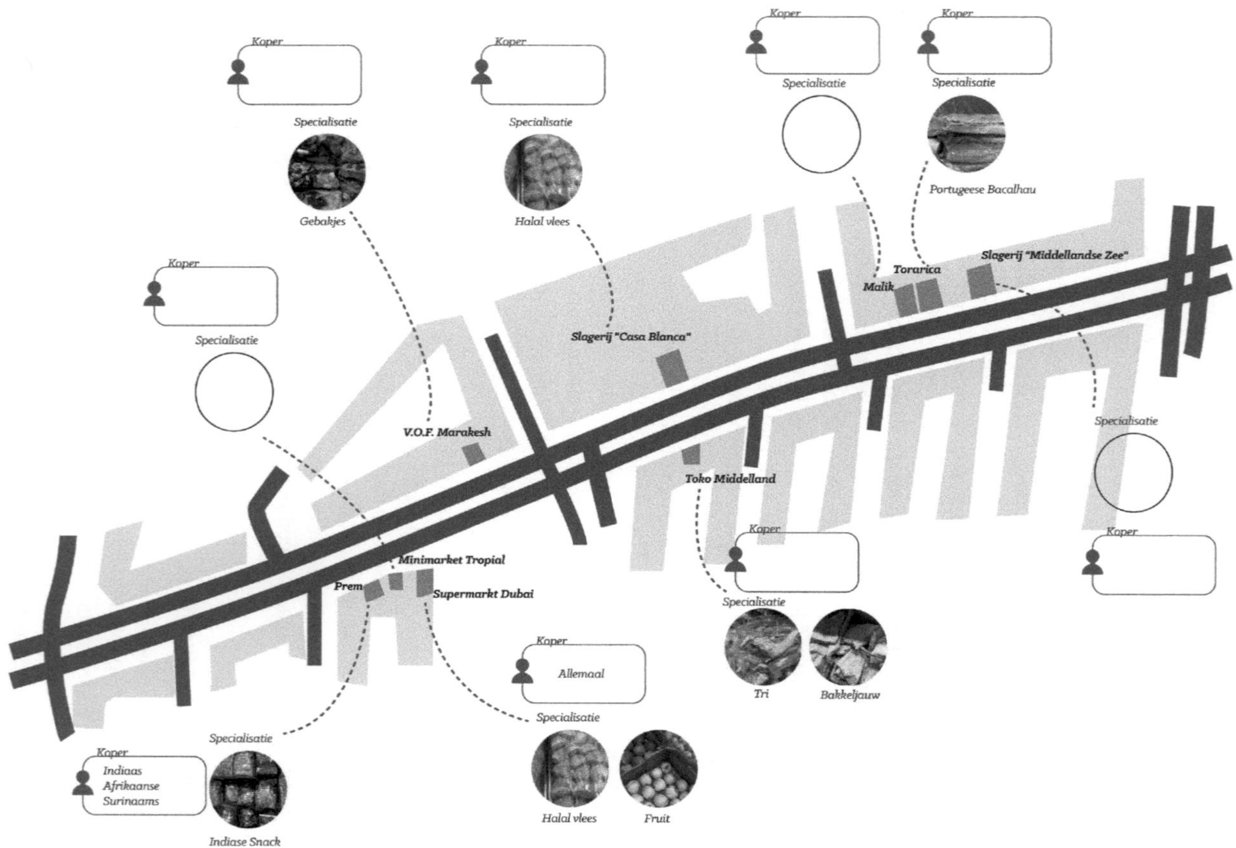

FIG. 6 Map of the area as a tool.

FIG. 7 Context mapping session and tools.

the store look. They would also worry that their existing customers, who were mostly from lower income families, would be discouraged from visiting the shops. Thirdly, they mainly relied on word of mouth to increase sales in the shops. They were proud of their promotion of organic products, but they did not try to improve the experience to enhance the word of mouth even more or use any digital marketing means to attract more customers. Some of the shop owners' reactions during the interviews were:

> *If I change something here to look modern, then my customers will be afraid, they will think I have raised the prices.*
>
> (Shop owner 1)

> *There was no result from the bazaar, it was a waste of time.*
>
> (Shop owner 2)

Potential visitors' perspective

Potential customers were discouraged to visit Middellandstraat, because in Rotterdam the area had an image of being dangerous and poorly managed. They felt insecure and tried to avoid the area. Moreover, they did not see the added value that made Middellandstraat worth visiting, because essentially it felt to them quite similar to other shopping streets in Rotterdam that had more to offer. Some of the reactions of potential visitors were:

> *I think Middellandstraat is messy, unsafe and the people there are like...how can I say.. a criminal....so I always walk fast if I need to pass through the street*
>
> (Potential visitor 1)

Middellandstraat and other multicultural streets? They are all the same

(Potential visitor 2)

Once inside the shops, it appeared that visitors could experience further problems. Some shops looked disorganized, and items were not easy to find. The fact that imported products had foreign language labels made unclear how they could be utilized, and their quality and value could not be easily understood, making shoppers uncertain. This usually made them prefer to buy in supermarkets. Furthermore, visitors felt uncertain about any shop's unwritten rules, such as how to buy and deal with plastic bags and whom to ask about unpackaged products (on sizes and amounts). Some shoppers felt rejected and unwelcome as not belonging to the "right" cultural group and because they were not part of the habitual visitors of the shop. One potential customer stated:

I will probably see a lot of foreign words and products, so I feel confused and perplexed and anxious because I don't know exactly what I can buy in there and what I can use and how to use it

(Potential visitor 3)

Potential visitors did not know much about other cultures' food, and they tended to buy only things they knew or had tried before or that had been recommended by someone else. Their lack of knowledge and awareness of different ethnic food undermined their interest as they did not know what they could get in those shops.

Any experiences of other cultures tended to be closely related to the use of digital communication. Useful information would be found online, for example, when searching for recipes or seeing recommendations. Online information was the base for many to get closer to a different culture. Personal relations could also be key in expanding horizons. Whenever potential customers had a friend who could accompany and introduce them to the shop and the products or provide a recommendation, this could break the ice. Once they felt familiar with the shop and the products, they felt confident enough to return directly to that specific shop.

Once I went to a Moroccan supermarket with Moroccan friends, and it felt very different visiting with local people who understand the culture. Because they will introduce a lot of different things to you, also they know the people in the shops, it was a really nice experience

(Visitor)

For the visitors who were outsiders (in terms of cultural background), the cultural diversity was interesting but unfamiliar. However, for visitors with a migration background, often mixed with Dutch roots, the cultures represented in Middellandstraat felt familiar but different for some residents and shop owners. The shops made they experience a form of nostalgia that came from the visitors' past experiences and memories of their country of origin, its food, and other aspects of the shop that they recognized from that culture. This experience was identified as a quality that should be taken into account when developing a design goal. Another important finding related to this quality was the unique and valuable architecture of the shopping street. Middellandstraat is one of the few streets in Rotterdam that was not destroyed by bombardments during World War II. It kept its original architecture with high ceilings, stained glass windows, and colors and building fronts typical of buildings designed before the war. This was a precious cultural asset in Rotterdam, which should be taken into account when designing to improve the place. This can also be part of a smart design approach, combining the old and unique architecture (hardware) of the Dutch with the new food culture of the immigrants to create new experiences for visitors and residents and set Middellandstraat apart from other shopping streets.

Accentuating the old architecture and bringing it into life was also one of the demands of the cocreation group, and the local government had a budget for it. Shop owners or property owners could get cofinance for 50% of the cost of renovation up to a certain amount if they agreed to renovate their building to show the uniqueness of the old architecture. In addition, the cocreation group decided to renovate the two squares and enhance the quality of the public space to attract visitors and new customers.

Another useful finding that showed the importance of understanding the local context was the following: West-Kruiskade, the neighboring street, had an alcohol-free restaurant, in accordance with Muslim practices. A new tenant in Middellandstraat wanted to start a restaurant with a Moroccan theme: Restaurant Mosaic. Similarly to the West-Kruiskade venue, the owner also wanted to serve nonalcoholic drinks only. However, in participatory sessions with local citizens, the attitudes toward alcohol were polarized. In the Netherlands, it is habitual to offer alcoholic drinks in any good restaurant. This practice is in line with the dominant culture and supported by an economic rationale indicating that venues serving alcohol have higher revenues. In the participatory sessions the owner expected a clear support for a nonalcohol policy, but opposite opinions and preferences were also represented. The coproduction method involved taking an inclusive approach, allowing all stakeholders in the group to voice their opinion, and this exposed those tensions. In the end, it was the restaurant owner who faced the final decision, choosing not to sell alcohol, despite some discontent. The coproduced solution was therefore informed by both the dominant Dutch culture and the Muslim tradition, resulting in the judgment that the restaurant could be run profitably even with a nonalcohol policy. The restaurant owner foresaw that the experience-related aspects of Moroccan culture such as food, tableware, service, programming of the exhibition of art, and the feel of being in the country while in the restaurant could offset alcohol as an economic and cultural factor and even attract Dutch customers. Culture could play a role in place and sense making but also as a means for achieving economic goals and contributing to the general goal of economic revitalization of the shopping street.

Define: Design direction

In the second phase the design direction was defined. This phase was based on gaining an understanding of the current situation of the street, the shops, and its relation to the visitors and other stakeholders. Insights from the discover phase were used to develop a "design goal." Based on strengths and opportunities found from different studies, the Search Areas method (van Boeijen et al., 2020) was used to find the most promising areas to be developed to a concept. The research work enabled us to identify opportunities and problems in the experience of multicultural shops. People could be stimulated to visit the street by providing them with information about the shops' identities and embedded values of the shop's food cultures and of the architecture. They could be helped by explaining the characteristics of the food items that are specific to the street. Designing to differentiate Middellandstraat from the nearby shopping streets was considered vital. We therefore identified several opportunities. The design of the physical space, such as the squares and the architecture, combined with the interior design and the design of the food stores should encourage people to explore shops on both sides of the street. To emphasize the differentiation of Middellandstraat compared with the other multicultural street nearby, it was important to make the visitor feel that its shops were part of a specific street culture. Potential visitors could be encouraged to go to the shops by giving them the food information in advance and inspire them with suggestions about how those ingredients could be used, which could lead to a more exciting journey. This could be encouraged through the online channel, as people already used it as a main way to learn about other cultures. The shops should also enhance a certain level of familiarity. This could be done, for example, through interior design by using design patterns to explain the countries of origin, through lighting, and through extra product information that would stimulate the exploration of new ingredients. Furthermore, new designs could support both any shop owners' and visitors' latent feeling of nostalgia. These opportunities for design could lead to a change of the structure and the visual appearance of the store interior.

The design opportunities and needs have been clustered into *three phases* of the shopping experience, which are *before* shopping, *while* shopping, and *after* shopping. For the before shopping phase, it was considered to be important to provide stimuli and inspiration for people to encourage them to visit the street. During the shopping, enough product information and support for exploration were needed to have a positive shopping experience. While after shopping, it was important that people were left with a good impression and feeling to make them come back again.

For the before shopping phase, our findings indicated that an integration of the online and the offline world would enhance the food-related experience. Online intervention was appropriate here because from our findings we knew that people were looking for information online and already tried to familiarize themselves with the culture digitally. Online intervention would act as a warming up before visiting the physical store. It would introduce the shops and the street and make these more familiar. It would allow shop owners, locals, and visitors to the area to interact more, becoming a platform to make the locals, who already knew about the shops, help visitors. It would also help clarifying any "unwritten rules" or terms of engagement to outsiders, so they would feel less puzzled with both the physical stores and the street.

The design direction was summarized in a "design goal," which was formulated as follows:

The design should enable people who usually do not visit Middellandstraat shops to get inspired and informed about food culture and shop identity, promote the authentic identity and historical value of the shops, and provide a comfortable atmosphere for the visitors to explore the multicultural shopping street, Middellandstraat. The design should support this envisioned experience in a sociable smart way, connecting both the online and offline world.

Develop: Cocreation

In the following phase of the work, ideas were generated, and through iterating on the insights gained in the previous stages, these were improved and further developed into concepts. Through brainstorm sessions, paper prototyping, idea evaluation with locals and shop owners, and creative sessions with design students, three concepts were cocreated and evaluated with a selected group of representative stakeholders. In this phase, solutions (as three concepts) were also evaluated with stakeholders. The concepts were tested with residents, visitors, and shop owners (see Fig. 8). Together with stakeholders a roadmap was then generated to help them start planning any possible preparation for implementation.

Designing to bring the old historic architecture back—high ceilings, typical pre–WWII storefront, and the stained glass windows—was not an integral part

FIG. 8 Feedback session with a resident.

of this assignment. Those would be carried out by others in the cocreation group of the stakeholders but were still necessary as they would further add value to the food experience and draw visitors and consumers to the shopping street.

Deliver: Resulting design

The designed answer to our question was *SmaakReis*. The name SmaakReis ("Taste Trip") refers to a visit to a transformed, more welcoming, and comfortable Middellandstraat to try different flavors from various cultures. SmaakReis was a service that enabled people to explore the food diversity on offer in the street. It inspired and informed visitors about the food culture and the shops' authentic values. It integrated the online and offline experiences and was articulated into different elements. The design elements were spread throughout the customer journey and served different goals to enable people to explore the food diversity in the street. Fig. 9 provides an overview of SmaakReis' visitors' journey and the designed elements.

The first goal was to get to know a particular shop, its products, and its owner. Visitors could experience this in advance of a physical visit through the website. Customers would be guided in the shop through the signage inside. If they were not sure what to do and what kind of products were available, the signage would help them in getting to know more about these. The second goal was to try and share the culture. There would be an open platform for visitors and residents to share recipes from their own culture (see Fig. 10; the visual shows a wireframe of the website in mobile view).

Fig. 10, left (1), shows how visitors could see information about a shop, its history, and its products. Visitors could further see a recipe made using a specific product in the wireframe number 2. They could share their own recipes, look at tips on the food and tutorials, and buy the ingredients. In this way, visitors could cocreate and share recipes to be featured in SmaakReis, getting involved and being part of Middellandstraat's diverse environment. Visitors could also join classes and workshops about Middellandstraat's cultures and foods to have a closer, more intimate cultural experience as shown in wireframe number 3 in Fig. 10.

Inside the stores, appropriate pointers for rules were displayed, helping customers reduce their anxiety and uncertainty while shopping. Information about unique ingredients and some tips on how to cook with them was also made available. See Fig. 11.

The third goal was to explore and immerse in the culture. Visitors could explore Middellandstraat's shops, foods, history, and the area with *a shopping buddy*. They could make an appointment with local people (as shown in numbers 6 and 7 in Fig. 12) to learn and exchange cultures and get richer experiences in the shops and in the street. Creating a more personal relationship with the area would encourage and attract customers. In this way, visitors (both local and nonlocal) would become part of the community, starting by interacting

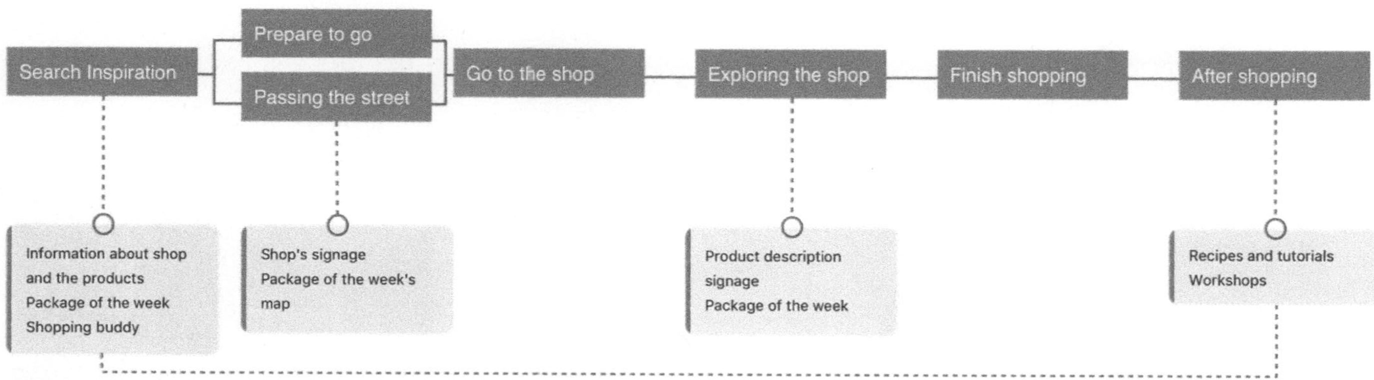

FIG. 9 Overview of the visitors' journey and the design elements in the service called SmaakReis.

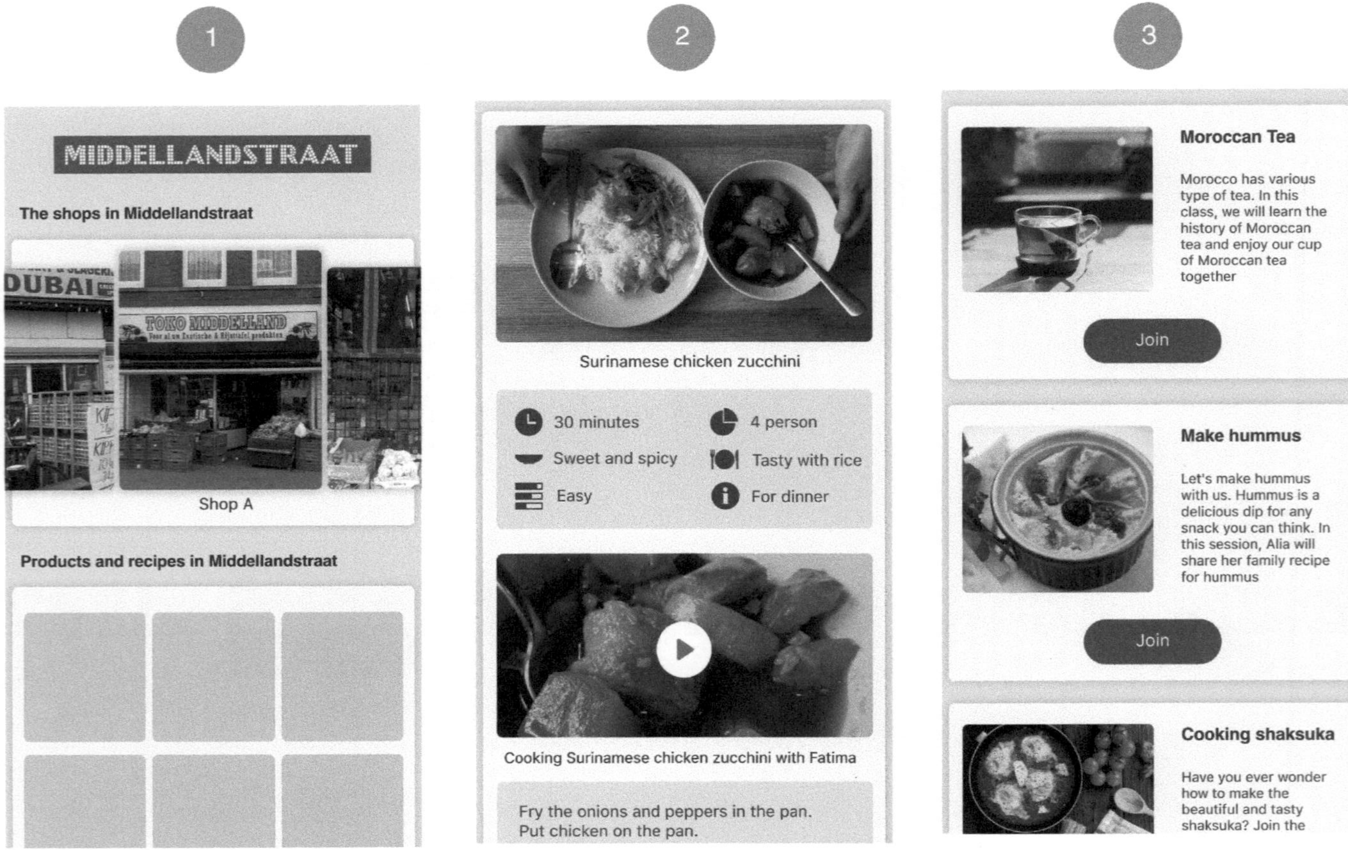

FIG. 10 Wireframe of the proposed website in mobile view. (1) Homepage with shop and product introduction. (2) Online recipe. (3) Workshops and events information.

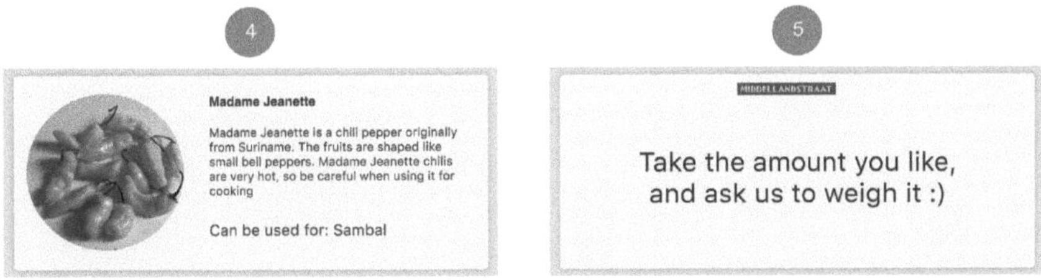

FIG. 11 Examples of pointers inside the shop. (4) Product description example. (5) Direction for rules inside the shop example.

with each other via the website and giving ratings and recommendations about recipes and stores. Packages of ingredients and guidance on where to get these were provided (as shown in numbers 8 and 9 in Fig. 12), and these packages would be accompanied with a special recipe by the shop owner or local people. Ingredients and ways to cook them were displayed visually to make visitors more familiar with them. All of this made exploring the area and the shop much more exciting. Yet, this was something that would not require a major effort for the community in Middellandstraat, because they already naturally had knowledge about ingredients and recipes.

The last aspect of the design intervention emphasizes visual unity in the street. The physical stores that were specialized in a particular country or products from a specific culture were also visually rebranded. The design intended to create some unity in how the various cultures were showcased. A mosaic shape with several colors served as Middellandstraat logo. The shops that were specialized in a specific product could use the mosaic to represent, for example, the country of origin. This helped the visitor to recognize the shops easily and looked more professional and credible, something that increased the feeling of being safe (see Fig. 13).

The SmaakReis service aimed to bring people to Middellandstraat and meet local residents, improving understanding between different cultures. The interaction between visitor and shop owner was transformed into a celebration of their differences and a way to connect with each other (see Fig. 14).

Visitors could also enjoy the unique architecture of the building in combination with the food and gain new experiences.

Discussion and conclusions

This chapter emphasized the specific values of a culture-sensitive design approach applied in the context of a hybrid smart urban intervention. Using such an approach the project enabled us to explore pathways to a sociable approach to the city to push the envelope of how smart urban design could evolve. In this section, we discuss barriers and enablers for shaping smart in urban contexts that we encountered through designing for the revitalization of the shopping

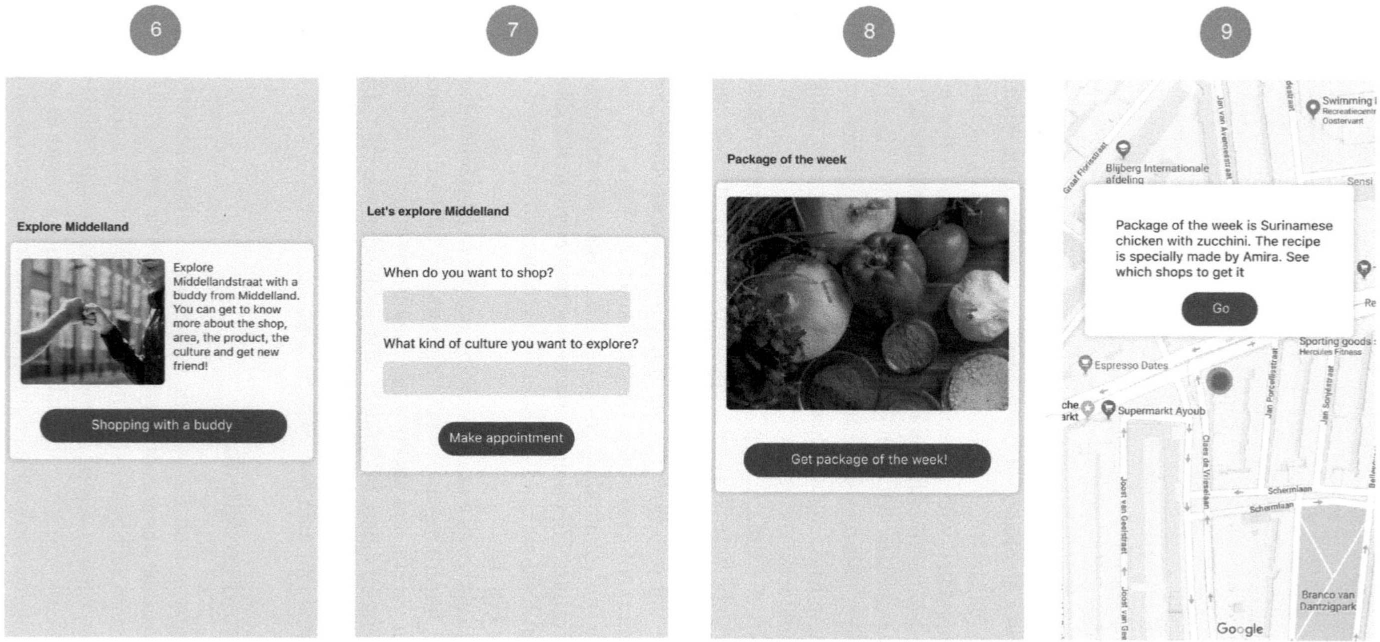

FIG. 12 Wireframes of the website in mobile view. (6, 7) Shopping buddy feature. (8, 9) Package of the week feature.

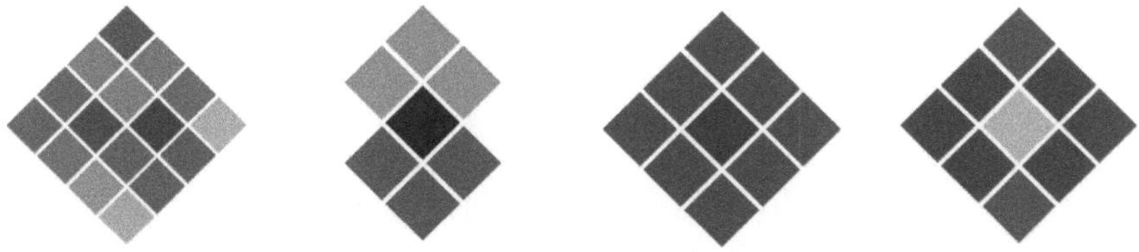

Middellandstraat Indian shop Moroccan shop Surinamese shop

FIG. 13 Examples of the visualization of shop logos to brand their identity.

street. More specifically, we reflect on the value of using a range of methods and techniques addressing inclusiveness and cocreation.

Delineation of the context

In general, it is difficult to delineate the context for which to design (van Boeijen, 2015; Hao, 2019). Many products and services are not bound to one place, and this raises the question on who is included as part of any design-related cultural study and who is not. The point of departure of this project is using a culture sensitive design approach focusing on the different cultural assets (Bates and Plog, 1976)—*shared beliefs, values, customs, behaviors, and artifacts*—of the different ethnic groups having a stake in the area. This aims at improving sense making in the shopping street, with shops attracting more customers and shoppers spending more time and money in the street, helping its economic regeneration. The economic dimension ends up being the guiding principle for the shopping street, as choices for certain aspects of the culture depend not only on the culture itself but also on whether these are a means to help generating revenues and improve the shops and the street. It is possible that in a residential area the culture would assume more a role on its own to represent and reinforce identities.

Culture in itself is rewarding. It represents and communicates certain values. In a cocreational setting involving different cultures, the beliefs, behaviors, values, and artifacts represented must be communicated so that people learn more from each other in developing a deeper mutual understanding and appreciation. The insight about West-Kruiskade, the neighboring street with an alcohol-free restaurant and the different tensions in Middellandstraat, showed that opinions could vary in nearby districts. An inclusivity goal implies that all opinions are needed. In this case, being culture-sensitive means there need to be an understanding and negotiation between the different (food) cultures at a local level. Differently put, no matter which topic, negotiation on a local level should be an essential part of a smart city design process. The cocreation process is a continuous negotiated setting where all the stakeholders during the whole process from idea conception to design and implementation are involved and communicate

FIG. 14 Examples of the kind of services offered through the Middellandstraat website.

their grievances and want to come to a better decision for a better place, in this case the shopping street. The choices whether to serve or not to serve alcoholic drinks in restaurant because of cultural consideration help to understand other choices that must be made and implemented. When it comes to the delineation, it is a matter of zooming in and zooming out, a dynamic process in which different perspectives help to understand the situation.

Top-down versus bottom-up

This design case illustrated the complexity of smart urban design in the context of a government in transition. The design of SmaakReis needs therefore to be seen in a broader governance perspective. A process that used to be top-down, where local government decides, makes, and enforces the change, has evolved into a more bottom-up and inclusive process, giving voice to a variety of stakeholders. Such a transformed process asks for continuous adjustment and adaptation, where time is an important factor. It shows that different processes can happen in parallel and, when organized in a transparent way, contribute to one another and construct a broader baseline for implementing a local smart city vision. There are some learning points that we can take from this project for the adoption of a smart service to leverage placemaking, that is, the revitalization of Middellandstraat. The first step is how to get the stakeholders (residents, property owners, and the shopkeepers) actively involved and engaged. It shows that an ongoing effort to build trust with the stakeholders by involving them in the process is key. As also argued by Hao and colleagues (Hao et al., 2017; Hao, 2019), it is important to consider how understanding the underlying problems implies understanding deeply the stakeholders themselves, their roles in the community, and their ultimate agendas.

In other words, understanding the specificities of the local context is crucial to make a "smart" design for an area truly smart. As illustrated earlier the differences between Middellandstraat and the first part of this street, the West-Kruiskade, are easy to observe, which stresses the value and necessity of culture-sensitive design even on a very local scale. In Middellandstraat, this sensitivity was also visible in the complexity in involvement and negotiation among the urban stakeholders and to guarantee an inclusive process.

Roles of stakeholders

A main challenge for achieving inclusivity was to deal with the stakeholders' different roles and bring together their different agendas along the design process. The local government was concerned with control and safety of the neighborhood and the equal or proportionate participation of all stakeholders (local democratization). But above all revitalization should result in the safety of the neighborhood. This stakeholder therefore thought that achieving this would involve a degree of gentrification, but property owners were critical. If shop owners were the ones who would be expected to pay for architectural renovation, they questioned whether the shopkeepers could afford to pay higher rents, making up for their investment. For local residents the quality of living was clearly the most important factor at play. They were more focused on aspects such as the quality of public spaces, playgrounds, roadside furniture, and objects that could make the environment more attractive and liveable. And the shop owners and their association acted as the promoters of economic revitalization, questioning whether any redesign would result in more visitors and sales, meaning higher revenues for their shops. The challenge for the design was to balance all these different interests, during both the design process and execution toward its final results, to keep the stakeholders committed and enhance the acceptability of the final scheme.

A central topic

For the design project, food was chosen as the central topic based on considering those design aspects that could more easily contribute to the economic revitalization of the shopping street. As mentioned before, food was also representing the different values of cultures and religions. This project showed the importance of identifying a key asset of an area, which can be used to push its revitalization through socially engaged means.

Role of the designer and the design process

The designer directly engaged in this project (at the time studying in the Netherlands, but with an Indonesian background) was familiar with the cultural and religious diversity in her country of origin. Coming from a culture that values a collectivistic way of dealing with each other, she could relate well with similar collectivist approaches in Middellandstraat. Consequently the final design also reflected this value and its related practices. On the one hand, this means that to a certain extent she could have played the role of an indigenous insider (Banks, 1998). However, she could not speak Dutch and needed to rely on using the English language or interpreters, and such limitations made her more of an outsider. It seems however that an outsider could also positively contribute to a contextual grounded design approach. The benefit of being an outsider (Banks, 1998) is that it is possible to identify things that may be missed by the established urban stakeholders in the project. These advantages have helped in the creation of the design; the designer was independent enough to see new possibilities, yet familiar enough to manage to attune the solutions to the different stakeholders. The question remains whether it was a good decision to leave the designer out of the implementation process and whether the implementation roadmap provides on its own enough guidance for the urban stakeholders to make the foreseen transition happen.

Designers should take into account the values, needs, and desires of the intended users and the stakeholders involved to be able to bridge the different cultures in the design context. At the same time the authors believe that designers in such a case should not (or are even not able to) ignore their own beliefs (that are partly based on their own cultural background) and therefore should be aware of their role in both the design process and the design result.

The design process, including research, conceptualization, and concept testing, was finely tuned to the local cultural context. For example, in the research phase, it was decided first to talk with the shop owners one by one, listening to their concerns and needs, and not in group sessions to avoid barriers due to different values and practices (as described by van Rijn et al. (2005), van Boeijen (2015), and Sun (2012)). Methods and tools that allow for understanding the cultural and local sensitivity were used not only to inform the design process but also supported the further communication and interaction. Arranging and forming questions in a way that matched the interviewee's cultural background was also found to be essential.

Smart design

The design incorporated collectivistic values, and these were materialized through the ideas of community recipes, one platform for all, and a buddy system for the shopping activity. However, at the same time, it was also accommodating the needs of the intended users, in this case the visitors, who were not familiar with the implicit way of communicating in the shops, by tuning it to a direct way of communication through straightforward design and signage. A considerable goal of smart city design is to utilize high technologies for the benefit of the residents. But the technology cannot be considered the only component of a smart city. Designers need to keep looking at the context, in terms of space, people, and local (sub)cultures. Designing for people in their cultural context is essential, and if the technology is not aligned to this, it will lose efficacy.

In this particular task the applied technology may not have been the latest one, or the most sophisticated, but its implementation was thought in close relation with the area and the residents. The technology could come into the project once the base of cooperation between residents, municipality, and all stakeholders involved had been established. We would argue that being smart does not just come from adopting technology, but from addressing the social aspects of a problem, being "socially smart." This means utilizing an area's resources for the benefit of the residents. It is equally essential and part of this to think about the acceptance of technology and readiness of the residents and other stakeholders. And all of this also needs time to be fully implemented, as a design is not a final solution but has an open end, developing further over time.

Guidelines

To conclude, we identified ten guidelines for smart urban design interventions focused on placemaking, and more specifically on the economic revitalization of multicultural shopping streets:

1. *Use the "sociable smart city" concept in terms of using the possibilities embedded in the targeted context and capabilities of the people.* Smart here does not primarily refer to the application of new technologies, but first of all to building an empathic understanding of the local cultural context with its subcultures.
2. *Consider how to delineate the local culture*, on the appropriate skill, depending on cultural variation within the city areas. Negotiation on a local level is an essential part of the smart city design process, and since the cultural context can highly differ from location to location, the scale for revitalization should be considered and chosen carefully.
3. *Identify a key asset of an area*, which can be used to push its revitalization through socially engaged means (such as "food" in our example).
4. *Integrate the virtual world with the physical world.* A shopping street is not only about the physical stores but also about their virtual and online presence. It is crucial to integrate the digital experience of visitors with the physical experience of the shopping street.
5. *Make use of the multicultural background of (local) people.* It is a strength that makes the area unique. Bringing together the whole experience of the culture and showcase it can be a way to do it.
6. *Adjust the design process with the local cultural context.* Participatory sessions should be attuned to the participants' backgrounds. For example, the used generative materials, selected participants, facilitators/design researchers, and language and style of communication should resonate with the people involved.
7. *Adjust the design with the local cultural context.* While embracing people's own cultural backgrounds, attune the design with the local culture (in our case this included the visitors). An example in this project is by providing ways to communicate in direct ways to cater to the audience in the Netherlands (dominant culture).
8. *Be aware of providing a sense of unity within the variety.* Different cultural concepts can be confusing for visitors, so a sense of unity in the midst of the differences should also be considered.
9. *Use the existing resources; facilitate people to enrich the experience.* The multicultural background of the people in the area should not be a wall to understanding each other, but it can be used as a resource to enrich the experience. An example from this project was the change of having a buddy from the area for shopping around. The buddy knew the area, the dominant culture, and other cultural groups well and was the best guidance for a clueless visitor. It helped them to have a more personal relationship with the area.

10. **And last but not least** *reflect on the researcher/designer's own culture* **to understand how that will influence the design process and the final result.**

Implementation of the concept as a learning process of capacity building

It can be concluded that SmaakReis was helpful in codesigning futures with the local stakeholders and enabled them to act within a smart city context by aligning strategic activities. Interestingly the guidelines also have been used in promoting further initiatives that contribute to the revitalization of the multicultural shopping street. Recently, several storefronts and shop windows (hardware) in Middellandstraat have been upgraded following the guidelines more or less intuitively. By collective and concerted effort of the various stakeholders, more than ten new food shops have been added to the street, making it a more attractive place with richer experiences for the visitors.

Further feedback regarding the implementation has recently emerged. One Moroccan foodstore, Restaurant Mosaic, followed the guidelines and the concept of SmaakReis, particularly for the cocreation at shop level while using the matrix of the design components and shopping street functions (see Fig. 4), and applied the design for economic revitalization more rigorously than others for its placemaking. The implementation of the concept was conducted with a team of designers, such as an architect, an interior designer, and a business consultant, with input from the cocreation team, including the owner and family members. The implementation of any complex product-service system as SmaakReis is often a slow process and can be seen as a journey toward a "sociable smart city." A cocreation process involving stakeholders with different interests, ideas of innovation and conceptual thinking, and economic agendas requires professional human capacity. In keeping with Kroesen et al. (2015), such capacity can be seen as a function of knowledge (information, experience, skills, and algorithm), culture (values, attitude, and beliefs), and social and institutional environment (enabling environment) and is indispensable to achieve results. Capacity can be further enhanced by bringing in experts.

The SmaakReis approach offered a useful framework to explore a sociable smart city concept. Taking the Restaurant Mosaic example, as this falls into the shopping category, implementation was focused on that function (see the design aspects such as hardware, software, and mindware in Fig. 4). Realization of the hardware aspects (building, furniture, lighting, terrace, and location) was relatively easy to achieve, and hardware capacity was built. The software however (business model, restaurant activities, food, and services) required more complex degrees of human influence and creativity. Consequently, it asked for a process of capacity building to help improve customers experience, which was further detailed through designing the mindware for the restaurant, for example, understanding why people are coming to the restaurant and why they consume (Underhill 2009; Lindstrom 2012).

Due to the different backgrounds of the shop owners, building the necessary capacity for implementation is not easy. Clearly no shop owner can be forced, and they only can be nudged into a certain direction. Improving the hardware, software, and mindware relating to the shops very often asks for serious investment and a learning attitude from the owners. One particularly weak spot is virtual placemaking for attracting potential buyers. Creating a website, using social media, and having a meaningful presence on the Internet all depend on the presence of a digital strategy. Mirroring the actual state of the physical shops takes time and is a continuous effort, and shop owners do not have the necessary capacity, nor do they often consider it as important. Another aspect is that the concept of SmaakReis focuses more on the hardware part of shopping function, while the software and mindware should be equally important.

The implementation of smart city-related designs takes place in a dynamic and complex environment. For example, the cocreation group has appointed two street managers to keep monitoring the process of the Mooi, Mooier, Middelland program. Every shop owner is trying to improve their own shop using their own entrepreneurial instincts. The shop owners' association has transformed itself into a Business Investment Zone (BIZ), an organization created by the government for the business community in an area to invest in the continual improvement of the business environment. The annual contribution for shop owners is mandatory; it is collected through the tax authorities and then made available to the BIZ. The association of property owners has also turned itself into a BIZ to allow for investment in the shopping street. Recently, Restaurant Mosaic has organized a workshop to look for frugal innovation, innovations where no huge sums of money are involved, but where the solution is nonetheless highly effective, with the support of the Centre for Frugal Innovation in Africa of the universities of Leiden, Delft, and Rotterdam. There are plans in the making for a structural internship program to support the improvement of shops, since it has been concluded that designing of and implementing for a smart city need continued attention in execution with the right capacity. The shop owners need support to use the guidelines and the concept of SmaakReis, and they likely deserve to be informed in benefiting from the concept as happened in the codesign process with Restaurant Mosaic. In addition to the food venues, there are also outlets selling groceries, health and beauty products and services, and domestic appliances. Similar action plans need to be developed for these units to turn the entire street into a "smart shopping street."

It can be concluded that Mooi, Mooier, Middelland as cocreation into sociable placemaking fostering a sociable smart city is a long road; it takes times, and it takes learning, commitment, and, apart from the financial investments, also social capital investment. The preparation of the project Mooi, Mooier, Middelland started in 2015 and was scheduled to end in 2018; the activities are, however, still ongoing in 2020.

Is there a philosophy of science to be distracted here designing smart in multicultural environment? Is there an epistemology, a metaphysics, or a methodology? Are there certain ethics and aesthetics? Is there a certain language? Is there some theory to be formulated?

Maybe, it is a too feeble basis to distract general (scientific) rules for multicultural design (different cultural designs side by side) or intercultural design (mixing different cultural designs by interrelation) based on one project. The first thing we can say is that they—multicultural and intercultural—are design and the epistemology (field) of design is applied to them; only you have to take the different cultures into consideration. The metaphysics, as well as their appearance in reality, can be researched. To find the true meaning, designers need to use ethnographic methods and must understand the language of the design and of the people from the different cultures. Ethical aspects can play a mixing of design patterns into multicultural or intercultural design, but they can also be used as innovation where different cultures living side by side or as one entity. Multi- and intercultural designs have a certain aesthetics (beauty) to it; in this project, we have used it to improve a location for economic goals.

References

Banks, J.A., 1998. The lives and values of researchers: implications for educating citizens in a multicultural society. Educ. Res. 27 (7), 4–17.

Bates, D.G., Plog, F., 1976. Cultural Anthropology, third ed. McGraw-Hill, New York.

Bennett, J.M., Bennett, M.J., 2004. Developing intercultural sensitivity: an integrative approach to global and domestic diversity. In: Landis, D., Bennett, J., Bennett, M. (Eds.), Handbook of Intercultural Training, third ed. Sage, Thousand Oaks, pp. 147–165.

Bessière, J., 1998. Local development and heritage: traditional food and cuisine as tourist attractions in rural areas. Sociol. Rural. 38 (1), 21–34.

Concilio, G., De Götzen, A., Molinari, F., Morelli, N., Mulder, I., Simeone, L., Tosoni, I., Van Dam, K., 2019. Chapter 4: Innovation and design. In: Concilio, G., Tosoni, I. (Eds.), Innovation Capacity and the City. The Enabling Role of Design. Springer Briefs in Applied Sciences and Technology. Springer, Cham, pp. 61–83, https://doi.org/10.1007/978-3-030-00123-0_4.

Darson, R., Veen, M., 2017. Het DNA van de Middellandstraat (The DNA of the Middellandstraat). Unpublished Report Written for the Shop Owners Association of Middellandstraat.

Design Council, 2005. The Design Process. UK design Council, London.

Foth, M., Brynskov, M., Ojala, T., 2015. Citizen's Right to the Digital City: Urban Interfaces, Activism, and Placemaking. Springer, https://doi.org/10.1007/978-981-287-919-6.

Google Maps (2017). Retrieved from: https://www.google.nl/maps/place/1e+Middellandstraat,+Rotterdam/.

Hao, C., 2019. Cultura – Achieving Intercultural Empathy through Contextual User Research in Design. Doctoral Thesis, Delft University of Technology, Delft.

Hao, C., van Boeijen, A.G.C., Stappers, P.J., 2017. 2017. Towards Cultura: a communication toolkit for designers to gain empathic insights across cultural boundaries. In: Proceedings of IASDR.

Kroesen, J.O., Darson, R., Ndegwah, D.J., 2015. Capacities, development and responsible innovation. In: Koops, B.J., et al. (Eds.), Responsible Innovation 2: Concepts, Approaches and Applications. Springer International Publishing, Switzerland.

Lindstrom, M., 2012. Buyology: How Everything We Believe About Why We Buy Is Wrong. Double Day, USA.

Mooi Mooier Middelland, 2016. Rotterdam. Retrieved from http://www.mooimooiermiddelland.nl/wp-content/uploads/2017/05/Mooi-Mooier-Middelland-Bewoners-ondernemers-en-gemeente-bundelen-krachten-in-co-creatie.pdf.

Mulder, I., 2014. Sociable smart cities: rethinking our future through co-creative partnerships. In: Distributed, Ambient, and Pervasive Interactions 2014. Springer, Cham, pp. 566–574, https://doi.org/10.1007/978-3-319-07788-8_52.

Mulder, I., 2015. Opening up: towards a sociable smart city. In: Foth, M., Brynskov, M., Ojala, T. (Eds.), Citizen's Right to the Digital City: Urban Interfaces, Activism, and Placemaking. Springer, pp. 161–173. https://doi.org/10.1007/978-981-287-919-6_9.

Municipality of Rotterdam, 2017. Middelland, Rotterdam Delfshaven. Retrieved from: http://www.rotterdam.nl/middelland.

Puerari, E., de Koning, J., Von Wirth, T., Karré, P., Mulder, I., Loorbach, D., 2018. Co-creation dynamics in urban living labs. Sustainability 10 (6), 1893. https://doi.org/10.3390/su10061893.

Razzaghi, M., Ramirez, M., Zehner, R., 2009. Cultural patterns in product design ideas: comparison between Australian and Iranian student concepts. Des. Stud. 30, 438–461. https://doi.org/10.1016/j.destud.2008.11.006.

Sanders, E.B.N., Stappers, P.J., 2012. Convivial Toolbox: Generative Research for the Front End of Design. BIS, Amsterdam.

Sleeswijk Visser, F., Stappers, P.J., van der Lugt, R., Sanders, E.B.N., 2005. Contextmapping: experiences from practice. CoDesign: Int. J. CoCreation Design Arts 1 (2), 119–149.

Sun, H., 2012. Cross-Cultural Technology Design: Creating Culture-Sensitive Technology for Local Users. Oxford University Press, New York.

Underhill, P., 2009. Why We Buy. The Science of Shopping. Simon & Schuster Paperbacks, New York.

van Boeijen, A.G.C., 2015. Crossing Cultural Chasms: Towards a Culture-Conscious Approach to Design. Doctoral Thesis, Delft University of Technology, Delft.

van Boeijen, A.G.C., 2020. Culture-Sensitive Design – A Guide to Culture in Practice. BIS Publishers, Amsterdam.

van Boeijen, A.G.C., Daalhuizen, J.J., Zijlstra, J.J. (Eds.), 2020. Delft Design Guide, second ed. BIS Publishers, Amsterdam.

van Rijn, H., Bahk, Y., Stappers, P.J., Lee, K.P., 2005. Three factors for context mapping in East Asia: trust, control and nunchi. CoDesign 2 (3), 157–177.

Author biography

Ummu Sakiinah (MSc) is a user researcher. She has an educational background in Industrial Product Design from the Institute Technology of Bandung. She continued her study at the Delft University of Technology in the Netherlands, majoring in Strategic Product Design as she wished to take design on a more strategic level to have more impact. She is interested in the creative research process and circular and regenerative economy. Currently, Sakiinah is working closely with millions of small business owners app users across Indonesia.

Dr. **Ingrid Mulder** is an Associate Professor at the Faculty of Industrial Design Engineering, Delft University of Technology. She has a background in policy and organization sciences (MA, University of Tilburg) and educational science and technology (Ph.D., University of Twente). In her current research, she studies the city as a space for transition while experimenting with participatory techniques and systemic design approaches for scaling and infrastructuring social change.

Annemiek van Boeijen (MSc/PhD) works fulltime at the Faculty of Industrial Design Engineering at the Delft University of Technology. Her research is focused on the role of culture in design processes, with the goal of designing methods geared to support designers in cultivating a culture-conscious approach. She is initiator and coeditor of the *Delft Design Guide* and author of *Culture Sensitive Design—A guide to culture in practice* and teaches a double-blended online course with the same name.

Rudi Darson (MSc) studied commercial sciences in Suriname and in the Netherlands and Public Administration at Leiden University. He has a degree in Change Management from the Maastricht University and he is an independent consultant on capacity building and entrepreneurship for creating economies. For more than 10 years, he is supervising multidisciplinary students teams from Delft University of Technology, doing projects in developing countries on entrepreneurship and development. Now he is focused on creating virtual and physical places to help revitalize economies.

Chapter 9

Affective technologies for enchanting spaces and cultivating places

Manuel Portela[a,*] and Carlos Granell-Canut[b]
[a]*Universitat Pompeu Fabra, Barcelona, Spain,* [b]*Universitat Jaume I, Castellón de la Plana, Spain*

Chapter outline

Introduction

The smart city discourse is informed in some of its most notable examples by a citizen-centric approach based primarily on the deployment of information and communication technologies (ICT). Its promises are often understood in terms of efficiency and speeding up of activities, services, and processes pertinent to people's daily routines and city functioning. Such promises are ways of generating fascination toward speed, political integration, and economic connectivity that provoke a "generalised sense of social good to which the majority of many people subscribe" (Harvey and Knox, 2012, p. 522). However, these promises driven by ICT are often materialized in urban spaces that lack spirit and liveness. Some authors have considered how modern urban infrastructures can cause a degradation of the "sense of place" (Relph, 1976). Moreover, routinized and optimized spaces can turn into meaningless places of monotony, poor

* Manuel Portela was affiliated with Universitat Jaume I when this research was conducted.

Shaping Smart for Better Cities. https://doi.org/10.1016/B978-0-12-818636-7.00011-1

existential experience, and alienation (Relph, 2016) or what Auge has defined as "nonplaces" (Auge, 1996). We can also relate this critique to the effect of disenchantment (Bennett, 2001), where spaces become empty containers that we fill with routine activities. For Bennett, this is a tale that predisposes ourselves toward rationalism, skepticism, and a culture of the meaningless. Authors like Lefebvre, Jacobs, and Harvey also echoed such criticism within discourses of politics and urban spaces (Harvey, 2014). Additionally, any fascination with technological solutions in the smart city is also criticized for potentially having adverse effects on people's liberties and democratic empowerment.

In this chapter, therefore, we address the importance of individual and collective meaning and how to design experiences, not just efficiencies, using digital technologies. In this context, we propose an alternative way to look at spaces and their infrastructure (digital and material), taking a different perspective that adopts the notion of enchantment.

Enchanted places, enchanting technologies

The concept of enchantment has a phenomenological origin and is entirely related to the mundane relationships with infrastructure. Being enchanted is to be struck and shaken by the extraordinary that lives amid the familiar and the everyday, or, in the words of Bennett:

> *[...] enchantment entails a state of wonder, and one of the distinctions of this state is the temporary suspension of chronological time and bodily movement. To be enchanted, then, is to participate in a momentarily immobilizing encounter; it is to be transfixed, spellbound.*
>
> Bennett (2001, p. 5)

Alfred Gell suggested that an "enchantment of technology is the power that technical processes have of casting a spell over us so that we see the real world in an enchanted form" (Gell, 1992, p. 44). He considered that all kinds of technical systems are immanently enchanted, producing an embodied change toward the world and the space we live in. Sengers et al. (2008) followed up from Gell's essay to suggest that technologies can be useful to reenchant ourselves in a new way that "inspires wonder, amazement, and emotion, and suggests magic and mystery" (Sengers et al., 2008, p. 347).

When looking at smart cities, we need to consider enchantment, developing an awareness of how technology affects toward fascination with the spaces that we inhabit. However, the current development of digital technologies can also be problematic in this regard, as we could be creating technologies of disenchantment where we treat emotions as data or computable information (Sengers et al., 2008). Bennett (2001) also suggested that we live in a moment of disenchantment when we rely on the modern disposition toward the rational and mundane.

Works in architecture and digital urbanism have tried to address the shortcomings of such dispositions through different approaches. For example, Caldwell and Foth (2014) discussed how "media architecture" can modify and engage citizens' attention within places, while Foth et al. (2015) addressed a series of urban interventions and their effects on urban infrastructures as a form of activism and a push toward the Right to the Digital City. Several examples of digital interventions can also be drawn within the concept of the *hackable city* (De Waal et al., 2017), a vision of a city that empowers people to exercise their right to participate in the creation of spaces through digital technologies. Empowerment and active participation are ways of embodying spatial transformation and contribute to enchanted placemaking.

Therefore, in this chapter, we consider the assumption that urban technology can undoubtedly play a pivotal role in enhancing enchanting experiences of daily life practices. Nevertheless, we propose a different viewpoint on how to use technology in spaces, where this becomes central and vital to create new tales and metaphors that can contribute to enchanting place in a form of collective appropriation. This viewpoint comes from our reflections and experience of place derived from experiments that we will introduce later in the chapter. Before discussing those experiments, it is important to look at some vital concepts underpinning them.

Designing experiences for enchantment

We argue that enchantment is not directly related to objects in a space, but to the experience that makes it meaningful: the tales and promises of being in a place. As designers, we can explore the opportunities to intervene in a space and create new forms of enchantment. Importantly in doing this, instead of envisaging special ad-hoc forms of intervention and original, brand-new places, we may consider operating on the mundane as our most recurrent scenario. Activities like taking a bus, having a dinner, or shopping at the supermarket are mostly routinized. However, Bennett (2001) suggested that the marvels of places are already there and we have to make use of the combination of the new artifice and the everyday, ordinary aspects of place in our daily life. Such combinations, we argue, can be enacted through technology.

The question therefore is, how can we contribute to the enchantment of mundane spaces and everyday activities using high technology? We follow a postphenomenological thinking (McCormack, 2017), critically adopting concepts and methods that will contribute to rethinking cities in terms of enchantment: place, affects, and atmospheres.

It has been argued that activities that connect humans to *places* are related to the symbolic and affective qualities of such environments (Najafi et al., 2011). For that reason, our approach involves exploring how to change spatial practices, reconfiguring space, from a perspective that considers materialities, affects, and emotions and the historical and cultural reasons for getting attached to it. As opposed to a more static or linear view of design practices, spatial reconfigurations are conceived as a critical approach over these. They introduce a perspective on space as relatively and constantly reconfigured by our readings and feelings, with the result of making the process of cocreation more open to others' feelings and perceptions.

The *affective* approach is inspired by Spinoza's affects as explained by Deleuze (Deleuze and Hurley, 1988; McCormack, 2008). Briefly, affects are manifestations of forces that are embodied and experienced as a particular feeling or state. It is a leading form of thought with broad presence in contemporary theories of space, place, and the city (Cadman, 2009; Duff, 2017). One of the implications of affective approaches is to discuss how the richness of daily activities creates meaningful relations to spaces.

A way in which affects can relate to a place and be experienced or felt collectively—though not necessarily openly shared—is framed by the concept of *atmospheres*. This definition was adopted in the fields of environmental psychology and cultural and human geography to study spatial practices with a perspective of the mundane and the performative. The concept can be used both to describe phenomena and as a method and approach to study and influence spatial configurations. Our hypothesis is that the concept and the methods of affective atmospheres can be key in the practice of enchanting spaces. This however will always be an open-ended approach, a matter of affordance more than cause and effect. Changing an atmosphere might and probably will affect people and their reactions and behaviors, though how cannot be ever fully determined.

Discussing place, space, and affects

The placemaking movement has been informed among others by the work of Jacobs (1961), Lynch (1960), Whyte (1980), and Debord (1955) and the radical urban planners of the 1960s and 1970s who focused on the full range of human senses, perceptions, and uncertainties in relation to the planning of urban spaces, as well as on the importance of the diversity of these within an apparently rationalized city. The discussion about how urban interactions take place and how spaces are configured has a long history, and we cannot cover it entirely in this chapter. In particular, Massey (2005) compared two contrasting visions and specific uses of the terms place and space. The first saw *place* as closed, coherent, integrated. This notion of place points at a secure retreat defined by tradition, identity, and immutable values as a politically conservative refuge. By contrast, she considered *space* as the product of interrelations, constituted through interactions. Spaces are also spheres of existence, the mess of social dynamics, in which distinct trajectories coexist in heterogeneity. Spaces, Massey continues, are products of material configurations that are always in the process of coming to be and are never finished or closed. Conceptual differences between place and space may be blurred, but Massey argues that the opening and closing of spaces do not reside in their opposition to places. Her point is to contest the political aspects of place through the open possibilities that spaces enable. Several urbanists such as Saskia Sassen and Jane Jacobs agreed with the idea of openness and contingency of spaces, due to the constantly changing everyday activities that take place in urban spaces and variations in agency related to these. The space and place terms can therefore be seen as physically inseparable in people's lived experiences and become vital to elaborating on a political argument about the capacities and empowerment of people.

In Portela et al. (2018), we studied how the description of place is reflected relationally between personal attachment and collective meaning. The purpose of that study was to understand the time-space dimensions of where the meaning of place arises. The case study was conducted in 2017 with master students from the Universidade NOVA de Lisboa (Portugal), where seven participants joined a workshop that included mapping activities, discussions, and a walking observation through the city to describe unique places that participants have previously pointed to on a map during the workshop.

One of the findings was that any place that is felt as central to our lives is perceived as having a spatial extension, an inside and outside, integrating elements from nature and culture, and being interconnected by a system of spatial interactions, which is also localized, with emergent historical components. This study helped us to explore and confirm the notions of genius loci and place attachment that arise from the work of Norberg-Schulz (1980) and Altman (1992), respectively.

Associating a space with individual feelings is only defined from a personal viewpoint and depends on specific life experiences. Nevertheless, a specific observation was that participants noted how monotony or familiarity could lead to a collective sense of boredom or frustration. That means that our participants could acknowledge both personal and collective feelings, identifying a degree of commonality in their reaction and a set of shared meanings to the places visited and referenced to. This aligned with the findings and reflections by Auge (1996) and Relph (1976) we were referring to earlier in the chapter.

Understanding affects

The notion of affects has its origin in Spinoza's *Ethics*, but it was more recently used by Deleuze and Hurley (1988). Later, geographers like Massumi (2002, 2015), McCormack (2003), Anderson and Harrison (2010), Griffero (2014), and Thrift (2008) used the concept to make sense of human actions and interactions in space.

Affects are a form of action potential or a force that pushes the body to act, relate, and feel. Affects are a nonconscious, nonintended phenomenon and can be understood as a bodily capacities or intensities. Examples range from the feelings that come to us by entering a train or a shop, or the collective adrenaline experienced when participating in a public demonstration, or the "weight" that our body feels when listening to a presidential speech in times of war.

That spark that makes our bodies react one way or another is shaped by subtle forces that are called affects. These are not inherently possessed by a subject (Anderson, 2006) or indeed intended by them and, can be embodied by many individuals who can also affect each other. For Anderson, affects are relational and embedded in processes of circulation, flow, transmission, or contagion between bodies.

Cultivating places

It has been said that affects exist in between bodies (Massumi, 2015) but not in specific spatial configurations. However, some geographers have focused their attention on the study of how theories on affects could be related to ones on space. The ideas of mediated spaces look at embodied experiences (Thibaud, 2015), which are characterized by their sensory and affective intensities (Casey, 1998). Contemporary ideas of place are built around this mediation, known as the "in-between" (Casey, 2001), and the cultivation of places through their intensities (Duff, 2010).

From the affective theory, we can argue that any space is part of the continuous intensities of daily life. Every encounter is a unique experience that conjures the lived, felt, and relational experience of a thinking, feeling body. Spaces can be enacted into *thick* places, transforming the "bodily capacities for moving, sensing, perceiving and attending" (Bissell, 2015, p. 132). According to Duff, places are *thick*when cultivated by local appropriations and through intimacy with the practices and encounters that take place in the construction of meaning and belonging to a place. On the contrary, *thin* places lack such a connection.

We argue that places are politically contested when we look at the in-between bodies and spaces by observing how affective forces alter the capacities between bodies, individuals, and collective practices. Therefore any space might struggle between being considered a place and a nonplace in a process of continuous negotiation through affective relations.

We consider affects of enchanting and feelings of fascination a possible vehicle for cultivating places. But how can we intervene in space in ways that would afford the embodiment of affects?

Atmospheres and spatial affective transformations

If a place is related to our own interactions and affective relationships, our perception of it is mediated by its atmospheres (Anderson, 2014; Böhme, 1993; Griffero, 2014) and rhythms (Edensor, 2010).

The concept of atmosphere was coined by the phenomenologist Gernot Böhme who called it "a new aesthetics." Böhme described these as "affective powers of feeling, spatial bearers of moods [...] experienced in bodily presence in relation to persons and things or in spaces" (Böhme, 1993, p. 117). Atmospheres can be seen as "a qualitative-sentimental *prius*, spatially poured out, of our sensible encounter with the world" (Griffero, 2014, p. 5). Sumartojo and Pink (2018) pointed out that atmospheres are in the midst of the affective and sensory experience and between memory and imagination. In simpler words, atmospheres are the way we "feel" the presence of a space and make sense of the place.

Atmospheres' affective forces evidence the spatial relationship with our surrounding environment and bodies. Atmospheres, however, are hard to acknowledge, to the point that we cannot define them with simple words. We can refer to them as "bubbles" (similar but not the same as Sloterdijk's spheres (Sloterdijk, 2002)) that are overlapped and appear and disappear in the concurrence of actions.

Atmospheres can describe the feelings about and within places but are also considered to have agency over other bodies (Duff, 2010). Atmospheres were also used as methodologies or mechanisms to being attentive to the emotions and sensations that are present in our body as it is involved in everyday practices (Anderson and Ash, 2015). For that reason, earlier practice-based works on light (Edensor, 2012) and sound (Gallagher et al., 2016), along with other sensory experiences (Middleton, 2010), can be used as an useful set of precedents in looking at atmospheres and space. Stemming from these, designing and analyzing a small-scale lighting intervention has allowed us to investigate how a space can be changed through illuminated atmospheres and the emotional traces of affective encounters they allow.

Light, darkness, and illuminated atmospheres

Visual and interactive arts can enable the transformation of spaces by using multimedia artifacts such as light, video projections, and sound. While these temporary transformations do not produce a different physical space, they produce perceptual differences by enabling new spatialities and through evoking sensory transformations (Edensor and Sumartojo, 2018). We chose to work with illuminated atmospheres, directing our inquiry toward the affective forces that exist in scenes in which the presence of artificial light is massively present, altering the way our senses are attuned to physical spaces (Anderson, 2014; Edensor and Sumartojo, 2015; Gandy, 2017; Healy, 2014; Sumartojo and Pink, 2018; Thibaud, 2015).

Light has always enchanted humans (Holloway, 2006; Sørensen, 2015), but, with the frantic pace of technological advancements, light technology also revealed and changed urban life forever (Purnell, 2017). Light as a phenomenon, implicated in atmospheric conditions, also received the attention of several scholars (Bille, 2017; Böhme, 2014).

Designed illuminated spaces and atmospheres were studied when it comes to specific settings that transform places for specific purposes such as the transformation of collective memory (Gandy, 2017, p. 366), to modify the character of urban atmospheres (Edensor, 2015), or the feelings associated with experiencing urban spaces (Sumartojo and Pink, 2017, p. 7). However, illuminated spaces and atmospheres can also be seen as a way to defamiliarize with habitual spaces (Edensor and Sumartojo, 2018), as they can affect us in ways that lead to feeling very differently about otherwise familiar settings.

Therefore we explored the opportunities that experimenting with the transformation of spaces bring. We intended to study how people's everyday activities are entangled with the forces of enchantment and how these forces can impact on the cultivation of places. For that, we did a spatial intervention involving the design of illuminated atmospheres.

We explored the opportunities to intervene in a space and grasp the existence and changes of atmospheres, recognize subjectivities, and follow the reactions in spatial behaviors of those who passed by. In the following case, we played with the idea that individuals can defamiliarize from the habitual and discover the atmospheric transformation as a form of enchantment.

Atmospheric methods to locate affects

Anderson and Ash (2015) claimed that atmospheres can be treated as objects of inquiry, attempting to separate the assembling of atmospheres from their effects. The atmospheric methods, as they called this way of studying the atmospheric presence and change, helped us to put in perspective our lighting intervention and to assess our design process to anticipate effects. Considering that atmospheres have effects on and emanate from multiple bodies that enter in relation to physical encounters, the approach to atmospheric methods is to focus on the organization of such encounters "where affects emerge when two beings or entities contact one another in some way" (Anderson and Ash, 2015, p. 35). The stages for studying atmospheres are identification, coexistence, causal powers, and transformation.

To do this, scholars have used different ethnographic strategies. They have not relied only on well-known approaches such as autoethnography and interviews but have also proposed new creative methods for qualitative study *in*, *about*, and *through* atmospheres (Edensor, 2012; Sumartojo and Pink, 2018). A way to understand the effects of atmospheres is that these produce a mute "attunement" to place. For Edensor, attunement means anticipation that "engenders the coproduction of an atmosphere by preparing visitors for an emotional and affective encounter with the resort in noncognitive and cognitive ways" (Edensor, 2012, p. 1115). Attunement to atmospheres can put the observer and participants in each other's shoes, even when considering human and nonhuman actors (Anderson and Ash, 2015; Schroer and Schmitt, 2017). Researching atmospheres can also be done through looking at visualizations of what is yet to come (Degen et al., 2017; Rose et al., 2014) or pictures and video recordings that capture atmospheres in different time-spaces (Edensor and Sumartojo, 2018; Sumartojo and Pink, 2017).

Designing the intervention

Our experiment aimed at observing how situated practices and behaviors mediated by atmospheric attunement help to understand place cultivation and how the specific space of the experiment was perceived through the changes that took place. We also addressed the issue of the temporary condition of the intervention.

Therefore we considered a series of questions that would enhance our understanding of how technology can promote the cultivation of places:

1. How do interactions generate new spatial reconfigurations?
2. How are interactions part of affective atmospheres?
3. Which feelings and emotions are expressed through the presence and absence of interactions?
4. How do spatial reconfigurations alter the activities and practices that take place?

We selected the bus station of our university campus at Universitat Jaume I in Castellón (Spain). It is an open structure made of glass, wood, and steel, which works as a public transportation hub. We attuned ourselves with atmospheres from the station by making video recordings and pictures to observe people behavior and activities, together with the spaces created by light and darkness. The station had luminaries placed in the ceiling covering only half of the surface, leaving some spaces in shadow. Therefore the effect of the buses entering the station generated not only a soundscape of engine and beeping sounds but also a projection of moving lights, which somehow produced a threatening atmosphere. The lack of soft lighting was aggravated by the fury of ambient noise and people's urge to get to their homes. This situation presented itself as a potential playground, where the effects of humans and machinery stimulated the senses.

Based on this scenario, we imagined an intervention that would allow both the people waiting for the bus and other passersby to change the lighting environment by interacting with the soundscape. We designed and developed a lighting system based on a set of RGB festoon-style light bulbs that were divided into two rows along the stations and controlled by a computer (see Fig. 1 and external video source). The light bulbs were programmed to change according to the ambient sound captured by two microphones. One microphone pointed in the direction of the bus bay, and the other one at the area where people usually sit waiting. As the system took both sound inputs, the feedback to users was in the form of variants of lights responding differently to each stimulus. Those perturbations (Ash, 2013) were meant to bring the potential of new atmospheres, generating different rhythms and spaces of darkness, together with glowing and colorful effects.

We complemented the lighting system with a wireless control/interactive system to let people actively change the lighting effect using their mobile

FIG. 1 Illuminated bus station.

phones by choosing between four different rhythms and reactions to the noise input. This capability was announced through posters within and around the station. This aspect is vital because the way elements and materials are spatially arranged also contributes to the atmospheric emergence (Bissell, 2010).

We designed the entire system getting inspired by different art installations whose information was available from the Internet, for example, the Llum BCN or Vivid Sydney light festivals. Such precedents suggested us some configurations outcomes, anticipating certain atmospheres in a realistic scenario. However, budget, technologies, and time constraints altered the final implementation, not entirely fulfilling our expectations. Adopting a flexible attitude is common when technological interventions are deployed to allow for the unexpected. We have professional experience in light installations, and we knew that the material qualities of our experimental intervention were not the same as it could be in a definitive, well-funded installation. We made our best efforts to make this attractive and compelling, while the unstable ecologies provoked mutability in the process (Rose et al., 2014), adapting our expectations of the process of material creation.

Additionally, we were influenced by people who asked us about the intervention while we were still working on it, changing our perspective regarding our goal. We had to make changes and adjustments to the system so that it worked together with their perception of it, in a way that it contributed to co-created atmospheres (Bille et al., 2015).

We ran the experiment from the second half of December 2017 to the end of January 2018. We made observations and took notes, pictures, and video recordings before, during, and after the intervention. We also interviewed 40 people who were around the station, including students, professors, administrative staff, cleaning and security staff, and bus drivers. These interviews were held in two periods—the first period during December and the second during the second week of January.

Guidelines to observe atmospheric effects

We experimented with an atmospheric transformation to modify the affective relations within a space and to create new ways in which infrastructures can enchant through bodily encounters (Harvey and Knox, 2012). Extending these atmospheric methods, we identified six dimensions to study how atmospheres mediate the space and the perceived effects through our intervention. These dimensions can be perceived as overlapping, which are encounters, spatial practices, materialities, senses, perception, and emotions. Reflecting on them can be used as a guide to perform atmospheric observations and interventions.

Effects on encounters

We know that previous encounters influence people (Edensor, 2012), so atmospheres are mediated through their anticipation, memory recollection, and individual dispositions to affects. The novelty of the first encounter with the lights provoked enthusiasm, but these quickly became perceived as a normal part of the infrastructure, losing their enchantment. Subjects and their practices get caught in affective relations and atmospheres.

Feelings toward spaces are also defined by previous encounters, where affects can be recollected by memory. Transforming spatial configurations means to alter the expected nature and outcomes of these previous encounters and potentially change the way we perceive them.

The lighting system was configurable by changing the behavior of lights using smartphones. Whenever someone made those changes, the perceptual space changed as well, recovering their promise and enchantment. These moments of enchantment became opportunities to challenge participants' memory and perception. Encountering colorful lights was always a temptation to be curious and get closer to see what was going on. The effects of novel light behaviors caused more visits and encounters with the station. But these quickly ended up being ephemeral, as the potential of enchantment dissipated with people becoming used to the novelty. The perception and feelings of the space turned then back into being mundane and everyday.

Effects on spatial practices

The way people moved around the space was entirely different depending on whether they were bus passengers, personnel, or other passersby. Some would stand, while others moved around the structure, and others liked to remain far from the crowd. These movements themselves caused changes in atmospheres, where affects kept circulating between them as an effect of continuity. Other affects arose depending on the moment of the day—the movements, or the number of people—influencing the emergence and collapse of atmospheres. This can be exemplified by referring to sudden and contrasting changes, for example, the physical arrival and presence of a larger group of people talking when the station ambient was calm and quiet, or, when a noisy bus approaches at night with head lights on, shifting the shadows in the station, or when a warm wind arrives in a cold night. These interchanges of forces not only make different atmospheres coexist, collide, or mix during some time but also make some disappear for longer periods.

Our exercise intended to anticipate and amplify some of the experiences of the space by translating the ambient acoustics into light expressions. Some of the interviewees told us that they do not use the bus, yet the lights made the stop more conspicuous (as it is placed in an open space, lights were visible from many points). Bissell (2010, 2015) noted that habits help people orientate themselves in how to organize their living space and their relations with other objects and subjects. The introduction of lights helped bring more life in the darkest time of the day, enabling new ways of using the space by the passengers waiting for the bus.

Effects on materialities

The ways that the affective power of infrastructure and materialities extend to other domains beyond bodies and practices are often predictable (Merriman, 2016). However, through the introduction of the new lights, we wanted to invite people to discover a different space. When approaching the station, they saw the effects, their brightness and glow, and felt a sense of curiosity. However, physically, the station is well separated from other sites and buildings around it by empty portions of space and does not easily attract attention. Without our lights the movement of buses would produce its only illuminated atmosphere. Through the intervention the character of the bus station changed; it acquired relevance and stood out over other parts of the university campus. The curiosity about it and the urge to discover it constituted a radical change of attitude and disposition toward this place. Even when it did not change any explicit pedestrians' behavior, it generated a feeling of unity and continuity with other nearby spaces that had been inexistent before. For example, two of the interviewees said that seeing the lights from far away made them curious and willing to visit the bus stop. A new and stronger sensorial connection with the space was established, making it more attractive and acknowledged from other buildings. The bus stop became more relevant, not because of the primary function it offered, but because it brought other new qualities to the urban space. Atmospheres are therefore always fluid and open to change, at both collective and individual levels.

Effects on senses

The introduction of different rhythms, modulations, and intensities stimulated the senses turning the previous atmosphere of boredom into a mood of excitement. People reacted to such stimuli changing their behavior into a more playful one, which in turn made the bus stop feel more attractive.

Effects on perception

The changes in how the space was organized and how it stimulated the senses gave us, as observers, an opportunity to perceive the materialization of atmospheres. Normally, it can be argued that feeling an atmosphere is a strongly subjective and difficult to describe phenomenon. Atmospheres became more clearly perceivable as they became manifested in the feelings expressed during interviews about how these were associated with the encounters with materialities and spatial configurations. Some participants, for instance, manifested the perception of safety brought upon by the lights during the night. This helped us to attune to the atmospheres generated by these brand-new spatialities.

Effects on emotions

Perceptions of the basic qualities of the station were also materialized as emotions and related to feelings of detachment and boredom, as often occurs in many habitual practices associated with urban commuting. However, boredom that presents itself as a state of stillness, or a static sensation, does not necessarily imply passivity. In Anderson (2004), boredom is represented as a form of practice implying relaxation and a preparation to disconnect from social life and/or return to it. Our lights interfered with this exercise of disconnection, demanding attention and a reactivation of the senses (especially with changing rhythms, glows, and shadows) and a more proactive mood. Some of the people interviewed stated that in the presence of our intervention, they felt less alone. Yet others claimed that the lights were an uncomfortable disruption. Those for whom the lights altered and disturbed habits denied the possibility of enjoying the installation.

Discussing the effects of atmospheres

Considering the observations so far, we can reflect on converging and diverging aspects of our analysis regarding the identification of atmospheres. This helps us discuss their coproduction, causal powers, and transformations:

- Change and transformation are vital characteristics in the identification of atmospheres. They are in continuous transformation and in affective relations with bodies. Therefore their fuzzy boundaries make them hard to identify and define.
- Even when atmospheres are perceived on an individual basis, there are common aspects that are repeatedly identified "with variations across sites, networks, or events" (Anderson and Ash, 2015, p. 49). So, despite all differences and individual characterization, their effects can in many cases be considered as shared.

- In the case of the bus stop, some of the feelings and emotions expressed by the interviewees remained unchanged. Atmospheric effects in participants were perceived only slightly, and the presence of atmospheres could just be grasped, as Lefebvre once described as a technique for Rhythmanalysis (Lefebvre, 2004).
- Movement and interactions between people and machines created new affective relations. In particular the interactive capability of our system allowed to create and change atmospheres in relation to other mechanic phenomena (such as buses' noise and wind). Such interactions acted more as a disturbance or, what Ash (2013) calls a *perturbation*, an affective change that affects technological systems rather than living beings.
- Finally the newly created atmospheres coexisted with others in time and space. We agree with Anderson that new arrangements between objects and people can prompt the transformation and change of atmospheres and the disruption of others.

Consequently the effects of atmospheric change and disruption are twofold. On the one hand, we must consider that atmospheres and affects are not strictly associated with, and limited by, their own structured spatial forms and materialities. For example, the illuminated atmospheres created by our intervention extended through the university campus much further away than the premises of the bus stop. Psychological, emotional, and sensory perceptions are always inscribed dealing with negotiated temporal and spatial experiences (Duff, 2010). Furthermore, affective conditions are always mediated and organized in "multiple, partially connected, apparatuses that attempt to structure the capacity to affect and be affected of people and groups" (Anderson, 2014, p. 36). In this sense, our spatial intervention opened opportunities for contesting the habitual, creating opportunities of enchantment. Consequently, atmospheric effects can be imprinted in memory and recollected (Bille et al., 2015) every time we remember the spatialized experience. It is possible, though not proven, that some of the people who experienced our installation will recall the feelings of the atmosphere even when it is no longer there, with all the possible individual and collective meanings and significances associated with it and the place.

On the other side, the ambivalence of atmospheres does not allow us to fully anticipate their discursive powers. Atmospheres emerge as affective forces that exceed our capacities to integrate the social and cultural in a simple overview. As Edensor pointed, atmospheres can only be anticipated thanks to our memories and past experiences, and we can attune with a place before being in it. But even when atmospheres are somehow collective, the way in which every individual is affected by the experience of an encounter will still vary, and it is important to challenge the idea that through an artistic installation everybody would or should experience exactly the same feelings. It is key, "departing from the assumption that contemporary, Western philosophy of atmosphere is universal" (Bille et al., 2015, p. 36), as this would only blind us from studying the mundane of everyday spaces. Thus, if we want to bring up new enchantments toward technological development, we suggest to explore spatial reconfigurations through the individual and collective disparities of atmospheric experiences.

Thibaud (2015) was concerned about how current urban design approaches were attempting to play with the senses. He stated a dual concern with contemporary urban policies, where a trend exists on "creating a festive spirit while integrating law and order". Because of the ambivalence of atmospheres and the disparity on bodily experiences, it is unrealistic to guarantee a particular effect, and we question whether atmospheres can be created to precisely and universally fulfill specific needs or intentions.

However, during our experiment, we accounted for many effects and materialities that generated specific arrangements that changed the ways in which we experienced the space, proving that atmospheres will have significant impacts, though these cannot be fully determined. Therefore we would recommend experimenting with interventions that can go from whole-designed purposes to smaller everyday changes in material and sensory configurations. In that sense, digital technologies could be considered as *ecstasies*, defined as things that articulate their presence through qualities (Böhme, 1993), in an ongoing spatial change and transformation. Reconfigurations are engaged in a constant coproduction of materialities, affects, and perturbations. As in our experiment, the changes in material objects can cause a direct spatial transformation but could also interplay with other environmental aspects that are out of our control.

Our ethnographic observations point toward the mediation of spatial arrangements through sensory embodiments, individual activities, and social practices. Through this experiment, we explored the possibilities of co-creating atmospheres, following a process of anticipation, design, contestation, and perception of them. It was evident that beyond the fussiness of atmospheres, their effects are not always perceived by everyone, but also that the participation in their creation is not limited to the designer or the researcher. On the contrary, its constitution is collective, shaped by the agency of humans and nonhumans. Even when material changes are subtle, they can reverberate into new relations. As a result, environmental changes and the atmosphere's affective forces are in direct relation to our habits and the everyday perception of certain spaces. Such transformations generate new orders and imply new affective relations, resulting in new place assemblages. Consequently, being guided by the atmospheric changes and their effects, it resulted that even in the most unexpected spaces, places are still cultivated through personal and collective configuration.

Conclusions

While exploring the idea of place, we have found that change and transformation are critical characteristics in the boundaries of atmospheres and their effects. Technologies can affectively alter our feelings anchored to a place by generating suspension or acceleration of the rhythms that mediate our experience (Edensor, 2010). Consequently, many versions of the same affects could radically change the experience of atmospheric qualities of the same physical space, leading to a subtle effect of enchantment.

We wonder whether places can be created for specific needs. We proposed a more critical approach on how we can observe spatial relationships regarding affects, feelings, and emotions. Our contribution to the exploration of how objects and artifacts conform to some practices lies in observing how those configurations become part of atmospheres. We followed the research questions that led us to grasp the effects of atmospheres and their relation to space. We delineated guidelines to follow the effects of atmospheres that can serve as a point of departure to attune the perception to spatial changes. Finally, we encourage designers to develop their attentiveness toward the affective relationships with the elements and the spaces that are designed and take the lead on using technology for enchantment.

For the understanding of a place, we emphasize the importance of the atmospheric qualities of certain spaces that are tightly tied to habits. At the same time, we also explain many of the effects and materialities that generate specific orders that drive our experience. We speculate about the place qualities that foresee the development of technologies to bring new, even temporal, *genius loci* to cities. Thus finding new ways to provide essential changes to urban environments requires understanding how technology provides a unique opportunity to change the material conditions and intensities through different atmospheric changes. As Bennett (2001) defines, enchantment is also an attitude toward the openness to the unusual, the captivating, and the disturbing in everyday life. This is why we introduced throughout this chapter the need to cultivate places as a form of enchantment. To develop potentialities of attachment to those places, reducing the effects of placelessness and detachment.

In conclusion, we argue that by stimulating the senses, playing with the materialities of the environment, and enhancing the affective capacities of bodies, there is a possibility to transform, reshape, mesh, destroy, or dislocate atmospheres. We can expect the unpredictable but new and diverse spatialities will definitively open the possibilities of enchanting, even if this is not free from negotiation, contention, and struggle.

In designing for smarter cities, we need to be critical to our practices, understanding the consequences of designing spaces and how these spaces affect people's lives. We must take specific care about how our cities, urban life, and spaces are organized and mediated by technology because a space in transformation will always remain a space of continuous and endless change on the collective and individual. Therefore the methodologies presented in this chapter are devices that can be used to design with care in a broad sense.

Acknowledgments

The authors gratefully acknowledge funding from the European Union through the GEO-C project (H2020-MSCA-ITN-2014, Grant Agreement Number 642332, http://www.geo-c.eu/). Carlos Granell is funded by the Ramón y Cajal Programme of the Spanish Government (RYC-2014-16913).

References

Altman, I., 1992. In: Altman, I., Low, S.M. (Eds.), Place Attachment. Springer US, Boston, MA, https://doi.org/10.1007/978-1-4684-8753-4.

Anderson, B., 2004. Time-stilled space-slowed: how boredom matters. Geoforum 35 (6 Special issue), 739–754. https://doi.org/10.1016/j.geoforum.2004.02.005.

Anderson, B., 2006. Becoming and being hopeful: towards a theory of affect. Environ. Plan. D Soc. Space 24 (5), 733–752. https://doi.org/10.1068/d393t.

Anderson, B., 2014. Encountering Affect: Capacities, Apparatuses, Conditions. Ashgate Publishing, Ltd. Retrieved from: http://www.tandfonline.com/doi/full/10.1080/14649365.2015.1078146.

Anderson, B., Ash, J., 2015. Atmospheric methods. In: Vannini, P. (Ed.), Nonrepresentational Methods: Re-Envisioning Research. Routledge, pp. 44–61. Retrieved from: http://www.academia.edu/10294576/Atmospheric_Methods.

Anderson, B., Harrison, P. (Eds.), 2010. Taking Place: Non-Representational Theories and Geography. Ashgate Publishing, Ltd.

Ash, J., 2013. Rethinking affective atmospheres: technology, perturbation and space times of the non-human. Geoforum 49, 20–28. https://doi.org/10.1016/j.geoforum.2013.05.006.

Auge, M., 1996. Non-places: introduction to an anthropology of supermodernity. Cap. Class 20. https://doi.org/10.1177/030981689606000111. Verso Books.

Bennett, J., 2001. The Enchantment of Modern Life: Attachments, Crossings, and Ethics. Princeton University Press, Princeton, New Jersey. Retrieved from: http://scholar.google.com/scholar?hl=en&q=The+Enchantment+of+Modern+Life:+Attachments,+Crossings,+and+Ethics&btnG=Search&as_sdt=0,5&as_ylo=&as_vis=0#1.

Bille, M., 2017. Ecstatic things: the power of light in shaping bedouin homes. Home Cult. 14 (1), 25–49. https://doi.org/10.1080/17406315.2017.1319533.

Bille, M., Bjerregaard, P., Sørensen, T.F., 2015. Staging atmospheres: materiality, culture, and the texture of the in-between. Emot. Space Soc. 15, 31–38. https://doi.org/10.1016/j.emospa.2014.11.002.

Bissell, D., 2010. Passenger mobilities: affective atmospheres and the sociality of public transport. Environ. Plan. D Soc. Space 28 (2), 270–289. https://doi.org/10.1068/d3909.

Bissell, D., 2015. Virtual infrastructures of habit: the changing intensities of habit through gracefulness, restlessness and clumsiness. Cult. Geogr. 22 (1), 127–146. https://doi.org/10.1177/1474474013482812.

Böhme, G., 1993. Atmosphere as the fundamental concept of new aesthetics. Thesis Eleven 36, 113–126. https://doi.org/10.1177/072551369303600107.

Böhme, G., 2014. Light and space. On the phenomenology of light. Dialogue and Universalism 4 (4), 62–73.

Cadman, L., 2009. Nonrepresentational theory/nonrepresentational geographies. In: Kitchin, R., Thrift, N. (Eds.), International Encyclopaedia of Human Geography. Elsevier, pp. 456–463.

Caldwell, G.A., Foth, M., 2014. DIY media architecture: open and participatory approaches to community engagement. In: Proceedings of the 2014 Media Architecture Biennale, pp. 1–10, https://doi.org/10.1145/2682884.2682893.

Casey, E.S., 1998. The Fate of Place: A Philosophical History. A Centenial Book. University of California Press.

Casey, E.S., 2001. Between geography and philosophy: what does it mean to be in the place-world? Ann. Assoc. Am. Geogr. 91 (4), 683–693. https://doi.org/10.1111/0004-5608.00266.

De Waal, M., De Lange, M., Bouw, M., 2017. The hackable city: citymaking in a platform society. Archit. Des. 87 (1), 50–57.

Debord, G., 1955. Introduction to a critique of urban geography. Les Lèvres Nues #6.

Degen, M., Melhuish, C., Rose, G., 2017. Producing place atmospheres digitally: architecture, digital visualisation practices and the experience economy. J. Consum. Cult. 17 (1), 1–25.

Deleuze, G., Hurley, R., 1988. Spinoza: Practical Philosophy. City Lights Publishers.

Duff, C., 2010. On the role of affect and practice in the production of place. Environ. Plan. D Soc. Space 28 (5), 881–895. https://doi.org/10.1068/d16209.

Duff, C., 2017. The affective right to the city. Trans. Inst. Br. Geogr. 42 (4), 516–529. https://doi.org/10.1111/tran.12190.

Edensor, T. (Ed.), 2010. Geographies of Rhythm: Nature, Place, Mobilities and Bodies. Ashgate, Manchester.

Edensor, T., 2012. Illuminated atmospheres: anticipating and reproducing the flow of affective experience in Blackpool. Environ. Plan. D Soc. Space 30 (6), 1103–1122. https://doi.org/10.1068/d12211.

Edensor, T., 2015. The gloomy city: rethinking the relationship between light and dark. Urban Stud. 52 (3), 422–438. https://doi.org/10.1177/0042098013504009.

Edensor, T., Sumartojo, S., 2015. Designing atmospheres: introduction to special issue. Vis. Commun. 14 (3), 251–265. https://doi.org/10.1177/1470357215582305.

Edensor, T., Sumartojo, S., 2018. Reconfiguring familiar worlds with light projection: the gertrude street projection festival, 2017. GeoHumanities 4 (1), 112–131. https://doi.org/10.1080/2373566X.2018.1446760.

Foth, M., Brynskov, M., Ojala, T., 2015. Citizen's Right to the Digital City: Urban Interfaces, Activism, and Placemaking. Springer, Berlin, pp. 978–981.

Gallagher, M., Kanngieser, A., Prior, J., 2016. Listening geographies. Prog. Hum. Geogr. 41 (5). https://doi.org/10.1177/0309132516652952.

Gandy, M., 2017. Urban atmospheres. Cult. Geogr. 24 (3), 353–374. https://doi.org/10.1177/1474474017712995.

Gell, A., 1992. The technology of enchantment and the enchantment of technology. In: Coote, J., Shelton, A. (Eds.), Anthropology, Art, and Aesthetics. Clarendon Press, pp. 40–63.

Griffero, T., 2014. Atmospheres: Aesthetics of Emotional Spaces. Ashgate Publishing, Ltd, Surrey.

Harvey, D., 2014. Rebel Cities From the Right to the City to the Right to the Urban Revolution. Eure. vol. 40 Verso, London; New York, https://doi.org/10.4067/S0250-71612014000100013.

Harvey, P., Knox, H., 2012. The enchantments of infrastructure. Mobilities 7 (4), 521–536. https://doi.org/10.1080/17450101.2012.718935.

Healy, S., 2014. Atmospheres of consumption: shopping as involuntary vulnerability. Emot. Space Soc. 10 (1), 35–43. https://doi.org/10.1016/j.emospa.2012.10.003.

Holloway, J., 2006. Enchanted spaces: the séance, affect and geographies of religion. Ann. Assoc. Am. Geogr. 96 (1), 182–187.

Jacobs, J., 1961. The Death and Life of Great American Cities. Vintage Books, New York, https://doi.org/10.2307/794509.

Lefebvre, H., 2004. Rhythmanalysis: space, time and everyday life (Stuart Elden and Gerald Moore trans.). Cap. Class. https://doi.org/10.1177/030981680508600110.

Lynch, K., 1960. The Image of the City. The MIT Press, Cambridge, https://doi.org/10.2307/427643.

Massey, D., 2005. For Space. SAGE Publications, London.

Massumi, B., 2002. Parables for the Virtual: Movement, Affect, Sensation. Postcontemporary Interventions. Duke University Press, https://doi.org/10.1215/9780822383574.

Massumi, B., 2015. Politics of Affect. Polity Press, Cambridge, UK.

McCormack, D.P., 2003. An event of geographical ethics in spaces of affect. Trans. Inst. Br. Geogr. 28 (4), 488–507. https://doi.org/10.1111/j.0020-2754.2003.00106.x.

McCormack, D.P., 2008. Engineering affective atmospheres on the moving geographies of the 1897 Andrée expedition. Cult. Geogr. 15 (4), 413–430. https://doi.org/10.1177/1474474008094314.

McCormack, D.P., 2017. The circumstances of post-phenomenological life worlds. Trans. Inst. Br. Geogr. 42 (1), 2–13. https://doi.org/10.1111/tran.12146.

Merriman, P., 2016. Mobility infrastructures: modern visions, affective environments and the problem of car parking. Mobilities 11 (1), 83–98. https://doi.org/10.1080/17450101.2015.1097036.

Middleton, J., 2010. Sense and the city: exploring the embodied geographies of urban walking. Soc. Cult. Geogr. 11 (6), 575–596. https://doi.org/10.1080/14649365.2010.497913.

Najafi, M., Kamal, M., Mohd, B., 2011. The concept of place and sense of place in architectural studies. Int. J. Hum. Soc. Sci. 6 (3), 187–193.

Norberg-Schulz, C., 1980. Genius loci: Towards a Phenomenology of Architecture. Rizzolli, New York.

Portela, M., Acedo, A., Granell-Canut, C., 2018. Looking for "in-between" places. Media Theory J. 2 (1), 108–133. Retrieved from: http://mediatheoryjournal.org/portela-et-al-looking-for-in-between-places/.

Purnell, C., 2017. The Sensational Past: How the Enlightenment Changed the Way We Use Our Senses. W. W. Norton. Retrieved from: https://books.google.pt/books?id=P0t8DAAAQBAJ.

Relph, E., 1976. Place and Placelessness. Pion Limited, London, England.

Relph, E., 2016. The paradox of place and the evolution of placelessness. In: Freestone, R., Liu, E. (Eds.), Place and Placelessness Revisited. Routledge, p. 278.

Rose, G., Degen, M., Melhuish, C., 2014. Networks, interfaces, and computer-generated images: learning from digital visualisations of urban redevelopment projects. Environ. Plan.D Soc. Space 32 (3), 386–403. https://doi.org/10.1068/d13113p.

Schroer, S.A., Schmitt, S.B., 2017. Exploring Atmospheres Ethnographically. Taylor & Francis. Retrieved from: https://books.google.es/books?id=b9NBDwAAQBAJ.

Sengers, P., Boehner, K., Mateas, M., Gay, G., 2008. The disenchantment of affect. Pers. Ubiquit. Comput. 12 (5), 347–358. https://doi.org/10.1007/s00779-007-0161-4.

Sloterdijk, P., 2002. Spheres, Volume I: Microspherology. *Pluriel*, 686.

Sørensen, T.F., 2015. More than a feeling: towards an archaeology of atmosphere. Emot. Space Soc. 15, 64–73. https://doi.org/10.1016/j.emospa.2013.12.009.

Sumartojo, S., Pink, S., 2017. Moving through the lit world: the emergent experience of urban paths. Space Cult. https://doi.org/10.1177/1206331217741079.

Sumartojo, S., Pink, S., 2018. Atmospheres and the Experiential World: Theory and Methods. Taylor & Francis. Retrieved from: https://books.google.es/books?id=KiBtDwAAQBAJ.

Thibaud, J.P., 2015. The backstage of urban ambiances: when atmospheres pervade everyday experience. Emot. Space Soc., 39–46. Retrieved from: http://www.sciencedirect.com/science/article/pii/S1755458614000371.

Thrift, N., 2008. Non-Representational Theory: Space, Politics, Affect. Routledge, London.

Whyte, W.H., 1980. Social Life of Small Urban Spaces. Project for Public Spaces, New York.

Author Biography

Manuel Portela is a multidisciplinary PostDoc researcher at Universitat Pompeu Fabra. He holds a degree in Graphic Design, an MSc in Urban Studies, and a PhD in GeoInformatics from Universitat Jaume I (UJI) of Castellón, Spain. Portela is interested in the impact of technology in society, including topics such as artificial intelligence, citizen science, and smart cities. Currently, he researches the development of Explainable and Participatory approaches to Artificial Intelligence.

Carlos Granell-Canut currently holds a 5-year Ramón y Cajal postdoctoral fellowship at the Universitat Jaume I (UJI) of Castellón, Spain. Before rejoining UJI in 2015, he worked 4 years as a post-doc in the European Commission's Joint Research Centre. His research interests lie in the multidisciplinary application of GIS (Science/Systems), spatial analysis and visualization of sensor- and user-generated geographic content, and reproducibility research practice.

Chapter 10

Smart engagement for smart cities: Design patterns for digitally augmented, situated community engagement

Callum Parker[a], Martin Tomitsch[a,b], and Joel Fredericks[a]

[a]*Design Lab, School of Architecture, Design and Planning, The University of Sydney, Sydney, NSW, Australia,* [b]*CAFA Beijing Visual Art Innovation Institute, China*

Chapter outline

Introduction

Public spaces within cities are heterogeneous social areas that are generally accessible to all people (Low and Smith, 2013). Around the world, cities and the public spaces within them are changing due to decreased hardware costs, smaller computing form factors, advances in digital technologies, and society's increasing dependence on them. These digital technologies create information layers that exist on top of the physical space, resulting in the space becoming augmented with dynamically changing information (Manovich, 2006). This digital augmentation of cities contributes to the notion of "smart cities," which has many competing definitions and is used for a wide variety of agendas.

Shaping Smart for Better Cities. https://doi.org/10.1016/B978-0-12-818636-7.00010-X

Broadly speaking, smart cities use embedded digital technologies to better optimize their resources, plan preventive maintenance activities, monitor security aspects while maximizing services to their citizens (Hall et al., 2000), and generally improve the lives of urban dwellers (Tomitsch, 2018). Hollands (2008) highlighted that the definition needs deeper consideration and a more critical perspective. They raised the risk of digital technologies causing socioeconomic divisions and emphasized that the importance of ensuring smart cities is progressive in nature, taking "much greater risks with technology, devolve power, tackle inequalities and redefine what they mean by smart itself." Others acknowledged that a smart city is one that should invest in human and social capital and achieve sustainable economic growth and a high quality of life through "participatory governance" (Caragliu et al., 2009). In other words, defining a smart city as "a city seeking to address public issues via ICT-based solutions on the basis of a multi-stakeholder, municipally-based partnership" (Manville et al., 2014).

This chapter focuses on *community engagement* as one of the elements that contribute to achieving participatory governance in smart cities. Within the context of the chapter, we refer to community engagement as the process of involving stakeholders, such as government agencies, private enterprise, community organizations, and everyday people in decisions that affect governance, policy, infrastructure, and lifestyle. Building on the notion of "augmented space," the chapter discusses the use of situated digital technologies, that is, technologies embedded within an urban space, for engaging urban dwellers around contextually related community topics. The term "community" has been criticized for being vague and value-laden, glossing over the complexity of urban demographics (Head, 2007). In the case studies presented in the chapter, we use the term to refer to people passing through or dwelling within an urban space. The chapter argues that situated digital technologies are well placed to engage a more relevant demographic compared with traditional community engagement initiatives. In the next section, we position the role of digital technologies for community engagement. This is followed by a discussion of three case studies, which informed the development of six design patterns for digitally augmented, situated community engagement. This contribution is important as it consolidates our understanding of digitally augmented, situated engagement in a form that researchers and practitioners can use to inform the creation and use of situated digital technologies.

Digital technologies and community engagement

Digital technologies are increasingly used for engaging communities through activities that augment or replace traditional engagement techniques, for example, through the use of urban screens, mobile devices, apps, and augmented reality (Gentile et al., 2017; Gordon et al., 2016; Lee et al., 2012). Engagement activities can take various participatory forms, for example, from informing people only through top-down mechanisms to a more collaborative approach that involves people throughout the engagement process (Fredericks and Foth, 2013). Building on participation literature, like early work by Arnstein (1969), the International Association for Public Participation[a] defines these forms of engagement according to five categories: informing, consulting, involving, collaborating, and empowering (Head, 2007). Digital technologies contribute to the implementation of "smart engagement" strategies by leveraging technology to meaningfully engage diverse groups of people across the city with the aim to deliver new shared decision-making models (The Committee for Sydney, 2018).

The notion of "augmented space" allows for digitally augmented community engagement within an urban space, therefore integrating digital community engagement tools with the physical space the people use and share on an everyday basis. In particular, two emerging technologies offer opportunities for engaging people within an urban space: public displays and augmented reality. Public displays, sometimes also referred to as digital signage, can take the form of low-resolution pixel displays (Hoggenmueller et al., 2018; Offenhuber and Seitinger, 2014) or high-definition screens (Huang et al., 2008; Parker et al., 2018b), ranging from small to large scale in their form factor. Public displays contribute toward the notions of a smart city and augmented space as they embed digital information within urban spaces. Due to hardware advances and decreasing costs, public displays have become ubiquitous in modern cities in recent years (Colangelo, 2016; Davies et al., 2017; Hastings, 2018). This has largely been driven by advertisers, who see public displays as an opportunity to reach out to the masses that pass through public spaces each day (Alt et al., 2012), such as train stations, shopping centers, and airports. The advantage of public displays over traditional static posters is that the content can be made more eye-catching, with videos and animations, and allows providers to schedule content, keeping it fresh and reducing clutter.

Previous research has demonstrated that in particular public interactive displays (PIDs) can be effective mediums for communities to have their say on relevant topics as they go beyond informing and also allow for people to contribute to the discussion. For example, Valkanova et al. (2014) designed and evaluated an interactive projected poll visualization allowing people to vote using one of four voting options for a topic by positioning themselves in front of their intended answer while raising their hand. Hespanhol et al. (2015) deployed a full-body voting application running on an existing public display as a mechanism to engage urban dwellers around local topics. Located in a busy civic square, the application could be used by passersby to vote on specific questions posed, by positioning themselves appropriately using the on-screen camera feed. Steinberger et al. (2014) enabled urban dwellers to interact and vote on yes/no questions by using a more discreet approach of stepping on buttons

a IAP2—https://www.iap2.org.

placed on the road below the public display at a bus stop. On a much larger scale, the Smart Citizen Sentiment Dashboard (Behrens et al., 2014) used an existing LED media façade on a building in Sao Paulo, Brazil, to visualize the aggregated mood of citizens in response to urban challenges, such as safety in the city. Citizens were able to submit an emotional response by swiping their RFID transport card at a terminal placed in front of the building. While the PID examples discussed have demonstrated benefits to their local communities, they may not be completely relevant on an individual level. Achieving this form of relevance is difficult to achieve with PIDs due to their "public" nature and the lack of information about the individual that is engaging with them (Davies et al., 2014).

In contrast to PIDs, augmented reality (AR) is not yet as widespread in terms of its application areas and penetration into everyday urban life. In cities, it has the potential to enable individualized experiences within public spaces, free from the spatial restrictions afforded by public displays. AR is commonly enabled through personal technologies like smartphones, which are already ingrained into the social fabric of modern society with most of the world's population owning one Taylor and Silver (2019). AR can bring people together, encouraging communities to form around virtual content, as demonstrated, for example, by the global success of Pokémon GO.[b] The mobile AR game resulted in players moving around the urban, physical space to capture digital monsters; collecting rewards by traveling to real-world points of interest (Pokéstops), and capturing locations by battling other players' Pokémons stationed there. During its peak usage in 2016, Pokémon GO created a ripple effect on public spaces as local businesses made efforts to market toward players (Frith, 2017) and brought communities of people together in certain Pokémon GO hot spots (Colley et al., 2017; Paasovaara et al., 2017).

These technological advances bring new opportunities for engaging citizens with local issues, which is now seen as central to the success of smart cities (The Committee for Sydney, 2018). However, careful attention needs to be paid toward the design of digital technologies to ensure they are connected to the local communities in which they are used or situated (Wouters et al., 2015). Therefore a growing number of research studies have investigated the viability of digital technologies for engaging communities on local issues that matter to them and to provide urban dwellers within these communities a public platform they can use to voice their opinions (Du et al., 2017; Hespanhol et al., 2015; Schroeter and Foth, 2009; Steinberger et al., 2014; Wouters et al., 2014).

The use of digital technologies for community engagement also addresses a commonly reported issue found with "traditional" community engagement methods, such as activities in form of workshops, focus groups, and online engagement platforms. These traditional methods typically reach only a certain demographic, which risks the exclusion of urban dwellers from other

b Pokémon GO—https://www.pokemongo.com/en-au/.

demographics, such as migrants, refugees, people who identify as LGBTQIA, and people with disabilities in the decision-making process (Darcy et al., 2019; Hespanhol et al., 2018). Previous research has found the use of situated digital technologies, such as large urban screens, to be effective for engaging a wider demographic. In particular the "walk-up-and-use" aspect of public interactive displays (Alt et al., 2011; Valkanova et al., 2014) and the use of multiple diverse channels offer advantages over traditional engagement methods (Fredericks et al., 2017a). In this chapter, we refer to such applications that support community engagement within an urban space as *smart engagement interface*.

Challenges in enhancing community engagement with digital technologies

Despite the potential of digital technologies and their smart engagement interfaces (SEIs) for enhancing community engagement, a number of challenges remain. Some of these challenges are found across public display and AR deployments, while others are specific to public displays only as there is more knowledge from previous studies about public displays and community engagement. We outline five challenges in the succeeding text, which provide the motivation for the remaining sections.

First the public nature of these systems may deter people from participating, potentially causing social embarrassment and linking the user to the interaction, which could affect participation when discussing controversial topics, like politics. Schroeter and Foth (2009) presented a more private method for giving opinions while at the same time enabling those opinions to be presented publicly. Utilizing an existing public display located in a busy urban space, passersby were encouraged to text, email, or tweet responses to localized questions displayed, such as questions related to new building developments in the local area. These different interaction options enabled a broader range of people to participate. The large screen gave individual members of the community a platform to voice their opinions about community issues. Baldauf et al. (2016) investigated the use of smartphones to interact with public displays, with the public display taking on the role of a companion device or broadcast medium. By studying mobile interaction techniques to access voting pages they showed that using mobile follow-up questions with multiple-choice answers were a suitable use for smartphone-enabled public displays.

Second, previous research has found that digital augmentations in public space in the form of public displays are often ignored by passersby, an effect coined as "display blindness" (Müller et al., 2009). This is a challenge for digitally situated community engagement activities, as people may ignore the public display and its content (Hespanhol et al., 2015). Display blindness has been linked to the observation that people fail to see the relevance of the displayed content to their particular situation and to the fact that society is becoming increasingly hyperconnected (Müller et al., 2009; Parker et al., 2018b), meaning that information access is ubiquitous through personal technologies, such as smartphones and smartwatches (Choi, 2014; Passarelli and Angeluci, 2017; Tragos et al., 2015. Furthermore, personal technologies play a key role in the communication of civic information to local communities, often used in conjunction with traditional media, such as newspapers and radio. Being constantly connected via personal devices has been found to have an effect on how people engage with digital technologies in public space (Parker et al., 2018b).

Third, making content relevant to individuals in a public space to overcome display blindness is challenging since the target audience becomes hard to define as it can consist of a broad demographic, incorporating people with different beliefs, cultures, occupations, and physical abilities. Unlike the web where services like Facebook and Google are able to create and access a profile of the user based on their history and interests, privacy could be impacted if people needed to sign in with a username and password on the screen. While this method of authentication works for public interfaces like ATMs, PIDs are normally of a larger scale and often utilize touchscreens for interaction, thus opening up the risk of shoulder surfing and smudge attacks. As smartphone usage has increased over the years, research has investigated smartphones as a medium to enable personalized content on PIDs. The benefit of a smartphone is that it is personal and already contains profiles from common apps, such as Facebook and Twitter, for user preferences. Davies et al. (2014) explored using a custom smartphone app for users to share preference information with nearby public displays. The information was implicitly sent to public displays that were in close proximity to the user, triggering the displays to show content based on the user's preference, such as the weather. Earlier work by Hosio et al. (2010) demonstrated a similar direct approach where users needed to manually launch the public display applications from their smartphone. However, using smartphones to make PIDs more relevant to individuals by personalizing their content should be approached carefully to ensure the information does not negatively harm users in any way, such as linking the content on the screen back to individuals (Munson et al., 2011). The Digifieds system (Alt et al., 2011), which harnessed the UbiOulu public display network, provided citizens with a digital version of a community noticeboard. Users could post content, such as advertisements, using the screen or their own personal devices. The content could then be read by other users. While the content may not be entirely relevant to individual users, it has a greater potential at being relevant to the local community as it is shaped around their collective interests.

Fourth the deployment of digital technologies in public space often takes a "one-size-fits-all" approach, with little regard for the differences of the individuals that use the spaces (Parker et al., 2017, 2018b). Ensuring that PIDs are relevant and accessible to everyone is especially critical in situations where the objective is to engage local communities around civic topics, such as the development of new infrastructures. Disregarding people that pass through public space risks the technology becoming unused and could exclude those who come from marginalized communities (Fredericks et al., 2017b; Parker et al., 2017; Schroeter and Foth, 2009), including lower socioeconomic backgrounds, culturally and linguistically diverse communities, and those with physical or cognitive impairments.

Fifth, deploying digital technologies in urban space is challenging due to costs, regulations, access for maintenance, etc. (Tomitsch, 2018). These challenges hinder the adoption of following a community-centered design process that involves community representatives and other stakeholders in the design process and evaluates the success of initiatives through testing with participants from the local community. At the same time, as identified in the aforementioned fourth challenge, it is not always possible to simply replicate initiatives from other parts of the world due to the unique requirements of the social, cultural, and physical character of urban space.

Common to the first four challenges is the question of relevance. To be successful, smart engagement needs to cater toward the particular needs of the local community within an urban space and consider how to deliver relevant information and engagement in a hyperconnected society where people can tap into information available via the Internet and to connect with each other from anywhere and at any time. This chapter investigates this question of relevance through an analysis of three studies, which involved the design and deployment of digital technologies in public space for different community engagement purposes. To address the fifth challenge, we present six design patterns to guide the use of digital technologies for smart engagement that is relevant to individuals. The design patterns provide guidance based on our empirical findings while offering sufficient flexibility to allow for a specific community engagement initiative to be customized to the specific needs of a community and local environment.

Designing smart engagement interfaces

To gain further insights into the design of SEIs, we reflect on three situated, digitally augmented community engagement studies. We selected these studies as representatives from our body of work as they demonstrate the use of public displays, augmented reality, and a hybrid combination of both digital technologies. Each case study is first introduced and then discussed in terms of three driving factors that were common across the three case studies (Table 1): user, locale, and time.

The *user* factor relates to an individual's interactions (Parker et al., 2018a), behavior (Ackad et al., 2013; Wang et al., 2012), preferences (Davies et al., 2014; Parker et al., 2016), and their physiological traits and characteristics, such as adapting the height of the content based on the user's height for easier reaching (Parker et al., 2017). The *locale* factor captures design concerns relating to space and place, as these concepts are intertwined and affect each other (Dourish, 2006; Kukka et al., 2014). Space describes the spatial characteristics

TABLE 1 The SEIs used in the community engagement case studies along with the type of digital technology employed and the factors addressed in each case study.

Case	Digital technology	Location	User	Locale	Time
Digital community noticeboard	Public display and AR	University cafeteria	✓	✓	
Transit selfie voting booth	Public display	University foyer	✓	✓	✓
Augmented reality tree trimming awareness game	AR	Park		✓	

of an environment, such as the physical objects and weather (Mäkelä et al., 2017). Place on the other hand relates to the social aspects of a space, such as the people and communities within it (Moere and Wouters, 2012; Parker and Tomitsch, 2017). Therefore this factor can affect the type of people that exist in a public space—workers, shoppers, tourists, etc. Both the user and locale factors are affected by the *time* factor. In terms of the user, their goals or needs may change over the course of the day, month, year, or season, thus affecting how and when they need to interact with an SEI. For instance, a user may not have much available time during the week as they need to work, however, during the weekend they may have more time to spare to participate in situated community engagement activities (Hespanhol et al., 2015). The *locale* factor is affected as space changes over time, for example, by activities, such as community-run events and shops closing.

Case A: Digital community noticeboard

To address the challenges discussed in the section "Challenges in enhancing community engagement with digital technologies" around personalization and privacy, we investigated using mobile augmented reality to provide a personalized layer to individuals that engage with digital community noticeboards. Specifically, this SEI was composed of two components:

Public display. The public display was essentially acting as a digital community noticeboard, where it displayed general content targeted toward students. The content included local news, university news, local buy and sell advertisements, and property rentals. The display could be tracked by smartphone devices with the AR app through its AR marker that was placed behind the content as a background image.

Mobile augmented reality app. The mobile AR app allowed users to view personalized content overlaid on the public display through their mobile device

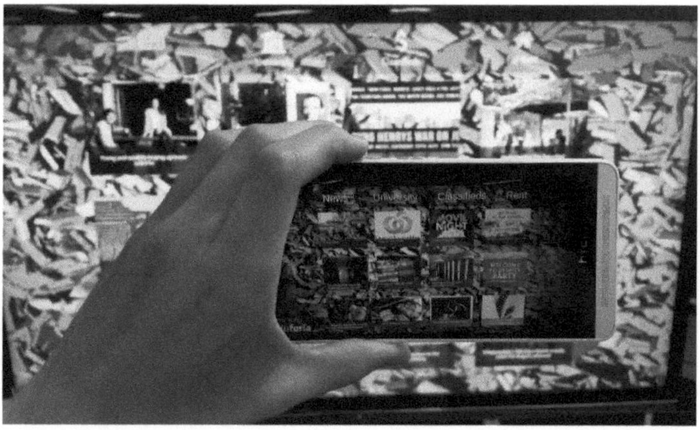

FIG. 1 Viewing a personalized augmented reality layer overlaid on a public display through a smartphone.

(Fig. 1). Virtual elements could be interacted with through the smartphone's screen using touch to select a tile and swipe gestures for navigation.

This SEI considered two factors in its design to make its content relevant to passersby: user and locale. Firstly the user factor is applicable as the SEI can adapt content to individuals based on preferences that users specify in their smartphone app, such as where a user lives and the degree they are studying. Based on this information the content in the augmented layer in the smartphone AR app will be tailored for each individual.

Secondly the locale factor influenced the general content displayed on the screen. As the touchscreen was situated within a main foyer area of our university campus, the content was tailored toward students—such as displaying affordable share houses for rent in nearby suburbs and news articles related to the local community.

Case B: Transit selfie voting booth

This SEI was composed of a touch-enabled public interactive display and a selfie-taking iPad booth, which were designed, employed, and evaluated as part of a community engagement pop-up study (Fredericks et al., 2017a). The larger aim of the study was to investigate the use of multiple physical and digital channels for the purpose of facilitating situated community participation on the topic of public transport options. The touch-enabled display and selfie booth are described further in the succeeding text.

Touchscreen public interactive display. The display (Fig. 2A) was 50 in. and in portrait orientation. The display had two functions: (1) display bus timetable information based on suburbs selected and (2) feature selfies taken from the selfie booth.

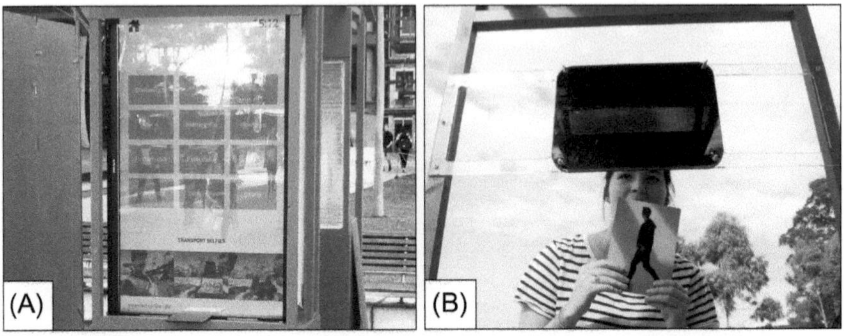

FIG. 2 Transit selfie voting booth composed of two modules: (A) touch-enabled public interactive display showing transit information and selfies and (B) selfie-taking iPad booth.

Selfie booth. The selfie module (Fig. 2B) contained a mounted iPad with a front-facing camera. The iPad featured a custom-developed software application that posed the question "Do you like the transport options to uni?" and allowed people to playfully respond to that question by taking a selfie while holding a card with the transport option they took to arrive at the university campus that day. After taking the photo, it would be automatically synced to the touchscreen and displayed in the lower portion of the screen, beneath the suburb buttons. Users of the touchscreen could then swipe horizontally on the photo area to see previous selfies.

This SEI addressed all three factors to make its content relevant to passersby within the urban space where it was located: user, locale, and time. The user factor was addressed as the display adapted to the person interacting depending on the destination suburb they chose. After selecting a destination suburb, the display would show three information ribbons containing information about the fastest transit options (walking, bus, or train) to the selected suburb at that time. Each information ribbon displayed the bus number or an icon for walking or train, departing time, transit information, and the closest departing stop. Below the transit information the display would show recommended places in that particular suburb that could be filtered by general shopping, bars, cafes, and restaurants.

The locale factor was addressed by showing a collection of selfies on the touchscreen from those who previously engaged, sparking the curiosity of passersby. This gradually built up a crowdsourced knowledge base of the transport options people use to travel to the university. This content changed over time and enabled passersby to get an idea of the transit options utilized at the time they viewed the screen. The transit display itself suggested places to visit according to the destination suburb the user selected and the time of day. For instance, if someone was interacting at night, they would get different local restaurant suggestions in their selected destination suburb compared with when they would interact in the morning.

Case C: Augmented reality tree trimming awareness game

This SEI was created for a study carried out in collaboration with a local electricity supplier, with the objective of raising awareness around the importance of tree trimming and to gain feedback from local communities about tree trimming. To support the community engagement initiative, we designed and developed an AR game that visualized the consequence of trees growing too close to powerlines. The aim of the AR game was to engage members of the local community in a discourse on tree trimming and powerlines while providing a playful, educational experience that illustrated why tree trimming was necessary.

The AR game (Fig. 3) was structured as a multiplayer activity, where one person would look through a head-mounted display (HMD) to see a video feed of the real world; other people would then move the AR markers, which were in the form of two 590 by 840 mm cubes. When the markers were in view, the person looking through the HMD would see 3-D models of a tree and powerlines appearing in front of the cubes. If the tree marker was moved too close to the powerlines marker, the 3-D model of the tree would catch on fire. The person looking through the HMD therefore had to guide the person moving the marker until they were far enough apart, which caused the 3-D tree to return to being a healthy tree. The system consisted of a pair of modified Google Cardboard HMD goggles with an HTC One Android smartphone. We cut a small hole into the Google Cardboard goggles where the smartphones' camera lens was located, so the camera feed was unobstructed. To enable walk-up-and-use interaction, the HMD was fixed onto a tripod. The HMD was mounted on a tripod with an adjustable strap. The AR markers were created as large cubes that could be

FIG. 3 People can visualize the effects of untrimmed trees growing being too close to powerlines.

easily moved around, built from Corflute, a lightweight building material. The cubes featured a black on white picture of a tree and a powerline, respectively. The colors and design were chosen to best stand out in outdoor conditions to ensure error-free tracking of the cubes within the AR application.

This SEI addressed the locale factor to make the content relevant to passersby. This was achieved by designing the app specifically to help members of the local community visualize and understand the effects of trees that grow too close to powerlines. The SEI was further designed specifically for the community engagement locations, which were public squares and parks across three Sydney suburbs.

Design patterns for digitally augmented, situated engagement

The concept of design patterns adopted for this chapter follows the definition of pattern languages outlined by Christopher Alexander and his colleagues in their two books *A Pattern Language* (Alexander, 1977) and *A Timeless Way of Building* (Alexander, 1979). Alexander developed the approach to capture and use design knowledge by establishing an extensive collection of pattern examples to assist architects and engineers in overcoming recurrent design limitations for buildings, towns, and other urban infrastructure.

Based on Alexander and his colleagues' work, other disciplines, such as the fields of information and communication technology (ICT) and human-computer interaction (HCI), have developed their own sets of pattern languages and design patterns based on their design experiences and limitations. Schuler (2008) developed a pattern language directed at social change and citizen activism for ICT design and social interaction. They developed 136 patterns created as an alternative design approach that involved a variety of stakeholders in the decision-making process. Similar to Alexander, for each pattern, Schuler provides a problem, context, and solution; however, each pattern incorporates detailed explanations through practical examples from previous literature and projects. Further building on Schuler's approach, Fredericks et al. (2017b) developed eight design patterns for designing and implementing digitally augmented pop-up interventions for localized community engagement. Each pattern specifies a problem, context, discussion, and solution through practical examples provided through their own research experiences and related projects.

From the discussion in the previous section, seven distinct but closely related design patterns emerged (Table 2). These patterns support the use of digital technologies in situated community engagement activities. Different to general design guidelines, which are of a prescriptive nature, patterns are constructive and "suggest how a problem can be solved" (Borchers, 2000). Furthermore the nature of design patterns allows them to provide empirically informed guidance while being flexible in their interpretation and application. This is important to meet the unique requirements of each community engagement initiative.

TABLE 2 Design patterns for SEIs that are relevant to individuals.

Theme	Design pattern	Description	Relevant work
User	1. Awareness of users' goals	How can an SEI be aware of an individual user's goals? What information is needed?	Cases A and B, Parker et al. (2017)
	2. Reacting appropriately to individuals	How can SEIs be designed so that they can detect and react accordingly to individual users and their specific physiological traits and characteristics?	Parker et al. (2017)
	3. Privacy and relevance trade-offs	Relevance often comes at the cost of privacy, while privacy comes at the cost of relevance. How can these two factors be managed? Furthermore, when should a trade-off be acceptable?	Cases A and B
Locale	4. Physical placement	SEIs can often end up being ignored by urban dwellers due to poor positioning. How can the ideal physical placement of an SEI be determined?	Case B
	5. Awareness of the community	Communities shape the places and spaces SEIs are deployed in. How can SEIs become aware of the community they are situated in? How can SEIs collect information about the local community in order to better respond to its needs?	Cases A, B, and C
Time	6. Changes over time	Certain content may only be relevant at certain times during the day, month, year, or season. How should SEIs change over time?	Case B

The format and the presentation of the patterns follow the approach used in previous work on patterns across the fields of ICT, citizen activism, urban design, and HCI (Fredericks et al., 2017b; Lyle et al., 2015; Schuler, 2008). For each design pattern, we present a description of the problem that the pattern addresses, the context in which the problem is placed, a discussion to ground the problem, and a possible approach. The patterns are organized according to three core factors that contribute to relevance introduced in the previous section: user, locale, and time. Ensuring that all three factors are covered addresses the commonly observed problem of community engagement activities and their effectiveness being limited by their ability to engage relevant audiences (Hespanhol et al., 2015; Steinberger et al., 2014; Valkanova et al., 2013) or in other words to clearly convey the activities' relevance to the local audience. The patterns can also be used to evaluate digitally augmented, situated community engagement activities before, during, or after their implementation. This is achieved by testing the digital technology intervention and its SEI against each of the patterns, informing iterative improvement and adjustment during and between activities.

User

Pattern 1: Awareness of users' goals

Problem

How can an SEI be aware of an individual user's goals? What information is needed?

Context

Individuals in public space all have different goals, which change over time. For example, a user's goal could be as simple as walking to the shop to buy milk. Therefore it is important that SEIs become aware of an individual user's goals, as this can enable them to automatically make judgments regarding the type of information that an individual user would be currently interested in. It is important to consider potential implications on privacy and surveillance, which is addressed through pattern 3.

Discussion

We explored the design of two SEIs to test two different approaches for making an SEI aware of a user's goals: profile based and device free, respectively. The profile-based approach implemented for the SEI in case A required the user to connect to a public display through a smartphone app with a prefilled user profile. The display would then determine the content that would be of interest to an individual, based on their interests and demographic information. The problem with this method is that it may not always be related to a user's current goal and may end up displaying content that is too personal. The other approach, device free utilized by the SEI in case B, tracked a user's explicit interactions and characteristics to recommend content they might be interested in. While this approach does not provide the SEI with detailed information about a user, it can be employed to determine the user's current goals while having little impact on a user's privacy and less of a barrier for entry as no external devices are required. For instance, the SEI in case B detected a user's goal by their interactions as, depending on the suburb selected, it would implicitly recommend places to visit. On the other hand, SEIs could implicitly detect a user's goals through built-in sensors, where they could detect if a user is having trouble engaging and respond with interface adjustments (Parker et al., 2017)—for instance, making buttons easier to reach for people in wheelchairs.

Possible approach

Designing an SEI that is aware of user goals and ideally their current goals should be achieved in a way that includes all potential users. Detection of a user's goals does not necessarily require detailed information about the user from a profile, much of the information can be implicitly gathered from their interactions, behavior, and physiological traits and characteristics. By designing SEIs in this way, it ensures that they are open to a broad demographic of potential users, which is inherently what the word "public" entails.

Pattern 2: Reacting appropriately to individuals

Problem

How can SEIs be designed so that they can detect and react accordingly to an individual user's physiological traits and characteristics?

Context

Urban environments can have a broad range of people passing through them and are therefore usually designed with affordances built-in to allow everyone access (Clarkson et al., 2013; Evans, 2014). For instance, pedestrian crossing gutter ramps that allow prams and wheelchairs easier and safer passage over gutters while still being accessible to everyone else. Similarly, SEIs need to respond to the unique characteristics of individuals, such as their height, to enable a broader demographic to engage with them.

Discussion

This pattern leads on from the previous one by discussing another aspect relating to the user—their physiological traits and characteristics. To tackle the problem of SEIs responding to different user characteristics, our earlier work (Parker et al., 2017) tested a PID that responds to the height of the user through four different height-aware modes. The main finding from our study suggested that height adjustments should be only made to the interface, drawing less attention than hardware adjustments, which could potentially result in social embarrassment for the user. Other findings indicated that PIDs should not rely on external personal devices or cards to identify users and instead use implicit methods, like cameras and interactions.

Possible approach

PIDs, as well as more broadly SEIs, need to be aware of and respond to the physical needs of users. This is becoming increasingly important as screens become larger (Huang et al., 2008; Parker et al., 2018b). The adjustments in response to the physical characteristics of the user need to be subtle, yet empowering, to ensure that attention is not drawn to an individual's physical capabilities. SEIs should also be designed so that they implicitly adjust to individual users while being accessible to a broad demographic—covering people of different capabilities.

Pattern 3: Privacy and relevance trade-offs

Problem

Relevance often comes at the cost of privacy, while privacy comes at the cost of relevance. How can these two factors be managed? When should a trade-off be acceptable?

Context

This problem is linked to patterns 1 and 2, as both require knowledge about certain aspects of individual users. While it is important to ensure SEIs are relevant to individuals, SEIs also need to be aware of privacy requirements to ensure they are not negatively impacting a user's privacy.

Discussion

From a reflection of our cases and other work, we found that privacy requirements among individuals can differ and change based on certain contexts. For instance, during the case B study, one of the participants mentioned that for the most part they were fine with the content displayed on the PID; however, if they were alone at night and they used the PID to find the next bus to their home suburb, they would be concerned about others seeing this information.

Therefore privacy requirements can differ depending on the context the SEI is being engaged in. In more privacy-sensitive situations, such as providing and viewing political opinions, the SEI from case A might be useful to protect an individual's privacy. The caveat to this method is that the user needs to provide data to a third party to enable the personalization, which can be influenced by how legitimate the SEI looks—as mentioned by participants in case B.

Possible approach

Designing an SEI that is relevant to individual users is a socially complex challenge. The more relevant an SEI is to an individual, the less private the experience becomes—which may not be desired in certain contexts, such as busy spaces. Therefore trade-offs need to be made, with compromises about both the user's privacy preferences and their relevance expectations. An individual's privacy preferences can potentially be discerned from the type of context and the people around them. For instance, an SEI in a busy space, such as a town square, should avoid displaying content that specifically identifies where a user lives or their private interests, as it is more likely that there will be onlookers.

An alternative is to explore methods to block onlookers. For instance, if the SEI was a PID, then it could have privacy protection features such as blacking out parts of the screen the user is not looking at (Brudy et al., 2014). Alternatively, mobile augmented reality could help preserve privacy by displaying personalized information on the smartphone screen (Baldauf et al., 2012; Parker et al., 2016). While these methods show potential, more work is needed to test their effectiveness in different real-world contexts, such as busy spaces and different social situations, for example, involving groups of users.

Locale

Pattern 4: Physical placement

Problem

SEIs, such as PIDs, can often end up being ignored by urban dwellers due to poor positioning. Therefore how can the ideal physical placement of an SEI be determined?

Context

This pattern can be used to inform the ideal positioning of SEIs and public displays.

Discussion

The position of an SEI needs to be considered carefully (in the case of AR content placement within the physical space is also important) aligning it with the engagement objectives and how the physical space is used. Despite the requirement that ensuring SEIs are positioned in a location that gets a lot of traffic being common knowledge, particularly in the public display research community (Huang et al., 2008; Williamson and Williamson, 2014; Wouters et al., 2016), it is not always followed in practice (Parker et al., 2018b). During case B, we carefully chose a location that was close to the entrance to the university and a nearby train station, resulting in people engaging with the SEI that recently completed their commute. The selfie booth and transit screen were both facing people as they were entering the university campus.

Possible approach

Before deciding on the placement of SEIs, it is important to first survey and observe the space in which the SEI will be deployed to see how urban dwellers use that space and move through it. Therefore the choice of an SEI's location should take into consideration the permeability of the space and how this changes over time (Montgomery, 1998). Additionally, SEIs should be positioned where there is a high likelihood of opportunistic interactions, such as nearby the entrance to the space or in an area where people are waiting, such as near café lines or escalators.

Pattern 5: Awareness of the community

Problem

Communities shape the places and spaces SEIs are deployed in. How can SEIs be designed so that they are aware of the community they are situated in? How can SEIs collect information about the local community to better respond to its needs?

Context

A challenge for SEIs is becoming integrated within the respective communities and gaining relevant information about them, such as local events being held and the demographics of the people that make up the communities. This pattern can be used to inform the design of SEIs that are more attuned to the people within the communities they are situated in.

Discussion

From reflecting on our first two cases A and B, it appears that while we had good intentions of making PIDs relevant to individuals, the content displayed did not necessarily reflect the local community (its needs and sentiment) despite being relevant to the location. Case C was designed to be an engagement activity linked to an issue affecting the local community, in this case tree trimming. This offered an experience that was not necessarily available via smartphones. Other research has shown that giving urban dwellers control over the content on SEIs can be an effective way to keep content relevant and give people in communities an accessible platform to reach out to one another. For instance, work by Alt et al. (2011) and Schroeter and Foth (2009) tested PIDs that allowed people to post content to them. The content could then be read by other users. What these examples show is that SEIs appropriated by their respective local communities can collect current, natural, and in-depth information about their communities—such data could be used to drive machine learning enabling SEIs to make inferences regarding the content they display.

Possible approach

To increase the relevance of an SEI and expand the data it can pull from, it needs to be strongly connected to the local communities within the location in which it is situated. Giving members of communities some control enables SEIs to have greater knowledge of the community, giving them a greater pool of data to draw from when tailoring content to individuals and empowering the people within their respective communities. To achieve this, researchers and practitioners could follow a codesign approach (Tomitsch et al., 2018), where stakeholders are actively involved in all stages of the design and development. Codesign needs to be a continuing process before and after the SEI is created and situated, which has been referred to as "middle-out design" within the context of community engagement (Fredericks et al., 2016).

Time

Pattern 6: Changes over time

Problem

Certain content may only be relevant at certain times during the day, month, year, or season. How should SEIs change over time?

Context

Time is an underlying factor that affects the other patterns. Therefore it is important to understand the impact of time on the physical space and place, the people that use the space, and the community (Moere and Wouters, 2012; Parker and Tomitsch, 2017).

Discussion

Despite the importance of factoring in the time when displaying content, not all SEIs change over time—for instance, commercial PIDs have been found to generally show the same content regardless of the time of the day or the day of the week (Parker et al., 2018b). Case B's PID module displayed bus times and recommended places of interest that were open at the particular time it was being interacted with. This meant that the PID was always up to date and allowed users to access the information at any time. However, the participants who were interviewed during this study mentioned that they would like to have the ability to see content related to events happening in the future, as they might already have a set plan in mind but would like to get inspiration for planning future activities. Therefore, depending of their particular nature, SEIs should consider offering temporal flexibility, to provide users with access to current and future information. SEIs should also change over time and, for example, respond to seasonal and cultural events celebrated by members of the local community. This may require ongoing curating and updating of content, which is an important consideration to keep in mind especially when implementing permanent SEIs. For example, the urban screen at Federation Square in Melbourne has been highlighted for its program, which constantly changes to augment local events and topics that are relevant to the local community (Tomitsch et al., 2015).

Possible approach

While showing current up-to-date information can be of relevance to individuals and communities, it is also important for SEIs to have temporal flexibility, where current and future information can be accessed. Permanently installed SEIs need to carefully consider how their content might have to adapt over time to remain relevant and to increase the connection with the local community.

Additionally, SEIs need to be aware of a user's time and adapt according to the time constraints the user has—for instance, if the user needs to be somewhere in a hurry, the SEI should optimize speed and present quick snippets of information that is relevant to the passerby at that moment in time. Conversely, users with time to spare, such as a traveler or retiree, can probably spend more time with the SEI and take in more information.

Implementing smart engagement

In this book chapter, we presented six interrelated design patterns situated across the three factors of user, local, and time, which are common factors identified from our own case studies. The specific intention of these patterns is to guide the design, development, and implementation of smart engagement interfaces (SEI) for public spaces. To exemplify how the patterns can be implemented, we presented a critical review of SEIs using PIDs and AR as examples of emerging digital technologies that offer new opportunities for engaging communities in a discourse of locally relevant topics. The review drew on previous work in this area and our three smart engagement case studies. The design patterns provide a reference for the design, deployment, and evaluation of SEIs as part of community engagement initiatives. As the field is constantly changing due to advances in technology and new methods, these patterns are intended to be dynamic, where they can be continually updated based on these changes. We hope these patterns contribute to the growing body of work and knowledge around the design of smart engagement interfaces.

To contribute toward the "smart" of smart engagement, we have highlighted in this chapter that digital technologies need to be designed so that they are responsive toward the needs of individuals who engage with them, the locale the technology is situated in, and how the users and locale change over time. It should be highlighted that smart engagement is not limited to being situated in public spaces, but more of a dynamic multidimensional engagement process that can be expressed through and may employ various media.

References

Ackad, C., Wasinger, R., Gluga, R., Kay, J., Tomitsch, M., 2013. Measuring interactivity at an interactive public information display. In: Proceedings of the 25th Australian Computer-Human Interaction Conference: Augmentation, Application, Innovation, Collaboration. ACM, pp. 329–332.

Alexander, C., 1977. A Pattern Language: Towns, Buildings, Construction. Oxford University Press.

Alexander, C., 1979. The Timeless Way of Building. vol. 1. Oxford University Press, New York.

Alt, F., Kubitza, T., Bial, D., Zaidan, F., Ortel, M., Zurmaar, B., Lewen, T., Shirazi, A.S., Schmidt, A., 2011. Digifieds: insights into deploying digital public notice areas in the wild. In: Proceedings of the 10th International Conference on Mobile and Ubiquitous Multimedia. ACM, pp. 165–174.

Alt, F., Muller, J., Schmidt, A., 2012. Advertising on public display networks. Computer 45 (5), 50–56.
Arnstein, S.R., 1969. A ladder of citizen participation. J. Am. Inst. Plann. 35 (4), 216–224.
Baldauf, M., Lasinger, K., Fröhlich, P., 2012. Private public screens: detached multi-user interaction with large displays through mobile augmented reality. In: Proceedings of the 11th International Conference on Mobile and Ubiquitous Multimedia. ACM, p. 27.
Baldauf, M., Reitberger, W., Güldenpfennig, F., 2016. Just one more thing!: investigating mobile follow-up questions for opinion polls on public displays. In: Proceedings of the 5th ACM International Symposium on Pervasive Displays. ACM, pp. 237–238.
Behrens, M., Valkanova, N., Brumby, D.P., et al., 2014. Smart citizen sentiment dashboard: a case study into media architectural interfaces. In: Proceedings of The International Symposium on Pervasive Displays. ACM, p. 19.
Borchers, J.O., 2000. A pattern approach to interaction design. In: Proceedings of the 3rd Conference on Designing Interactive Systems: Processes, Practices, Methods, and Techniques. ACM, pp. 369–378.
Brudy, F., Ledo, D., Greenberg, S., Butz, A., 2014. Is anyone looking? Mitigating shoulder surfing on public displays through awareness and protection. In: Proceedings of the International Symposium on Pervasive Displays. ACM, p. 1.
Caragliu, A., Del Bo, C., Nijkamp, P., 2009. Smart cities in Europe. In: Proceedings of the 3rd Central European Conference in Regional Science CERS 2009. ACM, pp. 49–59.
Choi, A.J., 2014. Internet of things: evolution towards a hyper-connected society. In: Solid-State Circuits Conference (A-SSCC), 2014 IEEE Asian. IEEE, pp. 5–8.
Clarkson, P.J., Coleman, R., Keates, S., Lebbon, C., 2013. Inclusive Design: Design for the Whole Population. Springer Science & Business Media.
Colangelo, D., 2016. Massive media: when cities become screens. In: Proceedings of the 3rd Conference on Media Architecture Biennale. ACM, p. 9.
Colley, A., Thebault-Spieker, J., Lin, A.Y., Degraen, D., Fischman, B., Häkkilä, J., Kuehl, K., Nisi, V., Nunes, N.J., Wenig, N., et al., 2017. The geography of Pokémon Go: beneficial and problematic effects on places and movement. In: Proceedings of the 2017 CHI Conference on Human Factors in Computing Systems. ACM, pp. 1179–1192.
Darcy, S., Yerbury, H., Maxwell, H., 2019. Disability citizenship and digital capital: the case of engagement with a social enterprise telco. Inform. Commun. Soc. 22 (4), 538–553.
Davies, N., Langheinrich, M., Clinch, S., Elhart, I., Friday, A., Kubitza, T., Surajbali, B., 2014. Personalisation and privacy in future pervasive display networks. In: Proceedings of the 32nd Annual ACM Conference on Human Factors in Computing Systems. ACM, pp. 2357–2366.
Davies, N., Clinch, S., Mikusz, M., Bates, O., Turner, H., Friday, A., 2017. Better off: when should pervasive displays be powered down? In: Proceedings of the 6th ACM International Symposium on Pervasive Displays. ACM, p. 19.
Dourish, P., 2006. Re-space-ing place: place and space ten years on. In: Proceedings of the 2006 20th Anniversary Conference on Computer Supported Cooperative Work. ACM, pp. 299–308.
Du, G., Degbelo, A., Kray, C., 2017. Public displays for public participation in urban settings: a survey. In: Proceedings of the 6th ACM International Symposium on Pervasive Displays. ACM, p. 17.
Evans, G., 2014. Accessibility and user needs: pedestrian mobility and urban design in the UK. Proc. ICE-Municipal Eng. 168 (1), 32–44.
Fredericks, J., Foth, M., 2013. Augmenting public participation: enhancing planning out-comes through the use of social media and web 2.0. Aust. Plan. 50 (3), 244–256.
Fredericks, J., Caldwell, G.A., Tomitsch, M., 2016. Middle-out design: collaborative community engagement in urban hci. In: Proceedings of the 28th Australian Conference on Computer-Human Interaction. ACM, pp. 200–204.

Fredericks, J., Hespanhol, L., Parker, C., Zhou, D., Tomitsch, M., 2017a. Blending pop-up urbanism and participatory technologies: challenges and opportunities for inclusive city making. City, Cult. Soc. 12, 44–53.

Fredericks, J., Tomitsch, M., Stewart, L., 2017b. Design patterns for integrating digitally augmented pop-ups with community engagement. Int. J. E-Plan. Res. 6 (3), 19–41.

Frith, J., 2017. The digital lure: small businesses and Pokémon Go. Mob. Media Commun. 5 (1), 51–54.

Gentile, V., Khamis, M., Sorce, S., Alt, F., 2017. They are looking at me!: understanding how audience presence impacts on public display users. In: Proceedings of the 6th ACM International Symposium on Pervasive Displays. ACM, p. 11.

Gordon, E., et al., 2016. Civic engagement. In: Dialogues on Mobile Communication. Routledge, pp. 178–193.

Hall, R.E., Bowerman, B., Braverman, J., Taylor, J., Todosow, H., Von Wimmersperg, U., 2000. The vision of a smart city. Technical report, Brookhaven National Lab., Upton, NY (US),.

Hastings, J., 2018. Digital signage adoption continues to surge forward. https://www.digitalsignagetoday.com/blogs/. digital-signage-adoption-continues-to-surge-forward.

Head, B.W., 2007. Community engagement: participation on whose terms? Aust. J. Polit. Sci. 42 (3), 441–454.

Hespanhol, L., Tomitsch, M., McArthur, I., Fredericks, J., Schroeter, R., Foth, M., 2015. Vote as you go: blending interfaces for community engagement into the urban space. In: Proceedings of the 7th International Conference on Communities and Technologies. ACM, pp. 29–37.

Hespanhol, L., Davis, H., Fredericks, J., Caldwell, G.A., Hoggenmüller, M., 2018. The digital fringe and social participation through interaction design. J. Commun. Inform. 14 (1), 4–16.

Hoggenmueller, M., Tomitsch, M., Wiethoff, A., 2018. Understanding artefact and process challenges for designing low-res lighting displays. In: Proceedings of the 2018 CHI Conference on Human Factors in Computing Systems. ACM, p. 259.

Hollands, R.G., 2008. Will the real smart city please stand up? Intelligent, progressive or entrepreneurial? City 12 (3), 303–320.

Hosio, S., Jurmu, M., Kukka, H., Riekki, J., Ojala, T., 2010. Supporting dis- tributed private and public user interfaces in urban environments. In: Proceedings of the Eleventh Workshop on Mobile Computing Systems & Applications. ACM, pp. 25–30.

Huang, E., Koster, A., Borchers, J., 2008. Overcoming assumptions and uncovering practices: when does the public really look at public displays? In: Pervasive Computing, pp. 228–243.

Kukka, H., Ylipulli, J., Luusua, A., Dey, A.K., 2014. Urban computing in theory and practice: towards a transdisciplinary approach. In: Proceedings of the 8th Nordic Conference on Human-Computer Interaction: Fun, Fast, Foundational. ACM, pp. 658–667.

Lee, G.A., Dünser, A., Kim, S., Billinghurst, M., 2012. CityViewAR: a mobile outdoor AR application for city visualization. In: Mixed and Augmented Reality (ISMAR-AMH), 2012 IEEE International Symposium on. IEEE, pp. 57–64.

Low, S., Smith, N., 2013. The Politics of Public Space. Routledge.

Lyle, P., Foth, M., Choi, J.H.-j., 2015. Design patterns for urban gardening. In: Citizens Right to the Digital City. Springer, pp. 79–98.

Mäkelä, V., Sharma, S., Hakulinen, J., Heimonen, T., Turunen, M., 2017. Challenges in public display deployments: a taxonomy of external factors. In: Proceedings of the 2017 CHI Conference on Human Factors in Computing Systems. ACM, pp. 3426–3475.

Manovich, L., 2006. The poetics of augmented space. Visual Commun. 5 (2), 219–240.

Manville, C., Cochrane, G., Cave, J., Millard, J., Pederson, J.K., Thaarup, R.K., Liebe, A., Wissner, M., Massink, R., Kotterink, B., 2014. Mapping smart cities in the eu. http://www.europarl.europa.eu/RegData/etudes/etudes/join/2014/507480/IPOL-ITRE_ET(2014)507480_EN.pdf.

Moere, A.V., Wouters, N., 2012. The role of context in media architecture. In: Proceedings of the 2012 International Symposium on Pervasive Displays. ACM, p. 12.

Montgomery, J., 1998. Making a city: urbanity, vitality and urban design. J. Urban Des. 3 (1), 93–116.

Müller, J., Wilmsmann, D., Exeler, J., Buzeck, M., Schmidt, A., Jay, T., Krüger, A., 2009. Display blindness: the effect of expectations on attention towards digital signage. In: Pervasive Computing, pp. 1–8.

Munson, S.A., Rosengren, E., Resnick, P., 2011. Thanks and tweets: comparing two public displays. In: Proceedings of the ACM 2011 Conference on Computer Supported Cooperative Work. ACM, pp. 331–340.

Offenhuber, D., Seitinger, S., 2014. Over the rainbow: information design for low-resolution urban displays. In: Proceedings of the 2nd Media Architecture Biennale Conference: World Cities. ACM, pp. 40–47.

Paasovaara, S., Jarusriboonchai, P., Olsson, T., 2017. Understanding collocated social interaction between Pokémon Go players. In: Proceedings of the 16th International Conference on Mobile and Ubiquitous Multimedia. ACM, pp. 151–163.

Parker, C., Tomitsch, M., 2017. Bridging the interaction gulf: understanding the factors that drive public interactive display usage. In: OzCHI'17: Proceedings of the 29th Australian Conference on Human-Computer Interaction (OZCHI 2017). ACM.

Parker, C., Kay, J., Baldauf, M., Tomitsch, M., 2016. Design implications for interacting with personalised public displays through mobile augmented reality. In: Proceedings of the 5th ACM International Symposium on Pervasive Displays. ACM, pp. 52–58.

Parker, C., Fredericks, J., Tomitsch, M., Yoo, S., 2017. Towards adaptive height-aware public interactive displays. In: Adjunct Publication of the 25th Conference on User Modeling, Adaptation and Personalization. ACM, pp. 257–260.

Parker, C., Kay, J., Tomitsch, M., 2018a. Device-free: An implicit personalisation approach for public interactive displays. In: ACSW'18: Proceedings of Australasian Computer Science Week 2018. ACM.

Parker, C., Tomitsch, M., Kay, J., 2018b. Does the public still look at public displays? A field observation of public displays in the wild. Proc. ACM Interact. Mob. Wearable Ubiquitous Technol. 2 (2), 1–24.

Passarelli, B., Angeluci, A.C.B., 2017. The hyperconnected contemporary society. In: Brazil: Media From the Country of the Future. Emerald Publishing Limited, pp. 343–362.

Schroeter, R., Foth, M., 2009. Discussions in space. In: Proceedings of the 21st Annual Conference of the Australian Computer-Human Interaction Special Interest Group: Design: Open 24/7. ACM, pp. 381–384.

Schuler, D., 2008. Liberating Voices: A Pattern Language for Communication Revolution. MIT Press.

Steinberger, F., Foth, M., Alt, F., 2014. Vote with your feet: local community polling on urban screens. In: Proceedings of the International Symposium on Pervasive Displays. ACM, p. 44.

Taylor, K., Silver, L., 2019. Smartphone ownership is growing rapidly around the world, but not always equally. https://www.pewglobal.org/2019/02/05/. smartphone-ownership-is-growing-rapidly-around-the-world-but-not-always-equally/.

The Committee for Sydney, 2018. Smart engagement: Leveraging technology for a more inclusive Sydney. https://www.sydney.org.au/wp-content/uploads/2015/10/Smart_Engagement_2018.pdf.

Tomitsch, M., 2018. Making Cities Smarter: Designing Interactive Urban Applications. Jovis Verlag GmbH, Berlin. ISBN 978-3-86859-492-812. 2017.

Tomitsch, M., McArthur, I., Haeusler, M.H., Foth, M., 2015. The role of digital screens in urban life: new opportunities for placemaking. In: Citizens Right to the Digital City. Springer, pp. 37–54.

Tomitsch, M., Wrigley, C., Borthwick, M., Ahmadpour, N., Frawley, J., Kocaballi, B., Nunez-Pacheco, C., Straker, K., Loke, L., 2018. Design. Think. Make. Break. Repeat. A Handbook of Methods. BIS Publishers.

Tragos, E.Z., Pöhls, H.C., Staudemeyer, R.C., Slamanig, D., Kapovits, A., Suppan, S., Fragkiadakis, A., Baldini, G., Neisse, R., Langendörfer, P., et al., 2015. Securing the internet of things security and privacy in a hyperconnected world. In: Building the Hyperconnected Society: Internet of Things Research and Innovation Value Chains, Ecosystems and Markets. River Publishers, pp. 189–219.

Valkanova, N., Jorda, S., Tomitsch, M., Moere, A.V., 2013. Reveal-it!: the impact of a social visualization projection on public awareness and discourse. In: Proceedings of the SIGCHI Conference on Human Factors in Computing Systems. ACM, pp. 3461–3470.

Valkanova, N., Walter, R., Moere, A.V., Müller, J., 2014. Myposition: sparking civic discourse by a public interactive poll visualization. In: Proceedings of the 17th ACM Conference on Computer Supported Cooperative Work & Social Computing. ACM, pp. 1323–1332.

Wang, M., Boring, S., Greenberg, S., 2012. Proxemic peddler: a public advertising display that captures and preserves the attention of a passerby. In: Proceedings of the 2012 International Symposium on Pervasive Displays. ACM, p. 3.

Williamson, J.R., Williamson, J., 2014. Analysing pedestrian traffic around public displays. In: Proceedings of the International Symposium on Pervasive Displays. ACM, p. 13.

Wouters, N., Huyghe, J., Moere, A.V., 2014. Streettalk: participative design of situated public displays for urban neighborhood interaction. In: Proceedings of the 8th Nordic Conference on Human-Computer Interaction: Fun, Fast, Foundational. ACM, pp. 747–756.

Wouters, N., Claes, S., Moere, A.V., 2015. Investigating the role of situated public displays and hyperlocal content on place-making. IxD&A 25, 60–72.

Wouters, N., Downs, J., Harrop, M., Cox, T., Oliveira, E., Webber, S., Vetere, F., Moere, A.V., 2016. Uncovering the honeypot effect: how audiences engage with public interactive systems. In: Proceedings of the 2016 ACM Conference on Designing Interactive Systems. ACM, pp. 5–16.

Author biographies

Dr. Callum Parker is a Lecturer in Interaction Design at The University of Sydney's School of Architecture, Design and Planning. Callum's research explores the potential of interactive augmented city interfaces, such as pervasive displays, {augmented, virtual, and mixed} realities, smart cities, and media architecture. He has compiled his research findings in books, journal articles, and conference papers.

Martin Tomitsch is an Associate Professor and Head of Design at the University of Sydney's School of Architecture, Design and Planning, visiting Professor at the Beijing Central Academy of Fine Arts Visual Art Innovation Institute, and founding member of the Media Architecture Institute. His research investigates the way new technologies will shape people's interactions in cities. His recent books include "*Design Think Make Break Repeat – A Handbook of Methods*" (BIS) and "*Making Cities Smarter – Designing Interactive Urban Applications*" (Jovis).

Dr. Joel Fredericks is a Lecturer in Design at The University of Sydney's School of Architecture, Design and Planning. Joel's research is transdisciplinary and sits across the domains of digital placemaking, media architecture, urbanism, smart cities, and immersive technologies. Joel has investigated collaborative and creative approaches to designing and deploying urban interventions that create curiosity and encourage people to playfully interact. He has authored and co-authored many publications in journals, edited books, and conference proceedings.

Section B

Co-producing smart places

Chapter 11

Platform urbanism and hybrid places in African cities

Nancy Odendaal
University of Cape Town, Cape Town, South Africa

Chapter outline

Introduction

> *... the most crucial task before us is not one of putting in*
> *place the digital plumbing of broadband communications links and*
> *associated electronic appliances (which we will certainly get anyway),*
> *nor even of producing electronically deliverable "content,"*
> *but rather one of imagining and creating digitally mediated environments*
> *for the kinds of lives that we will want to lead and the*
> *sorts of communities that we will want to have.*
>
> Mitchell (1996, p. 5)

The promotion of smart cities in corporate discourses, as well as some of the academic literature, focuses on technology as a solution to many of the intractable problems facing urban areas today. Underpinning many of these claims is a steady belief in the power of technology and data to solve the seemingly insurmountable problems that accompany rapid urbanization. Lately, this view has featured in local government strategies, particularly in the global South. The smart city phenomenon in Africa has been captured through proposed edge or satellite city developments, such as Luanda Satellite City (Watson, 2014), Konza Smart City outside Nairobi, Kenya, and Kigali in Rwanda. Visual representations of these initiatives show very little that links to the qualities of place.

Shaping Smart for Better Cities. https://doi.org/10.1016/B978-0-12-818636-7.00018-4

The impact of the digital age on African city streets displays a mix of old and new, far removed from the glossy visions depicted for master planned smart cities. The corporate branding of the contemporary telecommunications landscape is evident on billboards. The daily use of technology is also discernible in the motorcycle taxi (boda boda) riders (and drivers) clutching their cellular phones to their chins while negotiating Nairobi and Kampala's notorious traffic and potholes and the sale of airtime and associate services at street kiosks and bespoke container stores in South African townships. Fruit vendors and boda boda drivers compete for the same public roadside space, and being a pedestrian exposes you to many offers of contemporary mobility and street side nutrition. Technology appropriation is intrinsically linked to livelihood strategies and lifestyles, where disruption contributes to the ongoing unfolding of formal and informal activities in urban spaces. Social media, big data, the mobile phone, and the Internet of Things are all entangled with ongoing urban practices that respond to contemporary urban issues. African smart urbanism is a hybrid phenomenon that combines the old with the new, and this is not unique to the continent's cities.

In understanding what the place-based peculiarities are that are worthy of celebration or inclusion in an augmented urban design strategy, the first question should perhaps be: is it possible to do augment place making through effective urban design? Barns (2019), in the introduction to her book on "Platform Urbanism", draws attention to the ongoing tension between global standards and interface experiences of use of mobile phones, associated infrastructure, and advertising. It could be interpreted as a tension or a dance. I argue in this chapter that this is where the differentiated experiences of the urban are: the negotiation of this interface and how it relates to place. I reflect on my research experiences over the last decade in African cities to do so. The aim is not to exceptionalize the African experience, nor is it to generalize, but to use examples from marginal urban conditions to texturize the notion of smart places.

The relationship between "face to face" communication and online experience is informed by many technologies. Barns refers to McQuire (2008, in 2019) in reminding us that remote "being together" through radiotelegraphy, radiocommunications, and now electromagnetic spectrums have been part of the unfolding of industrial modernity since the 19th century. Either this separation has become entrenched, or we have become exceptionally good at traversing both. What draws people together in public space—if part of community, protest, storytelling, or mediated performance—are malleable themes threaded through many forms of media. Civic life has always been influenced by media and place. The more recent technological innovations have complicated and to some extent accelerated these relationships (Barns, 2019).

What is shaping cities though are not the technologies per se, despite claims and corporate visions to the contrary, but the interfaces between technology and users. These exchanges traverse scale where local actions are influenced by algorithms generated by global digital platforms, often outperforming locally bound operators. In Barns's (2019, p. 501) words "Scale, in other words, begets urban intelligence."

Practices evolve, imaginaries flood our expectations, as digital technologies are recombined and appropriated in the ongoing sociotechnical evolution of urban spaces.

Already in 1996, William Mitchell asserted the power of human agency in coproducing the "digital age"—almost quaint in today's times of "fake news" and online bullying. The application of this idea to the notion of co-created space is nevertheless still valuable. Practices, imaginaries, and expectations are negotiated on an ongoing basis as digital technologies are appropriated and recombined in ongoing urban transformation. When taking a symmetrical analytical approach, inspired by Science and Technology Studies (STS) and Actor Network Theory (ANT), we see technology simply as *part* of the story, not its defining feature.

This chapter is essentially a theoretical piece that uses empirical vignettes to further get under the skin of platform urbanism as a co-generator of place. The first part of this chapter provides a review of the literature on platform urbanism, with an emphasis on how the start-up and gig economy feature in African cities. Three themes on the relationship between platform urbanism and place are extracted from the literature review. Each of these is discussed using empirical vignettes to explore their implications for the shaping of space. There are emergent, somewhat contingent, and coproduction dynamics at work between platforms, urban life, and space (Barns, 2019). Hence, there is a need to have conversations about the impact of algorithms on behavior and decisions that shape such urban livelihoods (Barns, 2019; Leszczynski, 2016). The chapter concludes on how this may impact on future research and examination of urban practices.

Platform urbanism and hacking disruption

Over the last two decades, a scholarly interest has increased in the participatory cultures of networked mobile use as the smartphone has become increasingly embroiled in urban activism and community mobilization. Barns argues that this was an emergent form of platform urbanism, the term originating with her work. The contribution of urban informatics as providing more "fluid, mobile and networked imaginaries …" in understanding the makings of the smart city means that the user is not passive: "Engaging with platform services is today an integral part of being an urban citizen and as such involves many different kinds of value-sharing, not only the value extracted by technology companies" (Barns, 2019, p. 576). Thus what emerges from Barns's work is an architecture of interaction, a "read-write urbanism" that recognizes the agency generated through technosocial assemblages or participatory cultures of networked mobile use. The emphasis on big data has further emphasized the need to recognize this dynamic, rather than succumb to an automated reading of data-driven urbanism.

More recent work on platform urbanism has emphasized the implications for governance. An understanding of the notion of the platform is critical here. In what Fields et al. (2020) describe as a "black box," a technosocio framework enabled through algorithms and governed by intellectual property, platform applications have increasingly yielded influence on urban space. Unveiling the physicality of platforms starts poking holes at the mystique of the mythical digital other. Platforms constitute a technical architecture and are implicated in urban infrastructures. In what Barns refers to as the recombinant nature of platform architectures, their cross-operability is "… characterized by one platform underwriting a critical infrastructural function of another, in the way that Google Maps' digital map platform for instance serves to 'infrastructure' the visual interface experience for Uber riders" (Leszczynski, 2019, p. 191).

Thus the term "platform urbanism" builds on this functionality by recognizing the implications in terms of labor, the economy, governance, infrastructure maintenance and distribution, and the social and cultural practices that constitute urban life. This feature of contemporary urbanism has found its place in the global South and increasingly in African cities. How different styles of platform formation contribute to uneven geographies of saturation is beginning to emerge as studies of gentrification through AirBnB have shown (Wachsmuth and Weisler, 2018; Mermet, 2017). There is recent empirical interest in how publicly owned and cooperative platform mobilities could possibly make a more even impact (Stehlin et al., 2020). But are publicly owned platforms a possibility in urban contexts where institutions are weak and fiscal reach limited? A consideration of the winners and losers in the "sharing economy" in contexts with high-income inequality and structural legacies of sociotechnical change begs an examination into the hidden costs of the gig-economy and the impact that would have on place. In Africa countries, as in other parts of the global South, the global impacts of the digital revolution have coincided and—in some ways—contributed to increased labor insecurity, what Firmino et al. (2019, p. 206) refer to as "… the blurring of the boundaries between work and non-work – a distinction that has been rendered meaningless by the computer, the internet, and the smartphone." As in many parts of the global North also, the precarious nature of work is revealed through the disruptions that Uber and other ride-hailing apps have caused in the urban sphere. Of immediate concern in the documented research on ride-hailing apps such as Uber is the layering of informal economic arrangements on exploitative (informal) practices under a gaze of surveillance technology. The unlayering of assemblages that make up the platform economy in global South contexts is deeply entangled with disjunctures and tensions between economic empowerment and exploitation as the fluidity of such landings intersect with economic and labor histories, questioning its "normative, 'silicon-valleyan' rationalities" (Pollio, 2019, p. 762).

While the laudable focus on the exclusionary dynamics of the gig economy and platform urbanism is no doubt important, the point made that contextual differences produce very different experiences is an important one. These experiences are informed by colonial histories and postcolonial interpretations and relations. "In Brazil, we would say this is a move toward the tropicalization of global technologies, producing very specific ways in which these technologies are appropriated by local actors and resulting in unexpected situations" (Firmino et al., 2019, p. 207). This impacts the fine line between exploitation and entrepreneurialism as explored by the authors in Brazil. In particular, they focus on the implicit function of the smartphone as a form of surveillance through real-time monitoring, hence surveillance capitalism. Pollio raises a similar point in his work on Uber in Cape Town, South Africa. While drivers see themselves as entrepreneurs and welcome the temporal freedoms afforded by the ride-hailing apps, "a mediating technology of subordination" (Pollio, 2019, p. 767) nevertheless impacts their financial freedoms. Thinking infrastructurally about the use and influence of platforms goes beyond thinking digitally. It requires an engagement with the specificities of place and how it coincides with platform dynamics.

Universal rules of the market do not apply. The assumption that those entering the gig economy are entrepreneurs in the making is precarious at best. The sociocultural contexts within which economic relations are embedded have an impact on driver rights and targeting—the global entrepreneur does not exist—only local interpretations of worker access and rights (Firmino et al., 2019). These "entrepreneurs of sharing" are not necessarily equal partners (Pollio, 2019). But the benefits are important too: flexibility and yes some sharing of risk while earning an income. Pollio (2019), through a focus on Uber in Cape Town, shines a light on the transparency that enables drivers to escape corruption through "forms of technological liberation from existing labour arrangements" (p. 760). Drivers' opportunities to appropriate the means of labor through their own "hacks" speak to ongoing forms of capital mobilization and immobilization. "In fact, even the software-mediated, strict rules of ride-sharing uncannily allow for a series of manipulations and asymmetric conversions of values at the interface of other urban economies in Cape Town" (Pollio, 2019, p. 761).

Rather than interpreting technology disruption as new, as a rupture of sorts, I argue that some forms of continuity are present and that these are particularly pronounced and brittle in the African context. They relate to what exists geographically and materially and the social and institutional structures that form the backdrop to its socioeconomic urban landscape. The question begs: if public interventions are scarce, do homegrown platforms offer a closer attachment to place and therefore better opportunities for progressive sociotechnical evolution? I probe these possibilities under the following themes in examining examples of how start-ups and the gig economy evolve in the African city.

Theme 1: Sociotechnical agency and the remediation of public life

> *Returning again to our crowded train carriage, we might therefore see the daily habit of smartphone-enabled commuting as both a relatively new phenomenon, but also one that recombines and remediates centres of transportation, media and technology innovations, which have continuously acted to disrupt and retransform the experience of living together in cities. As citizens of an increasingly urbanised world, we continue to negotiate multiple historic layers of media technologies and scenic devices that shape our interactions with the world and with each other.*
>
> Barns (2019, p. 373)

As part of the assemblages of material, human, and data, digital platforms have become a banal part of everyday urban life, and their impacts are inherently spatial. How that agency is distributed cannot be assumed to be uniformly in favor of multinational companies such as Uber and AirBnB. Pollio's discussion of Uber in Cape Town reveals a more subtle play of power as the platform enables local narratives and economic assemblages to shape themselves in accordance with context and city aspirations. "Uber not only outsources some of its mechanisms of control that contribute to the subjectification of the drivers, but also better intersects local economic networks" (p. 768). The examination of how Uber's entrepreneurial rhetoric intersects with the City of Cape Town's emphasis on the startup industry as a core component of the local economy illustrates the articulation of specificities of place with algorithmic power. The recombination of software, digital platforms, and local infrastructures into endogenous responses to city issues is worthy of exploration, as a flip side of this coin.

SafeBoda is a Ugandan ride-hailing start-up company that has launched an App that serves motorcycle taxis, called boda-bodas, in East Africa (see Fig. 1). It was founded in Kampala by Ricky Thompson, a former boda rider, and two, other colleagues in 2015. Starting with 150 drivers at the end of 2017, there are now 19,000 drivers in Kampala alone in 2020. The issue of rider safety and the associated stigma associated with boda drivers, as reckless and unsafe, were the two motivating forces that led to the establishment of the app. No doubt informed by what anthropologist Jacob Doherty (2017, p. 195) refers to as the industry's allure of urbanity, mobility, and subaltern charisma, the SafeBoda app is intended to add a sheen of professionalism through technology. With "ride" being the core of the business, the SafeBoda platform has enabled diversification of its app-based services. These include package delivery, access to insurance and finance, and sale of airtime. Recently the expansion to food delivery has expanded the business model to include other forms of informal work: the preparation of food by local vendors. The latter includes the addition of two cloud kitchens to enable preparation by local cooks, as part of SafeBoda's network of service infrastructure. Initially titled Kya-kulya Chakula (Swahili

FIG. 1 SafeBoda riders in Kampala. *(Credit: Nancy Odendaal (Author).)*

term meaning food for everyday), the expansion to food services is significant. Expanding services to local food producers provides boda riders with additional income during off-peak hours. Enabling this continuous flow of business is essential to rider livelihoods (Personal Communication, Shekar). The digital platform enables partnering of complementary services in a quantifiable and formalized framework with some financial predictability. What may appear to be a professionalization of the informal economy is an actual construction of a frame for connections that enable livelihoods. The essential difference here is that the platform is homegrown and contextually embedded.

It is worth pausing on the contextual embeddedness of SafeBoda in relation to the qualities of place. My interviews with SafeBoda staff revealed a drive to articulate local sociotechnical conditions in their product offerings: access to local inexpensive food, enabling online payments for local services, and getting around town safely and inexpensively. As the COO Deepa Shekar put it: "We believe in what we are doing; Africa needs a localised option" (Personal Communication, 2020). Informal transport and food vending are common features of African city streets. What SafeBoda does is provide a platform for safe use of paratransit and a market for local cooks, together with an enhanced cloud infrastructure to enable that. There is a stated need to organize the informal, or as Ricky Thompson, CEO, puts it: "broaden the web for players that aren't connected to the web" (Personal Communication, 2020).

The recombination of infrastructural "bits" into a network of platformed mobility provision is bound to impact spatial relations. For example, a core part

FIG. 2 A boda boda Stage in Kampala. *(Credit: Nancy Odendaal (Author).)*

of the boda physical and institutional network is the "stage" (see Fig. 2). Stages are physical hubs that host 15–20 drivers. They represent catchment areas for boda riders, with the leadership of each stage elected by riders as part of a boda-boda association. The latter provides the connection between the riders and local authorities (Howe, 2003). A riding app disrupts this hub-based system, as riders are able to fill their time with rides and deliveries from anywhere, not only leading to a more efficient pricing system (Personal Communication, Shekar) but also undermining the functioning of stage catchment areas.

Stages are fairly ubiquitous across East African cities and provide some form of spatial reference for customers—walking to a stage and hitching a ride is straightforward and does not rely on the use of airtime. What SafeBoda introduces however, is a more individualized system that enables a broader catchment area, facilitating incomes on return rides also. The question then is does the platform entail an expansion and reordering of space? Thinking infrastructurally about the use and influence of platforms goes beyond thinking digitally. What distinguishes SafeBoda from mainstream hailing apps is that its intention is to grant more agency to riders. Speaking to staff, I was struck by the larger substantive values that guide them: enable mobility, deal with traffic congestion, and consolidate livelihoods. How this impacts on the spatial distribution of stages and their placemaking qualities remains to be seen, but it is clear that this evolved informal system of paratransit governance is seen to be in need of an upgrade.

Pollio (2019) shines a light on the transparency that enables drivers to escape the corruption inherent in Cape Town's taxi industry, through "… forms of technological liberation from existing labour arrangements" (p. 760). Drivers' opportunities to appropriate the means of labor through their own "hacks" speak to ongoing forms of capital mobilization and immobilization. In 2019, when visiting Nairobi, Kenya, I serendipitously spent time in the city's notorious traffic with an Uber driver, Joe,[a] who voiced his and many others' dissatisfaction that they were the losers in the technology innovation game. He articulated the desire of ride-hail drivers to have a sense of ownership and also take care of marketing their own platform. BebaBeba (meaning "carry carry" in Swahili), a ride-hailing service formed by the Drivers and Partners Association of Kenya (DPAK), was established in response to this need. Ironically, what was essentially a reaction to Uber was also inspired by the platform, or in Joe's words: "like Uber, we started on the street, not the office." BebaBeba was launched in December 2018 (2 months after our conversation) with the following statement by DPAK best encapsulating driver frustrations: "We have been subjected to losses each and every day, we have lost our vehicles through auction, because we can't afford to repay them and some of our friends have lost their lives because of long driving hours" (http://venturesafrica.com/bebabeba-enters-kenyas-taxi-industry/).

Joe explained how drivers can move between platforms as they fill their time and gently hinted at means whereby riders can be convinced to switch to platforms more amenable to driver needs. "In fact, even the software-mediated, strict rules of ride-sharing uncannily allow for a series of manipulations and asymmetric conversions of values at the interface of other urban economies in Cape Town" (Pollio, 2019, p. 761). Yet the commercial vestiges of investor capital is a substantive issue, reflected in recent posts on BebaBeba's Facebook page looking for investors for its "vehicle rent to own" program. The material elements of rider platforms not only extend to the physical impact *on* space, but beyond the algorithms are the logistics that require management of the network elements *across* space: the cars, the drivers, passengers, and payments. As pointed out by online Kenyan publication Techweez: "The smaller players will have to seek ways to manage their logistics much more efficiently. Typically, they should ask themselves how they would lower the arrival time once a rider hails a taxi? How do they ensure even distribution of their drivers? How do they make the process of payments faster and more seamless? What interesting technology will likely attract users to their service? A re-evaluation of these aspects will then likely inform their decision to continue seeking market share or opt out" (https://techweez.com/2016/07/13/kenyas-taxi-hailing-business-small-players-future/).

The material dimension is significant when considering the precarious nature of gig economy work in the mobility sector. As Doherty argues in his discussion of boda-bodas: "Boda bodies do not simply use infrastructure, but become a substantial component of it" (Doherty, 2017, p. 193). As a "critical supplement" to Simone's (2004) notion of "people as infrastructure", Doherty emphasizes the infrastructure violence boda riders are subjected to as their bodies, livelihoods, and labor coconstitute mobility systems. In the Uber example, explored by Pollio in Cape Town, the platform economy's self-employment regime essentially places drivers as supply infrastructure through the social that expand Uber's markets (Pollio, 2019, p. 766).

a. Name has been changed.

FIG. 3 BebaBeba promotion material. *(Source: https://bebabeba.ke/.)*

The corporate assemblages that have emerged from rider platforms hint at a localized set of practices becoming increasingly embedded in local commercial networks. The sociocultural contexts within which economic relations are embedded have an impact on driver rights and targeting—the global entrepreneur does not exist—only local interpretations of worker access and rights (Firmino et al., 2019). These “entrepreneurs of sharing” are not necessarily equal partners (Pollio, 2019). BebaBeba seeks to be the exception to that. While not the first Kenyan ride-hailing app—Little Cab, coowned by local technology company Safaricom (originator of mobile money app MPesa) was one of the first mobility platforms on the market—BebaBeba nevertheless places the driver at the center of its services (Fig. 3).

While the broadening of the ride-hailing market space is a formidable challenge for local start-ups, another form of disruptive hacking, I would argue, is the very diversification required to stay competitive in this growing market. These are often uniquely local and specific to place-based needs. In the popular press, they are listed as innovations. A timeous example (at the time of writing this during the COVID-19 pandemic) is Kenyan ride-hailing app, Little Cab’s offering of ambulance services. Partnering with St John’s ambulance services, the app offers the Little Rescue category within, which enables customers to get connected to a St John Ambulance dispatcher who will coordinate the emergency response with the ambulance crew. Much like their rides can be tracked in real time, the same can be done with ambulances. Speaking on the new partnership Fred Majiwa, Head of Program, Business Development and Communications at St John Ambulance Kenya said,

> *This is a timely partnership that will enable Kenyans in dire situations to receive quality emergency care and evacuation in record time. Customers can now track in real-time an ambulance that has been dispatched to help them, thanks to the live map that has been created. The crew will only need to use one live map to direct them to the emergency scene, which reduces time spent consulting the customer*

over the phone or use of different maps. This service also helps enhance our management of COVID-19 cases, which we are currently attending to free of charge, as part of our CSR.

(Kahende, 2020)

The quote earlier incorporates many aspects of platform urbanism as it relates to the temporal and spatial fixtures of cities. It speaks to technology convergence and also the centrality of platforms to urban life. As a "form of techno capitalism that entails a diverse, on-demand workforce" (Pollio, 2019, p. 761), the co-constitution of platform structures and urban space through a deepening of the geospatial dimensions of the platform economy (Söderström and Mermet, 2020) represents different styles of platform formation with geographically distinctive approaches within the same platform (Stehlin et al., 2020).

Theme 2: Entanglements and scale

The footloose nature of platform urbanism, or the ways through which data infrastructures are assimilated into our granular urban networks (Barns, 2019), reveals the potential for connections that traverse places yet maintain the same implementation values. How ideas travel across places and stay true to substantive goals with place-bound qualities is the emphasis of this section.

The genesis of SafeBoda and BebaBeba are examples of how informal networks gel with platforms to enable local interpretations of digital disruption. The beginnings of SafeBoda were seeded by friendship between the founder, a former Boda rider, and two friends connected into the start-up scene (Personal Communication, Thompson). In an online interview, Thompson referred to the "strategic partnerships" necessary to enable expansion, but then made an important point in the press, as well as in my interview with him, about the alignment of values. I am reluctant to attribute too much value to the emphasis on shared principles, but in my discussions in Kampala and Nairobi around SafeBoda and BebaBeba, it struck me that substantive issues were foregrounded and that they were directly related to issues related to place. They are also informed by people's lived experiences and how those manifest in particular locations as well as the regulatory environment that informs them. As eloquently illustrated by Doherty (2017), the life of the Boda rider is precarious and intimately enrolled in the material elements of degenerating road infrastructure, congestion, and associated dangers. The rideshare "partners" of Nairobi speak of livelihoods and day-to-day constraints in their oppositional practices to mainstream platforms.

The marketing tone of BebaBeba on social media emphasizes local pride and Kenyan innovation. Again, it is tempting to overromanticize these qualities of endogenous disruption, but they simply reflect a more granular engagement with local context. I recall listening to the country head of operations of Uber South Africa at an ICT summit in East London, South Africa, in late 2019, and while the presentation emphases on job creation and women's empowerment were contextually appropriate in the South African context, there as nevertheless a marketing spin that appeared somewhat contrived. The lived experiences of disruptor participants in Kampala and Nairobi was presented in an unrehearsed way, but what emerged strongly was a spontaneously sociotechnical spin whereby the technical, political, and material were interchangeable discussed. This was particularly discernible when discussing SafeBoda's expansion into the rest of Africa.

On the day I interviewed Deepa Shekar (in March 2020), SafeBoda was launched in Ibadan, Nigeria. When I probed how it may look different in the West African context, she spoke of the regulatory landscape being more complex in Nigeria, a local banking system that is sophisticated and allows for credit (important when designing the App), investor appetites enthused by the captive Nigerian market, and local smartphone use rates and skills. In Nairobi (launched earlier), users are more tech savvy, and the mobile money app M-Pesa is deeply integrated into commercial systems and lifestyles, while the uptake of e-commerce could push the continued rollout toward parcel delivery systems with local innovator SENDY being the main competition in this regard. Partnering with complementary services that respond more readily to the local character of place makes sense.

But there is a less articulated partnership that is more volatile that I became acutely aware of: the relationship between government regulation and local disruption. Again, this is central to how platform urbanism evolves and lands in different places. Digital platform-driven infrastructure such as AirBnB and Uber are difficult to regulate and predict. The inherent conflicts within platform urbanism are associated with the banality of platforms as they become a part of our daily lives, as well as the control platform firms have over data and code(Soderstrom, 2020). Technology innovation disrupt markets and regulatory regimes. In addressing mobility issues in these two examples, this issue cannot be neglected as duplication and upscaling evolve. Nairobi is a prime example, where rideshare drivers are motivating for the Kenyan government to declare ride-hailing companies as transport providers rather than technology companies in taxation and regulation, following steps in Europe to do so. But questions are often raised about the appropriateness of existing regulatory tools in doing so, as they relate to the qualities of place.

My discussion with SafeBoda revealed a more dynamic and interactive stance toward regulation: one based on ongoing interaction, proactive lobbying, and political pressure. The fact is that the boda rider community represents a formidable political constituency and leverage. Empirical work by Goodfellow and Titeca (2012) illustrate how "… market vendors and motorcycle taxi (boda boda) drivers showing how this engagement benefits both the informal groups and the president. Increased political competition has created an environment where informal groups seeking to protect their livelihoods can tactically leverage a presidential intervention in their favour, helping them evade the policies and regulations of the City Council."

Thus, tensions emerge between local vestiges of place-based regulation (often colonial remnants) and the increasing political and network capacities of what some would define "the informal." What McFarlane terms translocal learning (2011) are sociotechnical entanglements with strong connections to place, as platform urbanism recombines within local contexts of place-based dynamics and institutional frames.

Theme 3: The hybrid city

The relationship between the digital and material is hybrid and often described as continuously interconnected and coconstituted (Kitchin and Dodge, 2011; Leszczynski, 2019; Zook and Graham, 2007; Odendaal, 2014). The contextual depth afforded by a situated approach reveals how urban life is sometimes "rigged together from whatever is at hand" (Simone, 2011, p. 356) and then reassembled into a sociotechnical configuration that brings with it new (or renewal of old) spatial manifestations. The deployment of the SafeBoda app, for example, is foregrounded by safety concerns, mobility challenges, and marginal livelihoods that do not simply disappear once digital platform relations assemble. However, the relationships between these phenomena—material elements of the city, the relationships between interest groups, the exchange of information—are recombined into assemblages that may well disrupt existing relations. Boda riders are now in functional economic exchanges with food sellers, stages are disrupted by spatialities more under the control of the Boda driver, and the use of mobile money enables different forms of livelihood management as drivers are able to budget more carefully. Easterling (2014, p. 96) refers to infrastructure space becoming a "medium of information", with the platform element providing the software currency to reorganize and reshape space. In this reshaping, new "codes" evolve to contribute to a hybrid, coproduced urban form. In the example of Kenyan ride-share BebaBeba, the usual ride share network is expanded to include private car owners renting out their vehicles. The Little Cab example of ambulance services impacts (hopefully positively) the assemblage of healthcare provision elements that enable emergency care.

The concern that this cocreation of space is skewed toward the profit logics of multinational platform economy actors such as Uber and AirBnB is not entirely unfounded (Stehlin et al., 2020; Söderström and Mermet, 2020). Endogenous disruptive actors bring a different dimension to this dynamic that does speak to the generation and evolution of sociotechnical dynamics that are more contextually sensitive. I would argue that the literature on platform urbanism and indeed smart city writing that focuses on the dynamics of place generally lack this dimension of enquiry. In referring to the broader field of science and technology studies (STS), Philip et al. (2012) argue that the cultural dynamics of place are understudied in STS, while Furlong (2011) contents that privileging the technical does so at the expense of the more nuanced dynamics of sociopolitical processes. The "local" can be rendered quite abstract devoid of geographic and temporal specificity (Philip et al., 2012). Simply expanding the repertoire of STS case studies to the global South will not necessarily change that. Uncovering the dynamic of place required different questions to be asked, where technology appropriation is approached as a complex practice of translation and appropriation. Promoting blanket solutions and readings of infrastructure systems run the danger of ignoring the subtleties and dynamics of hybrid places. In fact, Philip et al. (2010, p. 9) refer to "hybrid knowledge practice" as a frame for understanding power, history, identity, and epistemology in telling sociotechnical stories.

Running a city on information, as envisaged by William Mitchell in 1996, ideally enables more democratic and citizen-oriented ways of placemaking. The emphasis on digital platforms, as initially sketched by Barns (2019), acknowledges the embodiment of technology through smart phones and its interface with platforms, as an extension or possibly, enhancement of contemporary urbanism where "... platform users will always be sharing, trading or 'co-producing' value" (2019, p. 121). The concern that spatial relations can be influenced from afar, given the debate on gentrification and AirBnB (Wachsmuth and Weisler, 2018; Mermet, 2017), thereby imposing a form of algorithmic governmentality, is a consideration. But so are the oppositional practices that have evolved in response to that. In Cape Town, the Reclaim the City campaign led by local advocacy group Ndufuna Ukwazi uses GIS data and social media to show how AirBnB has contributed to the lack of affordable housing in the central city (Odendaal, 2019). The campaign's strategies include online and offline strategies that range from populist representation of information to a technically astute interrogation of commonplace "truths" regarding property markets and the space economy of the city. A significant part of the campaign is the foregrounding of the "everyday" experiences of city dwellers in the face of gentrification and what some would argue state inaction. The enrollment of emotional, technical, and political "stories" into the campaign's narrative, together with the ongoing labor of legal, media, and policy engagement, gives an account of resistance to the externalities of platform urbanism. The cultural vertices of place are sometimes indiscernible but potentially powerful when articulated as part of a campaign of raising place-based awareness. I would argue this creates opportunities for progressive sociotechnical evolution that is more mindful of place.

Conclusion

In this chapter, I aimed to illustrate how local appropriation and design of platforms potentially yield a reconfiguration and contingency that is more attuned to context a more hopeful interpretation of digital urbanism (Leszczynski, 2019, p. 201). Rather than focus on the smart city as a standalone entity, I engaged the notion of platform urbanism as a continuous process of emergence and remaking. The focus on heterogeneity, everyday urbanisms as expressed in the empirical vignettes, was necessary to explore the role of place and human agency in ongoing sociotechnical change. Understanding the complexity and complexion of material-human entanglements reveals the coproduction of smart urbanism.

The importance of place is an important feature of a contextualized reading of smart places since it is so blatantly contrary to corporate representations of smart cities and speaks to embeddedness and situatedness. The examples show how the notion of place is foregrounded as a core reason for activism with smart tools used to enhance the experiential dimensions of residents' experiences of the city. The idea of surfaces: "exposures, folds, hooks, relays, hinges, soldering and shifting parts" (Simone, 2011, p. 364) are not necessarily all manifesting in the same geographic place. The contextualized interpretation of ride hailing, initiated by Uber and other multinational platforms in Kenya and then appropriated by BebaBeba, shows that these place-based manifestations are often tied to other spaces.

Whether localized ICT-informed interventions lead to meaningful spatial change is not conclusive. SafeBoda aims to professionalize the informal ride-hailing sector, but its destabilization of the stage system of evolved governance may bring its own externalities. Guma's work on Nairobi refers to "fractured constellations" as the spatial impacts of infrastructure are unevenly distributed (2019). The unevenness of service provision is informed by the nature of these assemblages and where they best gel or not. Differentiated services enabled through the input of mobile technologies have enhanced heterogeneity but not necessarily addressed structural inequalities. By emphasizing methodological approaches that do not privilege the revelation of visibility, this chapter sought to emphasize potentiality, slipperiness, and mobilities (Fields et al., 2020). The many unchartered futures of platform urbanism, I would argue, are functions of the relations between algorithms, people, and place. The empirical references discussed here show that how platforms evolve can manifest in unpredictable and generative ways. As Leszczynski (2020, p. 203) argues, revealing the glitches, or platform cracks, reveals possible urban futures, "... formed in more equitable ways via tactical manoeuvrings through strategically organized platform urban environments enacted in the form of everyday digital praxes at platform/city interfaces." In Kampala, Nairobi, and Cape Town, these futures are already here.

References

Barns, S., 2019. Platform Urbanism: Negotiating Platform Ecosystems in Connected Cities. Palgrave, Melbourne.

Doherty, J., 2017. Life (and limb) in the fast-lane: disposable people as infrastructure in Kampala's boda boda industry. Crit. Afr. Stud. 9 (2), 192–209. https://doi.org/10.1080/21681392.2017.1317457.

Easterling, K., 2014. Extrastatecraft: The Power of Infrastructure Space. Verso Books.

Fields, D., Bissell, D., Macrorie, R., 2020. Platform methods: studying platform urbanism outside the black box. Urban Geogr. 41 (3), 462–468.

Firmino, R.J., Cardoso, B.d.V., Evangelista, R., 2019. Hyperconnectivity and (im)mobility: Uber and Surveillance Capitalism by the Global South. Surveill. Soc. 17 (1/2), 205–212.

Furlong, K., 2011. Small technologies, big change: rethinking infrastructure through STS and geography. Prog. Hum. Geogr. 35 (4), 460–482.

Goodfellow, T., Titeca, K., 2012. Presidential intervention and the changing 'politics of survival' in Kampala's informal economy. Cities 29 (4), 264–270.

Howe, J., 2003. 'Filling the middle': Uganda's appropriate transport services. Transp. Rev. 23 (2), 161–176.

Kahende, K, 2020. Little launches ambulance services' (blog). https://www.little.bz/blog/little-launches-emergency-ambulance-services. (Accessed 20 June 2020).

Kitchin, R., Dodge, M., 2011. Code/Space: Software and Everyday Life. Mit Press.

Leszczynski, A., 2016. Speculative futures: cities, data, and governance beyond smart urbanism. Environ. Plan. A Econ. Space 48 (9), 1691–1708.

Leszczynski, A., 2019. Platform affects of geolocation. Geoforum 107, 207–215.

Leszczynski, A., 2020. Glitchy vignettes of platform urbanism. Environ. Plan. D Soc. Space 38 (2), 189–208.

Mermet, A.C., 2017. Airbnb and tourism gentrification: critical insights from the exploratory analysis of the 'Airbnb syndrome' in Reykjavik. In: Tourism and Gentrification in Contemporary Metropolises. Routledge, pp. 52–74.

Mitchell, W.J., 1996. City of Bits: Space, Place, and the Infobahn. MIT Press.

Odendaal, N., 2014. Space matters: the relational power of mobile technologies. urbe. Revista Brasileira de Gestão Urbana 6 (1), 31–45.

Odendaal, N., 2019. Appropriating "big data": exploring the emancipatory potential of the data strategies of civil society organizations in Cape Town, South Africa. In: The Right to the Smart City. Emerald Publishing Limited.

Philip, K, Irani, L., Dourish, P., 2012. Postcolonial computing: a tactical survey. Sci. Technol. Hum. Values 37 (1), 3–29.

Pollio, A., 2019. Forefronts of the sharing economy: Uber in Cape Town. Int. J. Urban Reg. Res. 43 (4), 760–775.

Simone, A., 2004. People as infrastructure: intersecting fragments in Johannesburg. Publ. Cult. 16 (3), 407–429.

Simone, M., 2011. The surfacing of urban life. City 15 (3–4), 355–364.

Soderstrom, 2020. The three modes of existence of the pandemic smart city. Urban Geogr. (forthcoming).

Söderström, O., Mermet, A.C., 2020. When Airbnb sits in the control room: platform urbanism as actually existing smart urbanism in Reykjavík. Front. Sustain. Cities 2, 15.

Stehlin, J., Hodson, M., McMeekin, A., 2020. Platform mobilities and the production of urban space: toward a typology of platformization trajectories. Environ. Plan. A Econ. Space. 0308518X19896801.

Wachsmuth, D., Weisler, A., 2018. Airbnb and the rent gap: gentrification through the sharing economy. Environ. Plan. A Econ. Space 50 (6), 1147–1170.

Watson, V., 2014. African urban fantasies: dreams or nightmares? Environ. Urban. 26 (1), 215–231.

Zook, M.A., Graham, M., 2007. Mapping DigiPlace: geocoded Internet data and the representation of place. Environ. Plan. B Plan. Design 34 (3), 466–482.

Author biography

Nancy Odendaal is an Associate Professor in City and Regional Planning at the University of Cape Town, in South Africa. Her research examines the intersection between urban infrastructure, human agency, and city spaces. Her work has primarily focused on the impact of digital infrastructure on cities of the global South. She is current chair of the Association of African Planning Schools.

Chapter 12

Learning lessons for avoiding the inadvertent exclusion of communities from smart city projects

Alan-Miguel Valdez[a], Edward Wigley[b], Oliver Zanetti[c], and Gillian Rose[c]
[a]*School of Engineering and Innovation Management, The Open University, Milton Keynes, United Kingdom,* [b]*Department of Geography, The Open University, Milton Keynes, United Kingdom,* [c]*School of Geography and the Environment, University of Oxford, Oxford, United Kingdom*

Chapter outline

Introduction

For some time now, it has been recognized by a wide range of smart city actors and commentators that the active participation of city residents, workers, and visitors in smart city activities is a necessity if smart technologies are to improve urban life. A number of justifications underpin this claim. There are those—particularly social scientists—who draw on the arguments of Henri Lefebvre to argue that there is an innate "right to the city" such that smart city technologies should enable meaningful participation by all people in a city (Balestrini et al., 2017; Cardullo and Kitchin, 2019a; Wiig, 2016). Others point to how widening understanding of the potential of smart technologies can increase both labor and consumer markets for commercial smart innovations (McNeill, 2017). Some enthuse about the "smart citizen" and how their active involvement can generate important data for smart city analytics (see Gooch et al., 2015; Granier and Kudo, 2016). Then there is the argument that local knowledge about local physical, social, and political infrastructure is crucial to the successful deployment of smart city technology (Saunders and Baeck, 2015; Tryfonas and Crick, 2015).

Shaping Smart for Better Cities. https://doi.org/10.1016/B978-0-12-818636-7.00003-2

All of these arguments are valid, although their implications will play out differently in different cities. This chapter takes as its starting point the slightly different claim that the institutions and organizations active in a city must participate as crucial partners if smart city projects are to enhance the quality of urban life. The chapter develops this claim by discussing a medium-sized town in the United Kingdom, which has been experimenting for some time with a range of smart city projects for some time. Milton Keynes (MK) is halfway between London and Birmingham and was the last of UK's postwar new towns to be built. It was incorporated in 1968 and now has a population of approximately 260,000. It is one of the UK's leading smart cities with a wide range of ongoing projects, and the chapter draws on data from a 2-year research project which analyzed several of these between 2016 and 2019.

While the first mention of "smart" in MK appears to be in a council transport strategy review in 2007, the development of MK as a smart city began around 2012, led by the council's Director of Strategy. He describes his strategy as one that brings smart city projects into the city to test things out:

> *often enterprises, organisations that were innovating and developing new approaches needed to do that somewhere and that gave me a bit of insight into something that has stood as well over time, which is thinking about our place as a place which can be welcoming and supportive and attractive and enticing for those sorts of projects (interview).*

MK is therefore an example of the urban laboratory model of a smart city (Evans et al., 2016; Karvonen and van Heur, 2014). There is no city control center and no oversight of smart activities by a council committee. Rather, smart in MK is "organic," to quote the Director, formed of whatever projects can be persuaded to land in, or emerge from, the city (cf Cowley and Caprotti, 2019; Taylor Buck and While, 2017).

The "urban lab MK" strategy has produced a number of successful bids for smart city projects. Examples include an £8 million grant for the deployment of charging infrastructure for electric vehicles, a £13 million OFGEM-supported smart grids trial, and £150 million for the operation of a transport innovation center, the Transport Systems Catapult (TSC). Another project, MK:Smart, was funded by the Higher Education Funding Council for England between 2014 and 2017. MK:Smart was led by computer scientists at The Open University with a number of partners: Milton Keynes Council, other universities (University of Cambridge and University of Bedfordshire), utility companies (HR Wallingford and e.on), and ICT providers (BT, Samsung, Huawei, and Tech Mahindra). The consortium was awarded £8 million, and match funding commitments resulted in an overall program value of £16.7 million (Valdez et al., 2018). Both the TSC and MK:Smart have facilitated many smaller projects, often in partnership with other funders and with local and national voluntary organizations. As a result, MK is a smart city full of relatively short-term projects with significant amounts of public and civic as well as corporate involvement and with a wide range of smart stakeholders, including local campaigners and MK:Smart and TSC partners.

None of these actors discuss their activities in the city in terms of the "right to the city," nor do they refer to the "smart citizen." Instead, our analysis suggests that they understand the city in terms of a complex network of organizations and groupings. And indeed, MK does have an extensive range of local social, voluntary, community, and other groupings, more or less formal, across a wide range of sectors. Many of these were facilitated as part of efforts to establish a sense of community in the new town as it was built and populated in the 1970s and 1980s, and that facilitation underpins ongoing forms of community organization in the city.

In this context the research project identified three particularly important actors in MK as a smart city: start-up businesses, the local council, and local voluntary groups. This chapter looks at the way in which these actors can enhance the participation in smart city projects of local groups and organizations. The chapter's discussion of two of these actors (the start-up and the voluntary group) are fictionalized. This is not only because of the ethics protocols of the research project but also because such a fictionalized discussion can merge insights from several real cases to better illustrate the key learning points with which each discussion concludes. The third case study (the council) is not anonymized. The chapter ends by discussing a tool used by many smart city organizations in the hope of engaging local communities: visual communications.

Start-up businesses

The commercial nature of many smart city projects has been well observed by social scientists, and many have critiqued the neoliberal logics that shape such smart city initiatives (Cardullo and Kitchin, 2019b; Grossi and Pianezzi, 2017). However, given the current political climate, there is a certain inevitability to those commercial logics, and the aim in this section is to examine how smart city projects emerging within that commercial framework can best ensure that as many people are involved as possible. There is an ethical imperative related to the Lefebvrian right to the city. But there is also a compelling commercial imperative, of course, as more involvement likely means more paying customers.

We conducted research with a range of start-ups in MK, as well as provide a fictionalized example, based on this research, described in the following section. We will call this example the Community Home Improvement Platform or CHIP. If it was real, the utility of CHIP might have been explained by its CEO in the following terms:

> *The UK's housing stock is in a bad state. We want to help building owners like householders, businesses and landlords get a better deal on the materials they need to make improvements. We can use big data to identify needs, and social tools to allow individuals to aggregate into groups. Like this, consumers will get a better deal from installers, installers will save on marketing costs, and we'll bring about positive change across the housing stock.*

Two relationships are central to the development and operation of successful smart city start-ups. First a productive relationship with one's client or user base is vital. Without this, no smart city project will be able to demonstrate either commercial success or positive social impact. Second, smart city projects very often entail building links beyond the business leading the development of the project, to groups or organizations working in the city itself. This often makes sound commercial sense. For instance, the extensive experience built up by a local council, voluntary group, or community organization can provide insights into the workings of the city that simple user research would be hard pressed to achieve. Moreover, funding from government business investment or accelerator programs are often key to the support of speculative new technologies whose potential is both great and unproven—in Connell (2014) terms, the act of "creating markets for things that don't exist"—and may require these links to be formed for funding to be awarded. This section will address each of those relationships in turn.

In working with the user base, in the early phases of the platform's development, there was some evidence of good practice. The platform set out to save its users money in two ways, first through the aggregation effect of having consumers negotiate with hardware sellers and builders in bulk and, second, through the savings that users might accumulate, for instance, through the installation of energy saving home improvements like insulation or replacement windows. Even so, home improvements are a costly investment. The platform's developers had thought of this and incorporated into their design a mechanism to help low-income homeowners access government grants to reduce their costs further. However, while income diversity had been considered, the developers had failed to account for other types of social difference. For instance, other social characteristics like age may affect motivation to have major building works done, with younger people put off by high costs and older people preferring to avoid disruption. However, the platform didn't look at these. Instead, it understood motivation as solely economic. In other words, people would only be interested if the works improved the value of their building or if the work saved them money. This approach could exclude people by not engaging with what matters to everyone. Furthermore, although usability was an aspiration for the platform at the outset, insufficient work was undertaken with user groups to test that usability. Ultimately, what was designed was too technically complex for all but the most tech savvy.

CHIP created a network of interested and relevant parties who contributed insights that were important to the development of the platform. One of their most significant links was with a local community group in the city, who acted as a voice for potential citizen users of the platform and provided access to publics for user testing activities. Representatives of the citizen group were enthusiastic about the platform, particularly in the early stages of their involvement with it, but frustrations did emerge. At the outset the community group's interest stemmed from their concern about city residents living in substandard accommodation, and they saw the platform as a tool that could help to improve lives for such people. Commercial pressures led to a shift in direction taken by the platform's designers, however. Their initial focus on community groups was replaced by an attention to commercial and social landlords who were able to operate at scale and therefore had higher spending power. Although the relationship did remain firm and optimism remained high for the potential benefit the project could bring to city residents in future, the change in immediate approach over which the community group had little control was straining. Ultimately, they felt disappointed and excluded from a project into which they felt they had invested a good deal.

Thus the following key lessons are offered to small businesses or businesses pivoting into the smart city domain:

- Build good relationships with local community representatives at the beginning of your project and be sure to work on them throughout.
- Think through all that motivates people to use your product or service. Remember social and cultural difference affects motivation, not just supposedly rational economic incentives.
- Don't overestimate your target audience's technical ability. If it's too complex, it won't get used.

Local government

Data have always been important to how cities in general and urban planning in particular are done, but new strategies and tools are needed to cope with the changing nature, volume, and speed of change introduced by the smart city (Kitchin et al., 2015; Marvin and Luque-Ayala, 2017). While the smart city concept is relatively recent, local authorities have been developing their own smart and inclusive data-driven practices for a long time. Local authorities often have intelligence units and well-established practices that use data in support of evidence-based policymaking. Moreover, city councils are well aware that community leadership is at the heart of modern local government. Councils work in partnership with local communities and organizations—including the public, voluntary, community, and private sectors—to develop a vision for their local area (Local Government Association, 2019). As result of this long-standing collaborative approach, councils may already have procedures to ensure inclusiveness in their data practices in place, even if they are not specifically framed as "smart city" policies (see, for instance, Local Government Association, 2018, p. 2, 3).

The following example (not fictionalized) draws on activities in MK to show that when introducing smart data infrastructure, there is a risk of losing valuable practices, networks, and communities that have previously developed and on which the effective use of data depends.

In 2003 MK Council decided to commission an "Intelligence Observatory": a data repository to make its own data and that of its partners more accessible. The

Observatory integrated many different datasets in a Geographical Information System (Fig. 1). Initially the investment was justified because the Observatory would improve the identification of health inequalities, which would allow healthcare provision to be focused on the areas most in need. The approach pursued by the council was not at all unusual, as various local authorities at the time were deploying similar data infrastructures and learning from each other. Identifying and addressing inequalities in health outcomes was a particularly strong driver for similar initiatives (Oliver et al., 2002; Hemmings and Wilkinson, 2003; Beaumont et al., 2005). The Observatory started operations 1 year after being approved by the council, and it rapidly became clear that delivery of the initial vision would require the establishment of a team of experts, to work and collaborate with partners, with a corporate brief for collecting, processing, and disseminating information about the town and the surrounding area to all interested parties. An intelligence team was created, which provided a full range of support and training to users of the site and partner organizations. The team also contacted potential partners to establish ways in which collaboration and information sharing would benefit them.

As it developed, data provided by the Observatory was used by a range of local organizations. The council constructed spatially specific indices of deprivation to apply for central government support for those neighborhoods, for example. The Observatory also provided baseline data for MK's planners. By

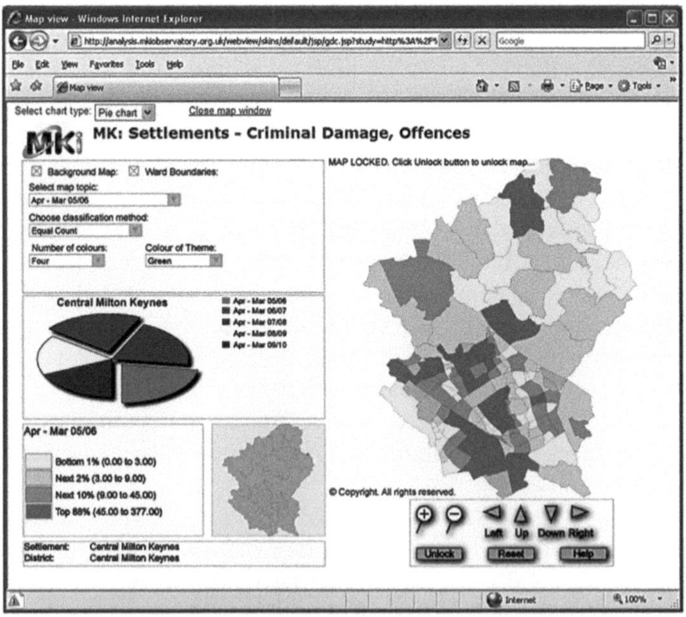

FIG. 1 A screenshot from the MK Intelligence Observatory. *(Source: Oakford, A., Williams, P., 2011. The use and value of local information systems: a case study of the Milton Keynes intelligence (MKi) Observatory. Aslib Proc. 63(5), 533–548. Emerald Group Publishing Limited.)*

2007 the Observatory had 500 registered users, and it was populated with data provided by the council itself and by a variety of agencies including the English Partnerships, the National Health Service, the Office of National Statistics, regional and national government, local partnerships, and emergency services. Data from the Observatory was also being used by a variety of voluntary sector organizations to concentrate their efforts in the areas that had a greater need. Voluntary sector organizations also relied on evidence from the Observatory to provide justification of need when requesting resources, lobbying for change, or responding to consultations related to their areas of expertise.

The MK Observatory operated for 15 years and was highly valued by a wide range of local actors. However, the Observatory ceased operation in 2017 owing to two somewhat related developments. Economic austerity made spending cuts necessary, and the Observatory represented a significant expense. Additionally, MK had developed its strong and largely privately funded smart city ecosystem, and data infrastructures that technically provided the same functionality of the Observatory were readily available. Data from the Observatory was transferred to a smart "data hub" at no cost to the council (Fig. 2).

However, the culture of inclusive, data-driven collaboration that had been developed over the last 10 years was put at risk through this "smart" development. Although all the original data were still available, in moving it from the council to the data hub, the relationships that had been developed by the council's intelligence team were temporarily severed. The new data hub had been developed with knowledge-economy businesses and start-ups in mind, and the

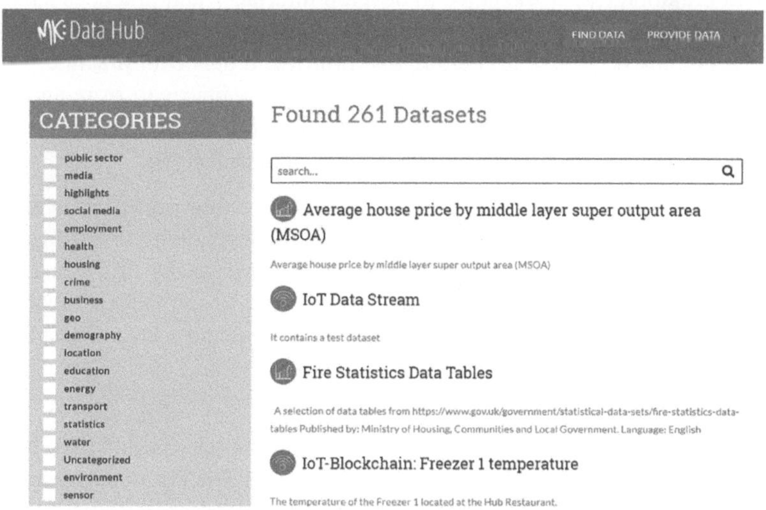

FIG. 2 The MK Smart Data Hub interface. *(Source: MK:Smart, 2019. MK:Data Hub Datasets. https://datahub.mksmart.org/ (retrieved 3 March 2019).)*

engineers in charge did not have the training or institutional culture required to communicate with local government officers and community groups.

Intelligence officers within the council identified the risk, however, and the council agreed to support a collaboration between experts from the council and the data scientists in charge of the smart city project. A budget scrutiny committee tasked with balancing the council's budget by "streamlining functions, working smarter and maximizing the use of new technology" welcomed this transfer of council intelligence capabilities to its smart city partners:

> *The Committee was concerned at the risk to the Council, both to its reputation and financially, which could arise from poor decision making due to Service Groups no longer being able to access the data interpretation and analysis skills currently provided by the Intelligence Observatory (MKC, 2017a). The Council, in association with the Open University, developed the MK Smart Data Hub which would continue to supply, via its website, the statistical data produced by the Research Team ... Staff around the Council have the necessary skills to produce and analyse such data and they should be encouraged to expand and share these capabilities (MKC, 2017b).*

If the city council had not made a point of remaining engaged through the transition, the culture of inclusive, data-driven collaboration that had been developed over 15 years could have been lost. Fortunately the council and the data hub team recognized the risk and brought on board a team of social scientists. The social scientists found that the original intelligence team at the council had produced a large volume of documentation related to their data practices and communities, which was available in the council's archives. This is now being used to get in contact with the former users of the Observatory and, through a series of collaborative workshops, to develop interfaces and practices suited to their needs. As such the new smart infrastructures are rebuilding a community that was put at risk by a smart culture that prioritized data over practice.

Thus new smart data infrastructures are not just a places to store and share data. They must also be embedded in formal and informal networks of users to be useful to a city. Key lessons for city councils are as follows:

- Cities have always been smart. Look into your own networks to identify the ways in which your city is already smart, and use technology to amplify them.
- Reflect on the city's information ecosystem. Who knows the city? Are they being engaged? Will they be able to engage with your smart platform? Work with your local communities and organizations to generate relevant and usable data.

Voluntary sector

This fictionalized example illustrates how neighborhoods do not have to wait and hope city authorities will include them some day. The Future Neighborhood Association, made up of representative organizations from across a Victorian town that was assimilated by the development of MK, took the initiative and deliberately aligned its plans to the sustainable innovation agenda developing around it.

Our imagined Future Neighborhood Association (FNA) was founded in 2002. Its aim was to seek the input of the community to ensure that the neighborhood would be able to fully participate of the economic development taking place around it, while protecting and enhancing its own heritage. The FNA's founding vision might have been expressed like this:

> *We are entering a period of rapid change in our town which is bound to have an impact on the community and businesses ... However, local organisations are determined to work together to ensure that local businesses thrive through this period of change and that the community are kept fully informed of what is happening.*

The visioning document was developed with the participation of people in the community. The association did not engage in visioning and engagement just for the sake of it, and the FNA's vision might also have included the following concrete aims:

- Ensure that community voices are heard in the development of its neighborhood through civic engagement with the local plans and those of the city council (Figs. 3 and 4).
- Develop and secure community facilities for residents to promote an active and involved community.
- Develop and support renewable energy and energy-consumption reduction projects.
- Support the conservation and community understanding of the neighborhood's unique heritage.
- Help to develop sustainable transport in and from the neighborhood.
- Be a well-run organization maximizing social impact and minimizing environmental impact.

FIG. 3 MK:Smart citizen lab on smart energy. *(Source: MK:Smart, 2014. MK:Smart News. http://www.mksmart.org/ (retrieved 3 March 2019).)*

FIG. 4 MK:Smart citizen lab on smart water. *(Source: MK:Smart, 2014. MK:Smart News. http://www.mksmart.org/ (retrieved 3 March 2019).)*

The FNA has periodically reviewed its vision, engaging with the community and with city authorities to update its plans and to ensure they are in line with the vision for the city, and that the neighborhood plans are respected and taken into account by city planners. By 2012 the efforts of the association had been instrumental in introducing various innovative and sustainable pilot programs that included an electric car-club program, a commercially successful electric bus service, and a program for increased energy efficiency and renewable energy production.

When smart city projects were initiated in MK, the FNA was therefore well positioned to participate. As several previous innovation projects in the neighborhood had become exemplars and were successfully implemented in other locales, the managers of smart city projects had a strong motivation to engage with the association and integrate its smart technologies to the ongoing sustainable innovation projects that were already taking place. Not all of the smart projects that were trialed in the neighborhood were successful, and in some instances the projects used the neighborhood as a testbed, but then they moved on to more affluent and profitable locales. Despite this the association persisted, became skilled at developing partnerships that would deliver lasting benefits for the community, and ultimately leveraged its reputation as a smart community to secure funding for renovations of its local center and its most run-down housing estates.

By having a long-term vision, understanding how it aligned with the broader city agenda, and engaging with local residents, businesses, and voluntary and community organizations, the FNA was instrumental in changing the way in which the Victorian neighborhood it represented was perceived within the city. Perception shifted from an area "down at heel" and in need of regeneration to an area that was leading the way on issues of sustainability and community cohesion. Key lessons for volunteer organizations and local community groups from this example are as follows:

- Have a vision for the future of your community and, if your local council has a smart city vision, develop an understanding of it too. Find ways to collaborate towards a joint vision.
- Work with institutions that can help you. Build partnerships with people in local government, education, and business who can guide you and help your community articulate how smart tech may be deployed in your neighborhood.
- Don't become a testbed for the sake of it. Question how potential smart solutions would (or would not!) make a lasting difference to people in your community.

Designing visual communications

Visual communications can be one of the most effective ways to inform and engage communities with smart city initiatives. Many smart city actors put considerable effort into picturing their products and projects (Rose and Willis, 2019). Visual communications are therefore an important aspect of inclusionary smart city projects. This section discusses how the most engaging visuals should relate to the experiences of smart city residents, workers, and visitors.

Because so many projects and products aspire to be used in many smart cities (Gardner and Hespanhol, 2018; Goodspeed, 2015; Söderström et al., 2014), images tend to play down the individuality or uniqueness of the places they picture, in favor of more generic and often more North American cityscapes. An autonomous vehicle that was demonstrated in MK, for example, was usually pictured in semideserted streets and pavements in the center (Fig. 5), and it was rarely pictured against any backgrounds that could identify the location, despite MK having a number of visually very distinctive sites. Such images allowed MK's smart products projects to be imagined both outside the local context and in other places globally.

As Vanolo (2016) notes, images produced by commercial vendors of smart technologies are often full of skyscrapers, wide streets, and streams of traffic. Images very often emphasize the smooth, frictionless, and flowing nature of smart cities (Rose, 2018). These commercial products may well want their viewers to imagine higher quality of infrastructures, networks, and services as a consequence of the market-led nature of this industry. Whether visualizing data packages transferred over wireless networks or the integration of transport systems through Mobility as a Service (MaaS) where smartphones allow contactless ticketless and synchronized travel patterns, images of the smart city overlook the daily frustrations of public transport users or the lack of digital connectivity. Instead, these visual communications imagine cities where crowds

FIG. 5 A typical publicity photograph of driverless vehicles in central Milton Keynes. *(Credit: UK Autodrive, Fabio De Paola/PA Wire.)*

of digitally connected people transit through urban space via faultless digital technologies and systems.

Many commercial smart city visuals are also very selective about who is and who is not pictured. While many smart city projects are embedded in cities inhabited by a range of people with diverse identities in terms of gender, age, ethnicity, nationality, and so on, much of the visual communication around these projects does not adequately represent this heterogeneity. Promotional videos focus on a narrow range of interviewees. These are often white, male, and middle class in appearance. A gendered split between science/engineering and social/caring roles is also not uncommon. For example, in a video for a community engagement group, men were visible in a number of roles demonstrating leadership abilities or talking about technology, while women were often seen doing the role of meeting and working with local community members, using social skills. Additionally, aside from the odd appreciative comment, the public's role in the smart city is often overlooked. Indeed, in a number of videos, the public merely form an audience for smart city speakers to talk to at meetings, presentations, and workshops.

This marginalization resonates with Vanolo's (2014, 2016) arguments that the public voice is silenced or "disciplined" in smart cities. Indeed, in the case of transport initiatives, the public are often reduced to manageable flows and units rather than complicated beings, with an emphasis on how to maintain the flows and remove frictions through the aforementioned MaaS-based systems. Moreover the narrative employed (along with the momentum affected through time-lapse photography of cityscapes and minimalist, driving music in the videos) is that these technologies and innovations *are* coming—without discussion of how appropriate they are for the city. Such visions raise concerns that instead of consulting, listening, and responding to individual city requirements, smart cities can only offer a range of preassembled solutions (Gardner and Hespanhol, 2018; Söderström et al., 2014).

In their efforts to imagine better future cities, this sort of smart city imagery seems all too often to lose touch entirely with the complex realities of actual places. They picture generic urban spaces, new and clean, gleaming in sunlight or dusk, saturated with smoothly flowing data streams, and unbreakable Wi-Fi signals, run by visionary political leaders or engineers (usually white men) and populated by passive crowds who exist only to be managed.

Discussions about this sort of vision of smart cities in MK generated what at first seems a rather counterintuitive conclusion: viewers are more convinced by the benefits of smart technologies when they are pictured as embedded in more realistic visions of city spaces. Visuals that picture local landmarks, that show smart city technologies occasionally glitching, that show suburbs and glossy city centers, and that show old places as well as new are all more striking to audiences, which is perhaps not surprising, when many people see digitally mediated city images in all sorts of other contexts, which they know are not real: computer games, estate agents' billboards, and sci-fi movies.

So the key lessons for the designers of visual communications in smart city contexts are as follows:

- Be true to a city, with realistic reference to images of the city. Avoid presenting a sterilized version of the urban environment or focusing on the technology instead of the city.
- Be sure your talking heads, animated personas, and people in continuity shots represent the diversity of contemporary society.
- Make sure the visuals, voice over, and interviewees speak in the language of the target audience. Engage people rather than trying to wow them with technology or saturate them with branding.

Conclusions

This chapter has reviewed four ways in which smart city activity, whether led by local companies, local government, or local voluntary sector organizations, can deepen the participation of local groups in smart city projects and visions. In all four areas of activity, a consistent theme has been the importance of engaging with local organizations, local priorities, and local senses of place. Local actors have long experience of their city and often have a great deal of insight into how it could be made better. Drawing on that expertise is surely a fundamental success factor that will ensure that the deployment of digital technologies will indeed make cities better places to live in.

It is important to note here what we mean by "local." Clearly, very little that happens in MK is confined to inside the city's boundaries. Every organization involved in smart city projects in the city is also embedded in institutional networks with regional, national, and in some cases international reach. Ideas and activities emerge from the city and flow out from it, and ideas and projects arrive in the city from elsewhere. Some of these connections and relations are deeply embedded, others provisional and temporary. The role of gatekeepers is also important: MK councilors and officials are frequently approached by smart tech vendors wanting to make their pitch, and over time the Director of Strategy has developed priorities and criteria to filter those requests.

However, places are also different. Their funding and governance structures are different, for example. MK itself has a particular history as a new town which, as this chapter has briefly noted, has contributed to a particular local social landscape of many quite active community groups and organizations. This has shaped this chapter's account of how to make smart projects work better. Also—and perhaps more importantly—the city has a hegemonic narrative that describes itself as forward-looking and innovative. Many of the various actors in many smart city projects in MK were very articulate about the distinctiveness of the city. They had a strong sense that MK from its foundation had been a sort of urban experiment and that it had a long history of trialing a range of innovative urban technologies, from solar-powered housing to cable television to smart dustbins. This sense of the particularity of MK, what one of our interviewees called its "sense of get up and go, we can do it," is also part of the why and how of local engagement in smart.

Finally, it is important to acknowledge that this particular configuration of civil society with the urban lab model of smart city development engaged in by the local council places another parameter on engagement in smart city activity in MK. As the business start-up case study suggests, much entrepreneurial smart city activity is more focused on turning a profit than addressing local problems, and not all local organizations agree with a local council's development policy. These conflicts of purpose and interest are perhaps less visible in a smart city like MK in which different projects appear and disappear generally rather quickly, very often dependent on short-lived funding or goodwill. They may be more evident—and challenging—in a smart city with a more centralized and strategic approach to smart city activity.

Acknowledgments

This paper is based on the research project "Smart Cities in the Making: Learning from Milton Keynes," funded by the Economic and Social Research Council grant ES/N014421/1. The principal investigator was Gillian Rose (University of Oxford), and the other team members were Nick Bingham, Matthew Cook, Parvati Raghuram, Sophie Watson, Edward Wigley, and Oliver Zanetti (all at The Open University).

References

Balestrini, M., Rogers, Y., Hassan, C., Creus, J., King, M., Marshall, P., 2017. A City in Common: A Framework to Orchestrate Large-Scale Citizen Engagement Around Urban Issues. ACM Press, pp. 2282–2294.

Beaumont, P., Longley, P.A., Maguire, D.J., 2005. Geographic information portals—a UK perspective. Comput. Environ. Urban. Syst. 29 (1), 49–69.

Cardullo, P., Kitchin, R., 2019a. Being a 'citizen' in the smart city: up and down the scaffold of smart citizen participation in Dublin, Ireland. GeoJournal 84 (1), 1–13.

Cardullo, P., Kitchin, R., 2019b. Smart urbanism and smart citizenship: the neoliberal logic of 'citizen-focused'smart cities in Europe. Environ. Plan. C Politics Space 37 (5), 813–830.

Connell, D., 2014. Creating Markets for Things That Don't Exist: The Truth About UK Government R&D and How the Success of SBRI Points the Way to a New Innovation Policy to Help Bridge the Valley of Death and Rebalance the UK Economy. Centre for Business Research, University of Cambridge. Available from: http://www.cc2-live.co.uk/davidconnell/docs/c%20dc-pub.pdf.

Cowley, R., Caprotti, F., 2019. Smart city as anti-planning in the UK. Environ. Plan. D Soc. Space 37 (3), 428–448.

Evans, J., Karvonen, A., Raven, R. (Eds.), 2016. The Experimental City. Routledge.

Gardner, N., Hespanhol, L., 2018. SMLXL: scaling the smart city, from metropolis to individual. City Cult. Soc. 12, 54–61.

Gooch, D., Wolff, A., Kortuem, G., Brown, R., 2015. Reimagining the role of citizens in smart city projects. In: Adjunct Proceedings of the 2015 ACM International Joint Conference on Pervasive and Ubiquitous Computing and Proceedings of the 2015 ACM International Symposium on Wearable Computers. ACM, pp. 1587–1594.

Goodspeed, R., 2015. Smart cities: moving beyond urban cybernetics to tackle wicked problems. Camb. J. Reg. Econ. Soc. 8, 79–92.

Granier, B., Kudo, H., 2016. How are citizens involved in smart cities? Analysing citizen participation in Japanese "smart communities". Inform. Polity 21, 61–76. https://doi.org/10.3233/IP-150367.

Grossi, G., Pianezzi, D., 2017. Smart cities: utopia or neoliberal ideology? Cities 69, 79–85. https://doi.org/10.1016/j.cities.2017.07.012.

Hemmings, J., Wilkinson, J., 2003. What is a public health observatory? J. Epidemiol. Community Health 57 (5), 324–326.

Karvonen, A., van Heur, B., 2014. Urban laboratories: experiments in reworking cities. Int. J. Urban Reg. Res. 38, 379–392. https://doi.org/10.1111/1468-2427.12075.

Kitchin, R., Lauriault, T.P., McArdle, G., 2015. Knowing and governing cities through urban indicators, city benchmarking and real-time dashboards. Reg. Stud. Reg. Sci. 2 (1), 6–28.

Local Government Association, 2018. The Councillor's Role. Local Government Association. https://www.local.gov.uk/our-support/guidance-and-resources/councillors-guide-201718/councillors-role (retrieved 3 March 2019).

Local Government Association, 2019. Equality Framework for Local Government (EFLG). https://www.local.gov.uk/sites/default/files/documents/guidance%20-%20equality%20frameworks%20-%20Equality%20Framework%20For%20Local%20Government%20%28EFLG%29.pdf (retrieved 3 March 2019).

Marvin, S., Luque-Ayala, A., 2017. Urban operating systems: diagramming the city. Int. J. Urban Reg. Res. 41 (1), 84–103.

McNeill, D., 2017. Start-ups and the entrepreneurial city. City 21, 232–239. https://doi.org/10.1080/13604813.2017.1353349.

Oliver, A., Healey, A., Le Grand, J., 2002. Addressing health inequalities. Lancet 360 (9332), 565–567.

Rose, G., 2018. Look InsideTM: visualising the smart city. In: Fast, K., Jansson, A., Lindell, J., Bengtsson, L.R., Tesfahuney, M. (Eds.), An Introduction to Geomedia: Spaces and Mobilities in Mediatized Worlds. Routledge, London, pp. 97–113.

Rose, G., Willis, A., 2019. Seeing the smart city on Twitter: colour and the affective territories of becoming smart. Environ. Plan. D Soc. Space 37 (3), 411–427.

Saunders, T., Baeck, P., 2015. Rethinking Smart Cities From the Ground Up. NESTA, London.

Söderström, O., Paasche, T., Klauser, F., 2014. Smart cities as corporate storytelling. City 18 (3), 307–320.

Taylor Buck, N., While, A., 2017. Competitive urbanism and the limits to smart city innovation: the UK Future Cities initiative. Urban Stud. 54, 501–519. https://doi.org/10.1177/0042098015597162.

Tryfonas, T., Crick, T., 2015. Smart Cities, Citizenship Skills and the Digital Agenda: The Grand Challenges of Preparing the Citizens of the Future. British Government Office for Science, London. https://www.gov.uk/government/publications/future-of-cities-smart-cities-citizenship-skills-and-the-digital-agenda.

Valdez, A.M., Cook, M., Potter, S., 2018. Roadmaps to utopia: tales of the smart city. Urban Stud. 55 (15), 3385–3403.

Vanolo, A., 2014. Smartmentality: the smart city as disciplinary strategy. Urban Stud. 51, 883–898. https://doi.org/10.1177/0042098013494427.

Vanolo, A., 2016. Is there anybody out there? The place and role of citizens in tomorrow's smart cities. Futures 82, 26–36. https://doi.org/10.1016/j.futures.2016.05.010.

Wiig, A., 2016. The empty rhetoric of the smart city: from digital inclusion to economic promotion in Philadelphia. Urban Geogr. 37, 535–553. https://doi.org/10.1080/02723638.2015.1065686.

Further reading

Oakford, A., Williams, P., 2011. The use and value of local information systems: a case study of the Milton Keynes intelligence (MKi) Observatory. Aslib Proc. 63 (5), 533–548. Emerald Group Publishing Limited.

Author biographies

Alan-Miguel Valdez is a lecturer in technology management and innovation at the Open University. Through his research, Miguel has interrogated how various publics as well as civic, industrial, and governmental bodies make sense of innovative technologies and collectively negotiate the future of their city. His research draws insights from urban geography and innovation studies to interrogate how the experimental spaces in cities may facilitate the development of new knowledges about sustainable innovation, and on how such knowledges coalesce and travel.

Edward Wigley is a staff tutor at the Open University, following a research post on the ESRC-funded project Smart Cities in the Making: Learning from Milton Keynes based at the Open University. Recent journal contributions have been published in Social and Cultural Geography (2016; 2018; 2019), Mobilities (2018), Globalizations (2018), Cultural Geographies (2019), and Geografiska Annaler (2020).

Oliver Zanetti is a senior researcher at the innovation foundation, Nesta, where he leads research into policy and practical interventions that can be made to support a thriving and inclusive innovation landscape in the UK's cities and regions. Prior to this, he was a postdoctoral researcher in the School of Geography and Environment at the University of Oxford, which he contributed to the ESRC funded project Smart Cities in the Making. Oliver has a PhD in human geography from The Open University.

Gillian Rose is a professor of Human Geography at the University of Oxford and a Fellow of the British Academy and of the Academy of Social Sciences. She is the author of Feminism and Geography (Polity, 1993), Doing Family Photography (Ashgate, 2010), and Visual Methodologies (Sage, fourth edition 2016), as well as many papers on images, visualizing technologies and ways of seeing in urban, domestic and archival spaces. Her current research interests focus on contemporary digital visual culture.

Chapter 13

Putting the people back into the "smart": Developing a middle-out framework for engaging citizens

Glenda Amayo Caldwell[a], Joel Fredericks[b], Luke Hespanhol[b], Marianella Chamorro-Koc[a], María José Sánchez Varela Barajas[c], and María José Castelazo André[d]

[a]*Queensland University of Technology, Brisbane, QLD, Australia,* [b]*Design Lab, School of Architecture, Design and Planning, The University of Sydney, Sydney, NSW, Australia,* [c]*Instituto Tecnológico y de Estudios Superiores de Occidente (ITESO) and SarapeLab, Guadalajara, México,* [d]*Infrarrojo Producciones, Guadalajara, México*

Chapter outline

Introduction

The greatest challenge within the smart cities arena is how can we put "people" back into the "smart." Research aligned under the smart cities umbrella is multifaceted and championed by movements of both inclusion and exclusion in countries throughout the world. The evolution of digital technologies over the past 30 years has created myriad "smart" technology trends that aim to enhance the "smartness" within urban environments to independently obtain and apply knowledge within these smart ecosystems (Ahmed et al., 2016; Shafique et al., 2020; Wathne and Haarstad, 2020). However, it can be challenging to deliver

Shaping Smart for Better Cities. https://doi.org/10.1016/B978-0-12-818636-7.00008-1

technology and citywide services that matter to all people from a variety of cultures, backgrounds, and beliefs.

Smart initiatives have been adopted by city governments and private enterprise to enhance services, improve public infrastructure, and foster efficiency for greater economic, social, and environmental benefits. However, it has been argued that these "smart" initiatives are techno-centric and driven top-down by stakeholders who often overlook human and more-than-human needs (Clarke et al., 2019). These top-down approaches to city-making struggle to engage and inspire communities in meaningful ways, particularly around smart cities (Anastasiu, 2019; Shelton and Lodato, 2019).

Conversely, grassroots movements championed by local communities are taking matters into their own hands with growing evidence of bottom-up approaches to city-making experienced around the world (Caldwell and Foth, 2014; Caragliu et al., 2011; Han et al., 2015). We define city-making as designing to create liveable and public spaces, foster cultural practices, and enhance community identity for urban dwellers across all demographics (Fredericks, 2020). Community organizations and individuals have taken it upon themselves to challenge top-down authoritarianism to test the needs, wants, and aspirations of urban spaces and technology in modern society. These contemporary approaches have led to bottom-up localized interventions in the form of DIY urbanism, guerilla urbanism, pop-up urbanism, tactical urbanism, temporary urbanism, and insurgent urbanism (Bisker et al., 2010; Hou, 2010, 2020; Finn, 2014).

Research traversing across political, epistemological, ecological, and spiritual dimensions is intrinsically complex in an age where people have competing interests, desires, and needs. In particular, research that is situated within the "smart city" and other techno-centric areas has predominantly been focused within the context of Western society. Other regions, such as the so-called global south, are often underrepresented within mainstream research due to lack of funding, or the research findings are primarily in languages other than English. In this chapter, we explore "middle-out engagement" in the contexts of non-Western cultures including Mexico and Thailand.

Following a case study methodology, we further expand and develop our concept of "middle-out" approaches to city-making (Fredericks et al., 2016a, 2019, 2016b). By employing these approaches, facilitated by digital and physical media, different actors at the "top" (decision or policy makers) and at the "bottom" (local citizens and community groups) can find common ground in the "middle." Although not necessarily always working together or collaborating, these actors can find novel forms of communicating with one another toward more inclusive outcomes, allowing city-making to move outward and forward.

We discuss specific projects from Mexico and Thailand. Each project has different stakeholders, such as academics, government, private entities, community groups, artists, and local citizens; with different levels of perception, affection, and engagement; employing different placemaking strategies; and across different scales of exhibitions, art festivals, and design workshops. The cases we discuss highlight the role that place and culture have on city-making initiatives, enabling us to propose a robust Middle-Out Engagement Framework to identify opportunities for smart cities initiatives that are socially and contextually relevant within emerging and Western countries. We discuss the implications to the cultural agenda, aesthetics, and social interactions played out in the city and in doing so contribute to contemporary urbanism with particular focus on the sociocultural agendas observed in highly populated, yet socioeconomically challenged, regions of the globe.

Middle-out engagement for collaboration

The term "middle-out" was coined to describe areas of knowledge coming from top-down or higher and bottom-up or lower information channels, which meet somewhere in the middle (Kinchla and Wolfe, 1979). The term has been applied in a range of disciplines including business, engineering, computing, and science (Fredericks et al., 2016a) and more recently in human computer interaction (Dow et al., 2019; Fredericks et al., 2019). Originating our interest to understand how top-down and bottom-up approaches could meet in the middle, our research initially focused on the creation of urban interventions for community engagement as seen in the InstaBooth project from Queensland University of Technology (Caldwell et al., 2015, 2016) and the Digital Pop-Up, Digitally Augmented Pop-Up, and Pop-Spot projects from the University of Sydney (Fredericks et al., 2015, 2017). As a result of these early works, we attempted to define middle-out design as a "process to draw on the collective knowledge of all actors to provide greater opportunities for more inclusive and collaborative community engagement processes" (Fredericks et al., 2016a, p. 2).

This approach to community engagement intended to integrate the objectives of top-down decision makers with the interests and views of those from the bottom, such as everyday people and community groups, to meet in the middle with the purpose to move outward and forward—hence middle-out. We acknowledge that these ideas are not novel and that strategic planning methodologies and engagement procedures promote this merging of interests with a desire to make change. Our particular interests as presented in this chapter are to further develop this approach into a research, design, and development method that acknowledges and encourages technical dimensions for application within urban development and urban HCI, city-making processes, architectural design, and smart city discourses. The mutual advantage for stakeholders of meeting in the middle is that with all collaborations and partnerships, there is a need to make compromises and to consider the interests of different parties—doing this creates a coalition of ideas and decision making necessary for collective action and universal buy-in.

By employing three phases of design, development, and deployment with a range of methods, middle-out community engagement seeks to gather the interests of different stakeholders from the outset and through the development of the engagement strategy. This approach relies on participatory or codesign methods (Bodker and Pekkola, 2010; Muller, 2003; Muller and Kuhn, 1993) through workshops, focus groups, questionnaires, and observations. The intention of participatory and codesign is to involve end users and stakeholders in the exploration of pertinent issues, the creation of new ideas through design developments, the opportunity to engage with the general public through eliciting responses to questions, and the analysis of results.

Brynskov et al. (2018) differentiate visions of smart cities as techno-centric versus citizen-centric. Although technology has a role to play in middle-out approaches, it is not the unique focus locating our discussion of middle-out as a balance between the citizen-centric and techno-centric contributions toward smart city developments. This addresses the need for sociopolitical and technological frameworks (Brynskov et al., 2018), which can be applied to address problems in a broad range of contexts, with different users and in different cities.

Middle-Out Engagement Framework

Further building on our middle-out design approach (Fredericks et al., 2016a) for the design, deployment, and implementation of urban interventions, we propose a Middle-Out Engagement Framework (Fig. 1) that can be applied to cultural agendas, community aesthetics, and social interactions played out in the city. This Middle-Out Engagement Framework therefore contributes to contemporary urbanism within city-making, community engagement, and urban technologies. The framework can be used individually or collectively among different groups, such as community groups, institutions, city councils, designers, or developers, as a mechanism for collaboration.

The Middle-Out Engagement Framework takes inspiration from the double-diamond design process model created by the Design Council UK in 2005.[a] The double-diamond design process model differentiates from other design models through its emphasis on divergent and convergent modes of thinking in the process of iteration required to make design solutions (Tschimmel, 2012). Also

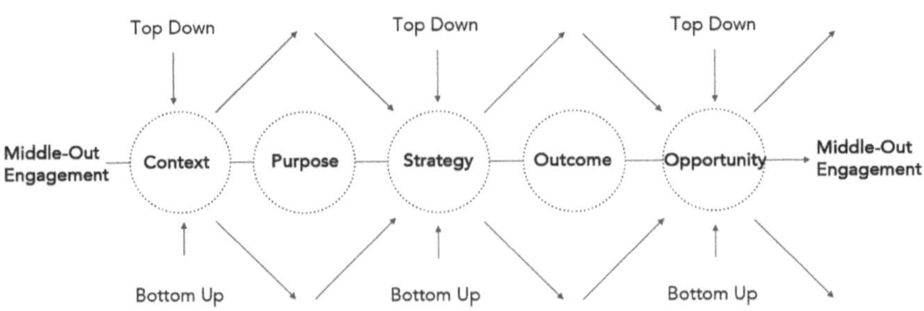

FIG. 1 Middle-Out Engagement Framework. *(Credit: Glenda Amayo Caldwell.)*

a. https://www.designcouncil.org.uk/news-opinion/design-process-what-double-diamond.

known as the 4D model, this process has four stages: discover, define, develop, and deliver.

The Middle-Out Engagement Framework has similar facets to the double-diamond design process in that it requires a design thinking approach to bring together the different stakeholders throughout the varied phases of the engagement process. There are three distinct roles that form part of our middle-out approach: (1) top-down stakeholders, which involve representatives from government organizations and private enterprise; (2) bottom-up stakeholders, which involve representatives from community groups, NGOs, and local communities; and (3) mediator(s) who acts as a facilitator and an independent entity and does not have a direct interest in the engagement process. With five phases, this framework is more complex and clearly focuses on bringing perspectives from different directions (top-down and bottom-up) into the middle part of the process. Each stage of the framework builds on the stage before it, which allows the stakeholders to progress to the next stage. The five phases operate as follows:

1. **Context**—The first stage of the Middle-Out Engagement Framework requires the stakeholders to decide on and review the *context* in which they operate in.
2. **Purpose**—The second stage allows the stakeholders to explore the goal of the engagement exercise and develop a common *purpose*.
3. **Strategy**—In the third stage, based on the identification of the context and purpose, the stakeholders create an engagement *strategy*.
4. **Outcomes**—The fourth stage implements the strategy within the given context to produce *outcomes*.
5. **Opportunities**—The fifth and final stage, *opportunities*, is critical to the Middle-Out Engagement Framework as it requires the stakeholders to reflect on the entire process and identify new opportunities that have arisen from the process. It is a future thinking stage that promotes the ability of the middle-out process to move forward and outward. This stage harnesses the desires and requirements from the stakeholders at the top and the bottom and empowers them to work together or acknowledge their different needs and to progress beyond the starting point.

The following sections describe the cases we have identified from two distinct parts of the globe. Each case is structured on the Middle-Out Engagement Framework to demonstrate the different stages of the process and how the stakeholder groups worked together.

Creative and smart city: Smart thinking from idea to strategy

This first case study is an account of the lived experience of two of the authors as part of the Civil Association in Guadalajara who act as the mediators in the middle-out approach. There are minimal accounts about this case published in

English, and to the best of our knowledge, these accounts have not been captured in academic publications regarding middle-out engagement and smart city agendas in Mexican contexts. There is information from the architects and designers about the Digital Creative City, or Cuidad Creativa Digital (CCD), masterplan, and other information derives from a master's thesis in Spanish, local newspapers, and social media sites (Fig. 2).

The top-down stakeholders that have been identified in this case are primarily from the Mexican government. They contracted international engineering company ARUP, world recognized university partner MIT, and Carlo Ratti Associates to design the CCD masterplan and associated buildings. The middle-out stakeholder—whom we refer to as the mediator—is an agency called the Civil Association who were created to mediate between government decision makers and local community members. The community who consists of the media sector including artists, actors, and local citizens are considered as the bottom-up stakeholders.

The following sections discuss the activities of Guadalajara and the Civil Association in the Middle-Out Engagement Framework as depicted in Fig. 3.

Context

Guadalajara is a city of contrasts that mixes modern buildings with colonial-style architecture, folklore, handicrafts, religious traditions, and gastronomy. In recent years the State of Jalisco has established itself as the most important technological hub in Mexico. Its capital city, Guadalajara, is recognized as the "Silicon Valley of Mexico" with numerous news articles having been written about the city's increasing technological production (Bokor, 2016). This in turn has generated economic activities around the research and development of specialized technology. In Guadalajara the digital creative industries have been identified as paving the way for the city to lead the country's smart city projects (Corona, 2017).

The first smart city project in Mexico, Guadalajara, was led by the government and labeled a Digital Creative City (CCD) (Cuidad Creativa, 2012). The CCD project aimed to establish Guadalajara as the first global node of digital creative production in the Spanish-speaking world. The CCD project was established through two key phases: (1) a sustainable urban development aimed to create a digital media hub in Mexico for the TV, film, animation, games, interactive media, e-learning, and advertising industries (Cuidad Creativa, 2012) and (2) initiatives to enact the CCD. The federal agency called ProMexico, a subdivision of the Mexican Secretariat of Economy that promotes international trade and investment, initiated the CCD project in 2009 and placed it at the center of attention of Mexico and the world's digital creatives (Navarrete, 2018). Architect and Professor Carlo Ratti, director of MIT's Senseable City Laboratory, was commissioned as the lead designer of the CCD urban development masterplan in 2012 (Bokor, 2016; Cuidad Creativa, 2012). Fig. 4 is an image as part of the CCD masterplan and brochure included on Professor Ratti's website.

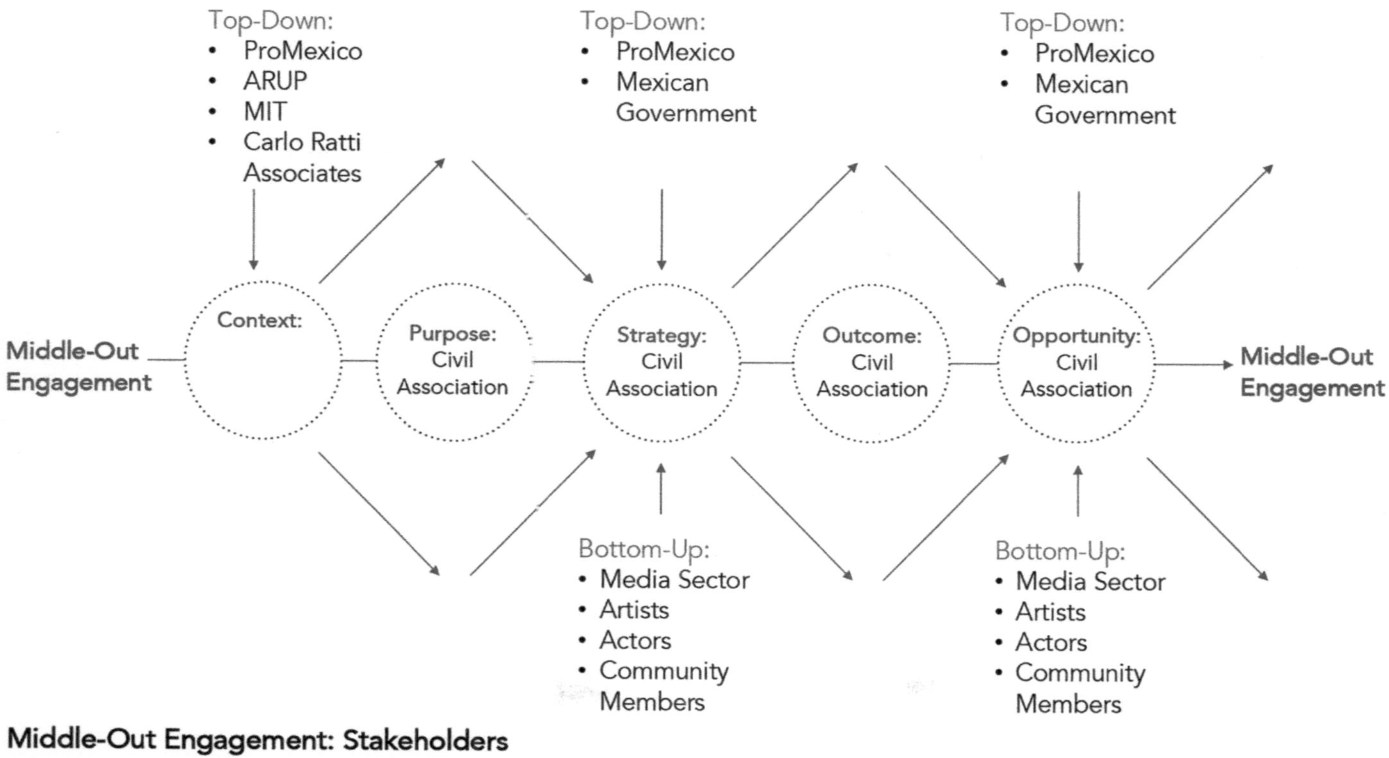

FIG. 2 Middle-Out Engagement Framework stakeholders for Guadalajara. *(Credit: Glenda Amayo Caldwell.)*

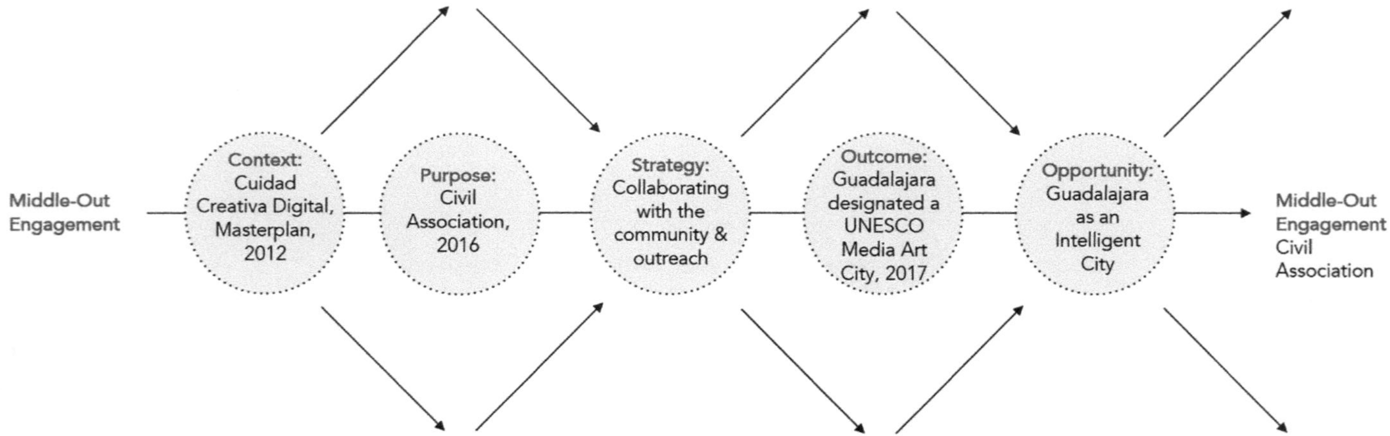

FIG. 3 Middle-Out Engagement Framework activities for Guadalajara. *(Credit: Glenda Amayo Caldwell.)*

FIG. 4 Image of the living lab as part of the CCD brochure. *(Credit: From http://www.carloratti.it/FTP/CCD/files/CCD_brochure.pdf.)*

The diagram of the middle-out engagement for Guadalajara in Fig. 2 shows how the CCD project involved stakeholders across government, industry, international design, and thought leaders in the context stage. At this stage the CCD was developed through a top-down approach, which included urban and architectural designs for key buildings within the masterplan. The diagram indicates no involvement from the bottom-up stakeholders in the development of the CCD masterplan as the project was largely driven by the Mexican government (Cuidad Creativa, 2012).

Purpose

Recent research acknowledges that Guadalajara is a city that historically has had strong activist movements that have been able to gain influence on political agendas (Acosta García et al., 2014; Haber, 2006). These activist efforts have been in response to violations against human rights, political corruption and mismanagement of public funds, and oppositions to urban planning (Garcia, 2018). Recent studies and research have focused on urban activism in Mexico where the growing middle class, educated and economically stable, have relied on their collective efforts to use public space and creative methods to communicate their claims to the city (Avalos, 2016; Garcia, 2018; de Padilla la Torre and Flores Márquez, 2011). These urban activist movements prevalent in Mexico and particularly in the city of Guadalajara are referred to by Tamayo as *movimiento ciudadano* (1999). After objecting to the lack of community engagement in the

CCD masterplan, in 2015, the creative media sector of Guadalajara demanded a change in the direction of the CCD project. They asked for more resources to be provided for local creative talent. In response to public outcry and demands, the local government established a Civil Association for the CCD project—this institution became in charge of operating the masterplan and project activities (Restrepo, 2018).

The middle-out aspect in this case is represented by the Civil Association. Fig. 2 shows the Civil Association positioned in the center of the first diamond under *purpose*, representing how this example of middle-out did not occur from the beginning of the CCD project but arose as the interests of those from the bottom needed to be incorporated as part of the outcome and progress of the masterplan.

The Civil Association consisted of a director and three coordinators who were tasked with developing the Guadalajara creative industries as part of the CCD project. The coordinators did not have political affiliations and were creatives and artists themselves, which is a critical point here, as they knew first-hand the needs of the sector. Their goal was to create strategies and projects focused on the development of the creative industries sector by linking talented creatives from local, national, and international levels.

As a result, Guadalajara adopted a cross-sectoral approach of media arts as an innovative and interactive tool that adds value not only to other creative fields but also to the way that urban dwellers experience the city. Employing a collaborative approach that involved both top-down and bottom-up stakeholders, initiatives were developed for linking the creative sectors with the objective to develop solutions for social impact. These initiatives were directed toward the people who lived and worked outside the perimeter of where the CCD masterplan was earmarked, which was classified as a vulnerable neighborhood within the city.

To reflect Guadalajara's commitment to the digital media and creative industries sectors and the community, the CCD, through the Civil Association, hosted a number of events, festivals, and exhibitions around digital media in Guadalajara. These took place on an annual basis to bring together projects combining creativity, art, and technology as a means to engage with a variety of top-down and bottom-up stakeholders, including government arts and cultural agencies, design practitioners, local artists, industry professionals, academics, and students.

As an example, Fig. 5 is a poster from an exhibition, *Circuitos Barriales*, demonstrating the outcomes from a series of community-based workshops hosted by CCD and directed and organized by the Civil Association. *Circuitos Barriales* was a program designed to connect and develop transdisciplinary approaches using art to bridge the gaps and create dialogue between people, science, society, and technology.

Fig. 6 is a tweet from @CCDJalisco promoting a free mixed reality workshop open to the public as part of the *Circuitos Barriales* media art program of events.

FIG. 5 CCD promotional poster of *Circuitos Barriales*. Director of the Program: María José Sánchez Varela Image. Program coordinators: Giovana Tovar and Ana Cristina Uralde e Irene Paez. *(Credit: Poster Design: Victor Razo & Sarape Social.)*

FIG. 6 Image of @CCDJalisco tweet promoting a public workshop. Director of the Program: María José Sánchez Varela Image. Program curator Jorge Ramírez by Anemonal. Program coordinators: Giovana Tovar and Ana Cristina Uralde e Irene Paez. *(Credit: Poster Design: Jorge Ramírez & Victor Razo.)*

Strategy

The Civil Association recognized that they were in a mediating, "middle" position between the decision makers at "the top" and the local people at "the bottom" of the CCD project. They had to make recommendations and advise the decision makers with strategies that were backed by solid economic outcomes while maintaining a strong social commitment to the locals. They were convinced that art and creativity were important factors that would contribute to social development as part of the CCD masterplan. Their strategy to progress and move forward with the masterplan was focused around the concept of "people" at the center of smart cities. Acknowledging that each member of society shapes their future, the Civil Association had to generate inspiration, learning, strategies, and tools that suited the individual while creating common benefit.

The Civil Association generated new strategies for positioning the creative ecosystem by including artists and actors at the center of innovation and collaboration. This was achieved through partnerships between local government authorities, universities, and local communities, artists, and actors. As far as we can ascertain, these strategies are not formally documented and are an interpretation of two of the authors who worked within the Civil Association. These strategies employed by the Civil Association team as identified by the authors are as follows:

1. Recognizing and mapping communities that are creative while acknowledging that communities and groups are already creative whether they are formally recognized or not.
2. Creating spaces of trust that cater for diverse creative and digital sectors of society, often resulting in the creation of concerts and other creative outcomes.
3. Validating a vision that represents present and short-term ideas as opposed to strategic long-term visions used during election cycles for political leverage.
4. Empowering and encouraging the local community through knowledge sharing with a variety of stakeholders (top-down and bottom-up) to enhance collaboration and project development.

In response to local concerns, the CCD project found itself needing to adopt a new way of communicating and moving from a top-down government-managed project to a project that is managed by different actors such as the Civil Association. Critical to its success was the development of strategies that included the locals' visions of the future. These strategies applied since 2016 have collectively served to visualize the value of the city's communities and creative actors. Through linking and mapping programs, the project has empowered and has given voice to different groups and new key agents for the CCD project.

Outcomes

Although the CCD project was initially driven by the top-down, it was evident that the middle-out approach as enacted by the Civil Association significantly contributed toward greater outcomes for the community. The implementation of the CCD masterplan and the distinction of Guadalajara as a creative and intelligent city have given rise to other important actors and actions that the plan itself did not initially contemplate. As a result, in 2017, Guadalajara was designated a UNESCO Media Arts City (UNESCO, 2017). This international recognition of the city is based on the response to the need for both public and private actors to have a common objective of community outreach. Through the expression of new digital media, the different sectors were able to converge and recognize that an intelligent city is created and configured by intelligent citizens. That assertion has many implications, part of which is aligning interests between the people who have the power and the urgent needs of the poorest who do not find value in an intelligent or smart city. This recognizes a different position of envisaging solutions and designing the smart city.

Since the creation of the CCD, companies from across Mexico and even from abroad have positioned themselves in Guadalajara to either prepare or work in different areas related to media art, generating more than 24,000 direct and indirect jobs in the city (UNESCO, 2017). The city of Guadalajara continues to grow in both technological and urban development. In recent years, there has been a rise in actors and organizations that have—from their trenches—acted more in the underground space. However, these actors are not less powerful. Rather, in a nonformal way, these underground or bottom-up approaches have generated networks beyond what the policy dictates or what the CCD masterplan intended to execute. These alternative groups have the vision that the city and collective intelligence is already there, within its existing communities, regardless of whether a government has a plan to name it. Artists, creatives, and "born ingenious" have been there and will continue to be there, without a masterplan. These sectors are assumed to be part of the conditions for Guadalajara to be recognized and an intelligent city to be named.

Since the new government took office in 2018, a forthcoming social development strategy is not yet clear. However, the idea of belonging to and being a creative in an intelligent city has a strong attraction for both companies and inhabitants. Private initiatives, underground groups, and agents of creative, digital, and technological ecosystems continue to organize and build new ways of thinking. They are able to rethink the future by establishing collaborative networks to achieve the best opportunities for their own development in a city that claims to be intelligent, whether it is part of the CCD government masterplan or not.

Opportunities and the role of champions in the middle

Although there are still many proposals and methodologies to apply to the concept of Guadalajara as a smart city, there is much more work that needs to be done beyond planning. Government structures and institutions cannot be expected to increase the involvement of the majority of citizens who, at least these days, are more concerned with their own safety or daily economic survival than having a smart city. The social, economic, and environmental benefits of smart cities are not immediately visible to these people; therefore there needs to be more effort in addressing their immediate survival needs and then highlighting how they can benefit from the smart city. The transformation of the city is not an easy task and involves a complex and time-consuming scenario.

Moving forward the "champions" who will manage the development of Guadalajara as a smart city will need to have new professional profiles. They will need training to be specialized in many fields and areas of activity that will be required in the coming years. This will enhance their capacity to respond to all of the challenges and transformations of urban environments, where talent, innovation, and creativity will intrinsically be at the forefront of community planning, collaboration, and action. Fostering higher levels of resilience and flexibility provides channels to adapt to change. The opportunity is to train these middle-out actors and give them the tools they need to continue to promote and generate their own vision of the future.

Nurturing a riverside creative ecosystem through community-driven placemaking

The second case study is an account of direct observations, fieldwork, walks, visits, and conversations conducted by one of the authors in Bangkok, Thailand, during two visits to the city and followed by subsequent research. Documentation of the activities described in this section is scarce in English and, to the best of our knowledge, has not yet been captured in academic venues or elsewhere—as is often the case for countries with large informal economy sectors. Other than in situ observation, most of the current information derives from interviews conducted by the author or the Thai press directly with the corporate project stakeholders responsible for enacting the middle-out approach with creative practitioners along the Chao Phraya River.

Fig. 7 depicts that the top-down stakeholders for this are the Bangkok Metropolitan authority and the Thailand Creative and Design Center (TCDC). The middle-out stakeholders are the Bangkok Creative District and the local leading architect Duangrit Bunnag. The bottom-up stakeholders come from the local community workers, shopkeepers, university students, and creative communities.

The following sections discuss the activities of the Bangkok case presented in Fig. 8.

Context

The Chao Phraya River is Bangkok's most important transportation artery, cradle of the local culture, and one of the historical centers of the Thai civilization. For centuries, communities have lived and traded along the river, yet Bangkok's rapid development in the past few decades has shifted much of the economic activity away from the riverside toward new developments in the

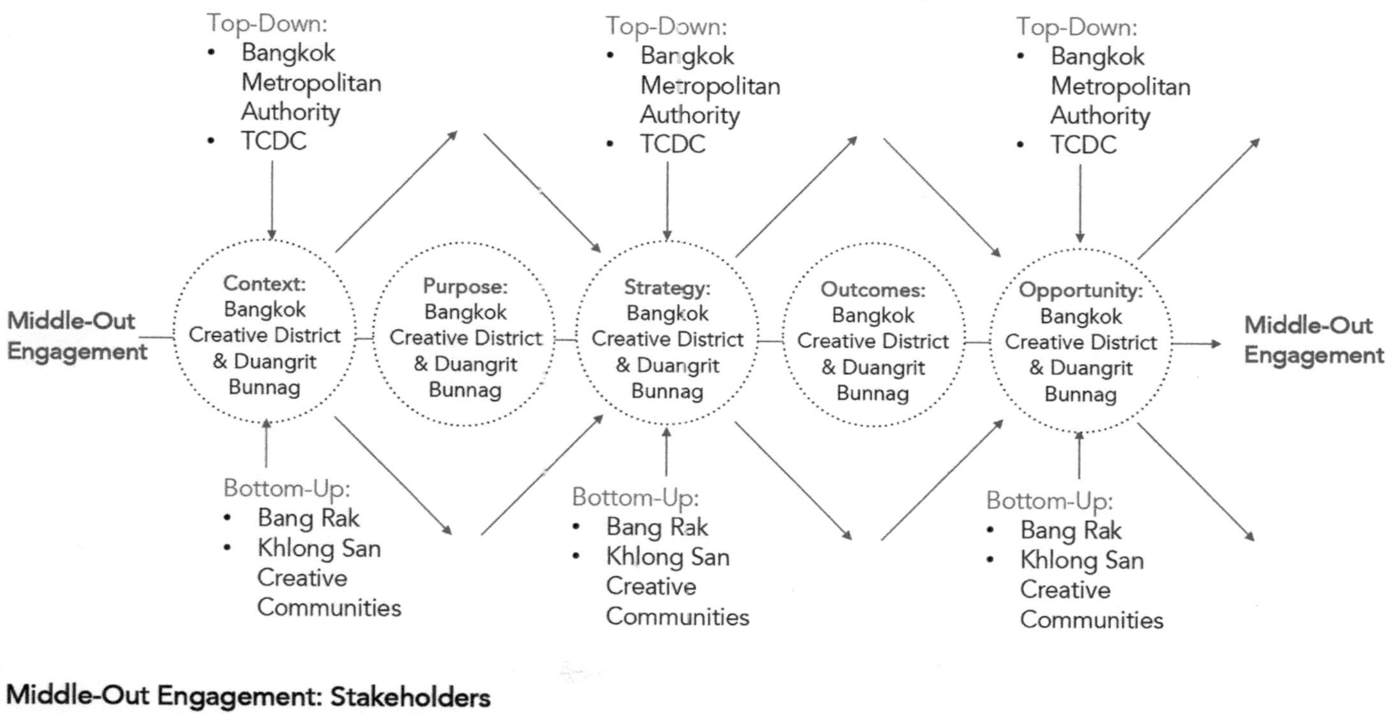

FIG. 7 Middle-Out Engagement Framework stakeholders for Bangkok. *(Credit: Glenda Amayo Caldwell.)*

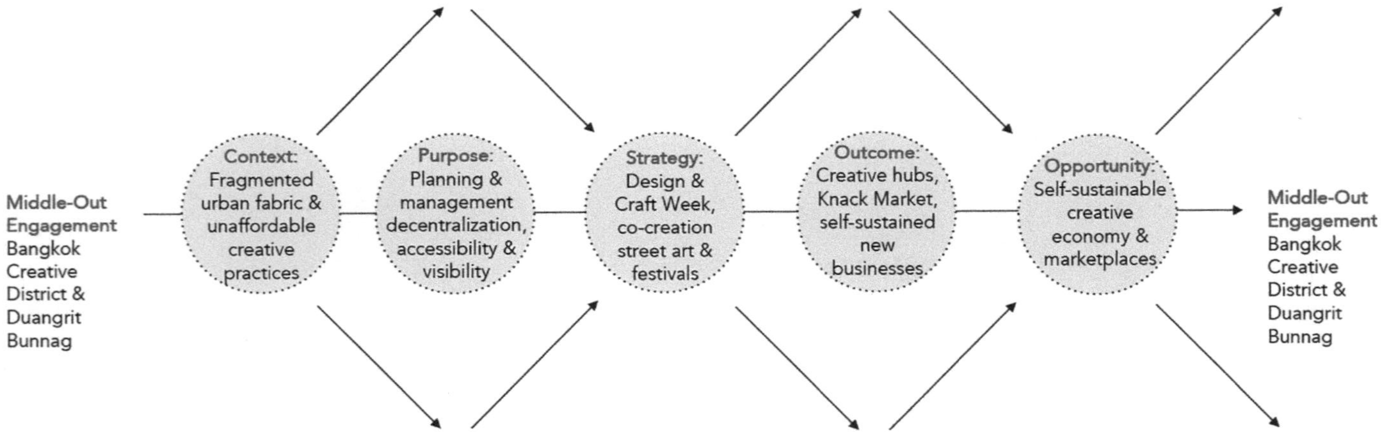

FIG. 8 Middle-Out Engagement Framework activities for Bangkok. *(Credit: Glenda Amayo Caldwell.)*

city's metropolitan region. In an effort to counteract the increasing decline and underutilization of the river as a cultural and commercial hub, a small group of business operators and property owners working along the Chao Phraya decided to join forces and found a new Organization. They started Bangkok River Partners, where they could collaborate to attract Bangkokians and foreigners alike back to the river (Fernandez, 2016a). Among other things, the group manages shared resources, organizes events and business meetings, and promotes general appreciation of the waterway, its surrounding neighborhoods, their history, and cultural activities (ibid).

Meanwhile, identifying the so-called creative economy as a substantial contributor to Thailand's economy, and following similar trends observed in other developing countries (from Brazil to China), in 2009, the Thai government launched an ambitious program for a Creative Thailand. That launch was followed up in 2010 with the establishment of the Thai Creative Economy Office (Howkins, 2011). Underpinning the strategy were the national policies on inclusion and equality. With greater funds allocated to creative sectors, and given the strength of Thailand's informal economy, the aim was to foster participation from a wide variety of agents across Thai society, forging a creative ecology where more people would "have a chance of expressing themselves, sharing ideas and working determinedly to succeed" (Howkins, 2011).

Both initiatives came together when members of the Bangkok River Partners saw the opportunity to reach out to both the government and local communities to promote decaying yet historically rich and culturally vibrant areas along the river, turning them into a combined creative precinct. A new public-private partnership was then devised to drive new cultural initiatives in the area, leading to the establishment of the Creative District Foundation in 2017.

Purpose

The Bangkok Creative District focuses on two of the city's oldest neighborhoods, located along the same section of the Chao Phraya River, south of the city center. To the east of the river, Bang Rak is one of Bangkok's oldest commercial hubs, home to Chinese, Christian, Muslim, and Thai communities (Fernandez, 2016b). It is also the site of the former British Legation and still retains traces of European design, including Charoen Krung Road, Bangkok's first paved road, cutting through it. It is also home to Thammasat University—a partner of the Creative District Foundation, contributing research and resources to the foundation and precinct. At the west margin is the Khlong San neighborhood, traditionally more residential and still largely underdeveloped, despite being also the site for ICONSIAM, a huge mixed-used development including a shopping mall and two of the tallest buildings in Bangkok (ICONSIAM, 2018).

The Creative District Foundation seeks to connect a wide range of stakeholders in Bang Rak and Khlong San, identifying opportunities for innovative approaches to culture and coordinating the resulting collaborations. It reaches out to Thammasat University to gather research knowledge capable of informing new projects. It liaises with the Bangkok Metropolitan Authority to canvass government support, ensure approvals, and align initiatives along government priorities. It advises entrepreneurs interested in developing projects in the precinct, and it engages with various community sectors, from business to street artists and students, to nurture codesign, open discussions, and realize events, enabling them to showcase their creative outputs.

Strategy

With the core goal of developing a creative ecosystem connecting the various stakeholders in Bang Rak and Khlong San, the Bangkok Creative District adopted a strategy of striking strategic partnerships that could lead to the establishment of permanent hubs in the district. The objective was to promote and coordinate regular creative activities in the district, as well as incubate new businesses, that is, supporting local traders and artisans to "set up shop." In particular the group shortlisted six tracks: art, design and digital, property preservation, urban planning, community, and food (Fernandez, 2016a).

As in any middle-out engagement initiative, the first challenge faced by the Bangkok Creative District was to gain the trust of potential participants and prove the strength of the concept. To do so the group has regularly held town hall meetings inviting audience with a wide range of profiles, from local community workers, shopkeepers, and university students to foreign embassy staff (Fernandez, 2016a).

Two key enablers of the Bangkok Creative District's activities at a community level are the Thailand Creative and Design Center (TCDC) and the architect and entrepreneur Duangrit Bunnag. Founded in 2005, TCDC's mission is to integrate traditional Thai culture with modern knowledge and technology (TCDC, 2017). The center also oversees the Creative Economy Agency, established in 2018 with the mission of promoting creative sectors as the "driving force to a balanced and sustainable economy" (ibid). Duangrit Bunnag is one of Thailand's leading architects and original cofounder of TCDC. He is also the founder and director of two of the main incubation hubs in the district—The Jam Factory and Warehouse 30—which we further discuss later. Another strategic partnership is with Thammasat University, which provides the ecosystem with research resources and support.

The key partnerships previously mentioned also led to the establishment of permanent creative hubs in the district, driving ongoing design innovation. In early 2017, TDCD moved its headquarters to the Grand Postal Building in Bang Rak, backed by Thailand's Department for Architecture. The goal was "to create at TCDC a space that felt open and facilitated the exchange of ideas, playing up the permeable boundaries of disciplines within design and the arts" (Phelps, 2017).

The Jam Factory (Fig. 9) was founded at Khlong San in 2014 by architect Duangrit Bunnag as the new office for his DBALP architectural firm. The site, at the margins of the Chao Phraya River, was a former site of a battery factory,

FIG. 9 The Jam Factory in Bangkok. *(Credit: Luke Hespanhol.)*

ice mill, and medicine plant, with abandoned warehouses and wide, open space. Bunnag spotted an opportunity to develop the space further as a creative incubator by inviting young local artists to occupy the space while he, in return, would offer industry advice. The initiative rapidly evolved into a multidisciplinary hub housing not only the DBALP offices but also restaurants, a bookstore, a café, a boutique furniture shop, a fashion label, and an art gallery (The Jam Factory, 2019).

Following the success of The Jam Factory, Duangrit Bunnag expanded the model to Bang Rak, on the other side of the river, transforming a 4000-square-meter space of seven abandoned World War II–era warehouses into a new creative hub: Warehouse 30 (Fig. 10). Opened in 2017 the place is now a mixed-used space, home to coworking spaces, cafés, artisan shops, and a regular cultural program including film screenings, yoga classes, and TEDxCharoenkrung (Bangkok River, 2019).

Outcome

In addition to the actively creative businesses incubated by The Jam Factory and Warehouse 30 hubs, a program of regular events has taken roots in the Creative District, ensuring a continuous community engagement and nurture of local talent. The TCDC, in particular, has been instrumental in leading the way as a cultural center by grassroots events and codesign workshops. For example, in January 2016, the *Bukruk Urban Arts Festival*[b] brought to the communities a 10-day event featuring public wall paintings, art exhibitions, workshops, artist talks, live music, animations, and projection mappings. Another initiative,

b. http://bukruk.com/festival/.

FIG. 10 Warehouse 30. *(Credit: Luke Hespanhol.)*

Co-Create Charoenkrung, involved a series of workshops run from December 2015 to June 2016, led by researchers from Thammasat University, among others. It involved interviews with members of the community, followed up by focus groups and workshops where participants engaged in participatory design activities with the goal of identifying issues and elaborating initial approaches to them. Subsequent iterations of the workshops refined the ideas further, evolving them to models and eventually proposals for turning Charoenkrung into a creative precinct. In addition to brainstorming, discussions, and ideation, a "co-create test day" was organized in May 2016, including urban prototyping and art in the form of live music, square dancing, bespoke street signage and visual identity, street performances, outdoor movie sessions, projection mapping on buildings, and interactive light installations (TCDC, 2017).

The Bangkok Creative District is now a well-established ecosystem in the emerging creative economy in Thailand (Fig. 11). While grassroots programs and incubation continues to take place—such as *The Knack Market*,[c] run monthly at The Jam Factory with over 100 local shops—some more high-profile events are gaining traction.

In January 2018 the TCDC ran the first Bangkok Design Week,[d] following the successful model adopted by many other global cities worldwide. The event consolidated its presence in a second edition in 2019, following the launch of Bangkok to World Design Capital in 2022. Likewise, new partners have joined and invested in the Creative District—such as the large development ICOSIAM, which, while transforming Khlong San, is also promoting local communities

c. http://theknackmarket.com/.

d. http://www.bangkokdesignweek.com/.

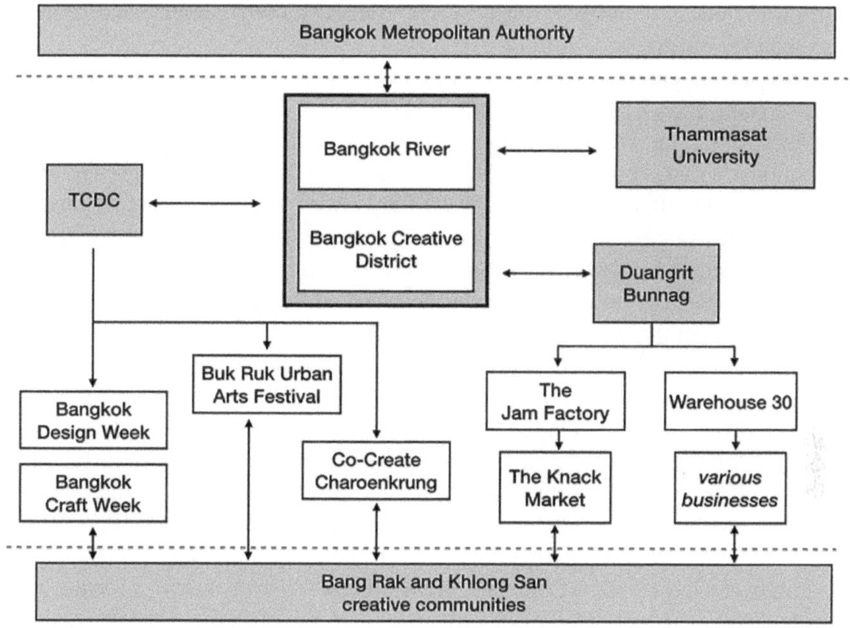

FIG. 11 Bangkok Creative District ecosystem. *(Credit: Luke Hespanhol.)*

through the realization of the inaugural Bangkok Craft Week in 2019. This was inspired by a similar (and arguably highly successful) event elsewhere, the London Craft Week.

Opportunities

The very successful rejuvenation of Bang Rak and Khlong San through creative city-making points to an increasing awareness across the ASEAN (Association of Southeast Asian Nations) region about smarter ways of developing cities. Like elsewhere, approaches to date seem to echo a tension between two complementary, yet very distinct paradigms: (a) placemaking, often articulated at street level by community actors, and (b) the smart city, generally involving top-down government masterplanning and infrastructure rollout.

In 2016 the Thai national government announced *Thailand 4.0*, an ambitious 20-year program to turn the country into a value-based economy driven by creative innovation, and based on smart industry, in smart cities with smart people (Languepin, 2016). In line with that initiative, Bangkok was inducted in April 2018 into the first cohort of the ASEAN Smart Cities Network (ASCN) (Gnanasagaran, 2018). Likewise, new top-down greenfield smart city pilot projects are currently underway, such as the One Bangkok[e] "city-within-a-city" district, which lists as its core values the integration of smart technologies into the

e. onebangkok.com.

district, including smart parking, traffic and energy management, and centralized security, safety, and cooling.

The site for One Bangkok is in a relatively central area of Bangkok, very close to Bang Rak. The fact that it has been carved from within the existing city is reflective of some of the challenges in upgrading existing infrastructure, such as Bangkok's outdated electricity network. Rolling out smart solutions to older neighborhoods, often culturally rich yet very poor in resources and infrastructure, becomes largely impossible without large urban redevelopment—which, while offering the possibility of "restarting planning from scratch," also risk the disappearance of preexisting cultural, social, and economic networks forged over centuries in the downtown areas along the Chao Phraya River.

Symptomatic of these growing concerns, ASEAN's first placemaking conference, Placemaker Week ASEAN, was held in November 2019 in Kuala Lumpur, cohosted by the United Nations Human Settlements Programme (UN-Habitat). Among the conference's core concerns were "how best to create a sense of place and belonging in cities, how to effectively forge social connections beyond homes and how to develop meaningful experiences with others as well as with our surroundings" (Thomas, 2019). In the context of Bangkok, the initiatives led by the TCDC and the Bangkok Creative District for the past decade stand as clear examples of successful creative placemaking rejuvenating a large historical, yet previously neglected area.

We argue that one factor particularly crucial to their success—and one that should be nurtured further—is the wide support network of collaborators from government, academia, local businesses, and community representatives and centered around entrepreneurs such as Duangrit Bunnag and key champion organizations such as TCDC and the Bangkok Creative District. Their role as mediators allows them to become central delegates discussing with government and academia on behalf of the local communities while also leveraging on community support to enact actual changes on the ground. More importantly, it enables significant improvements to the livelihood of those communities through the development of economic and cultural capital tapped from skills already existing in the community, in preparation for the (much slower) overhaul of public infrastructure and deployment of inclusive digital platforms. Combining local action with global strategy is, after all, a smart way of optimizing the wide range of resources available in the city, both tangible and intangible.

Toward middle-out outcomes

Applying a middle-out approach to facilitate collaboration between top-down and bottom-up stakeholders is fundamental to putting a diverse range of people back into the smart. The more we can see collaborative community-led or middle-out cases, the more we need to observe and track the power relationships happening between top-down and bottom-up stakeholders through our presented Middle-Out Engagement Framework. By acknowledging and defining the roles through the framework, we can aim to improve the relationships developed between stakeholders. This includes the flow of information between all the parties involved; then the workflow required to address common objectives; and continuously measuring the impacts. The cases of Guadalajara and Bangkok reveal that this process can be traced and it is not repetitive nor straightforward. It emerges according to the context and the needs of the different stakeholders.

In the case of Guadalajara, the middle-out approach was not included in the CCD masterplan in the initial phase of the project due to top-down decision making. However, it was incorporated at a later stage in response to public backlash from bottom-up stakeholders. The Civil Association was established by the Mexican government to assume the role of the mediator to facilitate a middle-out approach and by incorporating employees who identified with the bottom-up stakeholders, such as artists themselves. This shift in the engagement approach enabled the Civil Association to create credibility and establish trust with the wider community. Therefore its efforts were directed at generating common goals and outcomes beneficial for all stakeholders involved.

In the case of Bangkok, the middle-out stakeholders were the ones driving the engagement process from the beginning by coming together as a collective to create the Bangkok Creative District. They saw the opportunity to align the government agenda with the needs of the local area and its people. By carefully balancing these interests, the Creative District worked with the renowned architect, Duangrit Bunnag, to develop creative hubs that were central in bringing people together.

By clearly articulating the role of the mediator, of middle-out stakeholders, and of the middle-out engagement approach in the framework, we can see and share the benefits associated with these actors who actively bring the needs of the people from the "bottom" with the interests of the decision makers at the "top." The two cases reveal that the middle-out approach used by the Civil Authority in Guadalajara and the Bangkok Creative District involved strong levels of collaboration, including the creation of events and platforms that provided creatives outlets to share their work, bring people together, and learn from one another. This process allowed the community to engage in dialogues about the smart city and opportunities for digital technologies that were important to them while simultaneously establishing stronger presence and momentum around the creative industries—an interest of the Mexican and Thai governments as part of their smart city agendas.

Implementing middle-out engagement

Collaboration, cocreation, and collective knowledge are essential for future practice in designing, developing, and deploying "smart" initiatives, policies, and technologies. Through increased globalization and connectivity, contemporary society continues to explore "smart" opportunities, such as artificial intelligence, machine learning, and Industry 4.0. However, it is important to recognize the key to a smart future is the human factor by celebrating people, culture, and traditions.

The Middle-Out Engagement Framework has the potential to be applied across various contexts and scales beyond city-making within the smart city. This may include any type of engagement that involves mediation between different groups of people in the decision-making process; this would be most needed, for example, in projects or endeavors where mediation between local and external stakeholders requires the building of rapport and trust around the project goals.

Our previous work in developing the Middle-Out Engagement Framework was situated within the Australian context (a developed country). We successfully demonstrated that middle-out engagement has the potential to play a significant role in enhancing collaboration in the city-making process, regardless of geographic location, socioeconomic circumstances, and political influences. It was our intention to further explore its application within the so-called global south (developing countries) to bridge the gap between spheres of discourse, knowledge, and culture regardless of global circumstances and status.

The researchers and designers involved in these cases have acted as the mediators between top-down and bottom-up stakeholders. Through our analysis, we have examined the benefits and challenges that they have faced when designing, developing, and deploying what we refer to as "middle-out" approaches and the valuable—yet often overlooked—role they play in the city-making process. In the presented cases, we have discussed the application of the Middle-Out Engagement Framework. This has assisted us in determining key aspects of the approach with the aim of developing it from a "concept," to a "research method," to "everyday practice."

References

Acosta García, R.G., Larrosa-Fuentes, J., Paláu Cardona, M.M.S., 2014. Decisiones públicas sin diálogo público: Análisis de los argumentos sobre el caso de la Vía Exprés vertidos en la prensa de Guadalajara. Comun. Soc. 21, 139–159.

Ahmed, E., Yaqoob, I., Gani, A., Imran, M., Guizani, M., 2016. Internet-of-things-based smart environments: state of the art, taxonomy, and open research challenges. IEEE Wirel. Commun. 23 (5), 10–16.

Anastasiu, I., 2019. Unpacking the smart city through the lens of the right to the city: a taxonomy as a way forward in participatory city-making. In: de Lange, M., de Waal, M. (Eds.), The Hackable City. Springer, Singapore, pp. 239–260.

Avalos, J., 2016. El sujeto político juvenil: prácticas technopolíticas y saberes en la experiencia de activism de jóvenes en México, 2012–2016. Argumentos 18, 120–148.

Bangkok River, 2019. Warehouse 30. Retrieved from: http://www.bangkokriver.com/place/warehouse-30/.

Bisker, S., Gross, M., Carter, D., Paulos, E., Kuznetsov, S., 2010. Personal, public: using DIY to explore citizen-led efforts in urban computing. In: CHI'10 Extended Abstracts on Human Factors in Computing Systems.

Bodker, S., Pekkola, S., 2010. A short review to the past and present of participatory design. Scand. J. Inform. Syst. 22 (1), 45–48.

Bokor, M., 2016. Digital Transformation Underway in Mexico's Tech Hub. Retrieved from: https://newsroom.cisco.com/feature-content?type=webcontent&articleId=175240.

Brynskov, M., Heijnen, A., Balestrini, M., Raetzsch, C., 2018. Experimentation at scale: challenges for making urban informatics work. Smart Sustain. Built Environ. 7 (1), 150–163.

Caldwell, G.A., Foth, M., 2014. DIY media architecture: open and participatory approaches to community engagement. In: Proceedings of the 2nd Media Architecture Biennale. ACM, pp. 1–10.

Caldwell, G.A., Osborne, L., Mewburn, I., Crowther, P., 2015. Guerrillas in the [urban] midst: developing and using creative research methods-guerrilla research tactics. J. Urban Technol. 22 (3), 21–26.

Caldwell, G.A., Guaralda, M., Donovan, J., Rittenbruch, M., 2016. The InstaBooth: making common ground for media architectural design. In: Proceedings of the 3rd Conference on Media Architecture Biennale. ACM, p. 3.

Caragliu, A., Del Bo, C., Nijkamp, P., 2011. Smart cities in Europe. J. Urban Technol. 18 (2), 65–82.

Clarke, R., Heitlinger, S., Light, A., Forlano, L., Foth, M., DiSalvo, C., 2019. More-than-human participation: design for sustainable smart city futures. Interactions 26 (3), 60–63.

Corona, S., 2017. El Silicon Valley Mexicano está en Jalisco. El país. Retrieved from: https://elpais.com/tecnologia/2017/03/12/actualidad/1489275848_767120.htm.

Cuidad Creativa, 2012. Digital Brochure. Retrieved from: http://www.carloratti.it/FTP/CCD/files/CCD_brochure.pdf.

de Padilla la Torre, M.R., Flores Márquez, D., 2011. El studio de las prácticas políticas de los jóvenes en internet. Communición y Sociedad 15, 101–122.

Dow, A., Comber, R., Vines, J., 2019. Communities to the left of me, bureaucrats to the right: here I am, stuck in the middle. Interactions 26 (5), 27–33.

Fernandez, D., 2016a. Bangkok Creative District: Interview With David Robinson. Retrieved from: https://culture360.asef.org/.

Fernandez, D., 2016b. Bangrak—Klongsan Creative District Bangkok. Retrieved from: https://culture360.asef.org/.

Finn, D., 2014. DIY urbanism: implications for cities. J. Urban. 7 (4), 381–397.

Fredericks, J., 2020. From smart city to smart engagement: exploring digital and physical interactions for playful city-making. In: Nijholt, A. (Ed.), Making Smart Cities More Playable. Springer, Singapore, pp. 107–128.

Fredericks, J., Tomitsch, M., Hespanhol, L., McArthur, I., 2015. Digital pop-up: investigating bespoke community engagement in public spaces. In: Proceedings of the Annual Meeting of the Australian Special Interest Group for Computer Human Interaction. OzCHI.

Fredericks, J., Caldwell, G.A., Tomitsch, M., 2016a. Middle-out design: collaborative community engagement in urban HCI. In: Proceedings of the Annual Conference of the Australian Computer-Human-Interaction Special Interest Group. ACM.

Fredericks, J., Hespanhol, L., Tomitsch, M., 2016b. Not just pretty lights: using digital technologies to inform city making. In: Proceedings of the 2016 Media Architecture Biennale. ACM.

Fredericks, J., Hespanhol, L., Parker, C., Zhou, D., Tomitsch, M., 2017. Blending pop-up urbanism and participatory technologies: challenges and opportunities for inclusive city making. City Cult. Soc. 12, 44–53.

Fredericks, J., Caldwell, G., Foth, M., Tomitsch, M., 2019. The city as perpetual beta: fostering systemic urban acupuncture. In: de Waal, M., de Lange, M. (Eds.), The Hackable City. Springer, Singapore, pp. 67–92.

Garcia, R.A., 2018. Toma-la ciudad: intersubjective activism in Guadalajara's stresses and city museum. J. Latin Am. Caribb. Anthropol. 42 (1), 221–241. https://doi.org/10.1111/jlca.12386.

Gnanasagaran, A., 2018. Smart City Spotlight: Bangkok. Retrieved from: https://theaseanpost.com/article/smart-city-spotlight-bangkok.

Haber, P.L., 2006. Power From Experience: Urban Popular Movements in Late Twentieth-Century Mexico. Penn State Press, University Park.

Han, H., Hawken, S., Williams, A., 2015. SMART CCTV and the management of urban space. In: Harrison, D. (Ed.), Handbook of Research on Digital Media and Creative Technologies. Information Science Reference, Hershey, pp. 430–447.

Hou, J., 2010. Insurgent Public Space: Guerrilla Urbanism and the Remaking of Contemporary Cities. Routledge, Abingdon.

Hou, J., 2020. Guerrilla urbanism: urban design and the practices of resistance. Urban Des. Int. 25, 117–125. https://doi.org/10.1057/s41289.

Howkins, J., 2011. Developing Thailand's Creative Economy. Retrieved from: https://www.nesdb.go.th/ewt_dl_link.php?nid=3789.

ICONSIAM, 2018. About Us. Retrieved from: https://www.iconsiam.com/en/about_us.

Kinchla, R., Wolfe, J., 1979. The order of visual processing: "top-down," "bottom-up," or "middle-out". Percept. Psychophys. 25 (3), 225–231.

Languepin, O., 2016. Thailand 4.0, What Do You Need to Know? Retrieved from: https://www.thailand-business-news.com/economics/54286-thailand-4-0-need-know.html.

Muller, M.J., 2003. Participatory design: the third space in HCI. Hum.-Comput. Interact. Dev. Process 4235, 165–185.

Muller, M.J., Kuhn, S., 1993. Special issue on participatory design. Commun. ACM 36 (6), 24–28.

Navarrete, C., 2018. Gobernanza de distritos creativos y culturales en metrópolis latinoamericanas. Universidad de Guadalajara, Guadalajara.

Phelps, L., 2017. TCDC and the Bangkok 'Creative District'. The Artling. Retrieved from: https://theartling.com/en/artzine/tcdc-and-the-bangkok-creative-district/.

Restrepo, J., 2018. Nuevos modelos de desarrollo urbano: Medellin cuidad digital, Guadalajara cuidad creativa digital: Una mirada desde las expresiones asociativas juveniles. Colegio de la Frontera Norte.

Shafique, K., Khawaja, B.A., Sabir, F., Qazi, S., Mustaqim, M., 2020. Internet of things (IoT) for next-generation smart systems: a review of current challenges, future trends and prospects for emerging 5G-IoT scenarios. IEEE Access 8, 23022–23040.

Shelton, T., Lodato, T., 2019. Actually existing smart citizens: expertise and (non) participation in the making of the smart city. City 23 (1), 35–52.

TCDC, 2017. About Us: Philosophy. Retrieved from: https://web.tcdc.or.th/en/aboutus.

The Jam Factory, 2019. Space. Retrieved from: https://thejamfactory.life/space/.

Thomas, J., 2019. Placemaking: Creating Better Cities for All. The Asean Post. Retrieved from: https://theaseanpost.com/article/placemaking-creating-better-cities-all.

Tschimmel, K., 2012. Design thinking as an effective toolkit for innovation. In: ISPIM Conference Proceedings. The International Society for Professional Innovation Management (ISPIM), p. 1.

UNESCO, 2017. Creative Cities Network: Guadalajara. Retrieved from: https://en.unesco.org/creative-cities/guadalajara-0.

Wathne, M.W., Haarstad, H., 2020. The smart city as mobile policy: insights on contemporary urbanism. Geoforum 108, 130–138.

Author biographies

Dr. Glenda Amayo Caldwell is Associate Professor in Architecture at the School of Design, Creative Industries Faculty, Queensland University of Technology (QUT). She is an architecture and design scholar with expertise in physical, digital, and robotic fabrication, leading Industry 4.0 innovation through human-centered research in design robotics and media architecture. Caldwell leads the QUT Design Lab's Designing Creative and Resilient Communities Program and is the author of numerous publications in the areas of media architecture, urban informatics, and design robotics.

Dr. Joel Fredericks is a Lecturer in Design at the University of Sydney's School of Architecture, Design, and Planning. Fredericks's research is transdisciplinary and sits across the domains of digital placemaking, media architecture, urbanism, smart cities, and immersive technologies. Fredericks has investigated collaborative and creative approaches to designing and deploying urban interventions that create curiosity and encourage people to playfully interact. He has authored and coauthored many publications in journals, edited books, and conference proceedings.

Dr. Luke Hespanhol is an interaction designer, lecturer, researcher, media artist, and producer. His research ranges from creative technologies, digital storytelling, digital placemaking, and technology-mediated social interactions, to urban robotics, media architecture, and smart cities. He is the lead designer and curator of the Footbridge Gallery at the University of Sydney, and a member of the Media Architecture Institute.

Associate Professor Marianella Chamorro-Koc is an Industrial Design Researcher and Educator. Her work aims to contribute to the design of enhanced people interactions with everyday technologies. Chamorro-Koc investigates the experiential knowledge embedded in people's activities and the contextual aspects shaping them, and her applied research is positioned in the design of technologies applications to help people manage their health and wellbeing. Chamorro-Koc leads the QUT Design Lab Designing for Health Program and is a member of the Design Research Society (DRS).

María José Sánchez Varela Barajas is a Mexican philosopher, dancer, entrepreneur, and researcher. She has a concentrated interdisciplinary work portfolio and publications in the field of artistic embodied experience and phenomenology grounded in the philosophy of Edmund Husserl and Merleau-Ponty. Barajas led the successful bid of Guadalajara as UNESCO city of Media Arts, involving a unique collaboration across Academia, Industry, Governance, and citizens. She mentors emerging entrepreneurs at several universities across Mexico. Barajas also cofounded Culturista, a digital culture, and tourism startup. This innovative proposal continues its mission to research about the future of intelligent cities and the interrelationship of how we cohabit spaces.

María José Castelazo André is born on May 23, 1983 in Cuernavaca, Mexico. The André family migrated to Guadalajara, the city where she grew up, studied and worked for the next few years. Seeking justice, she studied law, but following Oscar Wilde's steps, she participated in more than 30 theatrical productions. With the conviction to learn from other cultures, she engages with tech-related communities, and participates all over the world in workshops specialized in Smart Cities as a solution that combines with the arts in efforts to make the world better.

Chapter 14

Digital twins of cities and evasive futures

Paul Cureton and Nick Dunn
Imagination Lancaster, Lancaster University, Lancaster, United Kingdom

Chapter outline

Smart cities evolution(s)

Digital twins of cities are sensing systems enabled by the IoT, data analytics, and machine learning. They are typically managed via control rooms or cloud and mobile open data platforms, with interactive virtual models that enable procedural testing and scenarios. Situated as an element of smart cities, digital twins are precise simulations of a physical asset providing an "ability to simulate the behaviour of the system in digital form is a quantum leap in discovering and understanding emergent behaviour" (Grieves and Vickers, 2017, p. 90). A digital twin is a "mirror" that connects "real space, virtual space, the link for data flow from real space to virtual space, the link for information flow from virtual space to real space and virtual sub-spaces" (ibid, p. 85). This reflects the predictions of computer representations proposed by (Gelernter (1993)), in the creation of "mirror worlds," where particular physical information becomes delocalized and data are held in a cloud server.

The general nonurban concept of digital twins was initially developed by NASA to create technology roadmaps along with the Defense Advanced Research Projects Agency (DARPA) for emerging military technology. They are intended for computational simulation of materials and their complexity, with particular application in manufacturing to enable failure detection within systems primarily on aerospace vehicles. Developed by large corporations such as IBM, CISCO, ANSYS, GE Digital, Microsoft, and Siemens, among others, digital twins run simulations of physical assets and assess their performance allowing "what-if"

Shaping Smart for Better Cities. https://doi.org/10.1016/B978-0-12-818636-7.00017-2

scenarios. The prototype or enhancement to a physical asset can be designed in virtual space; thus a product or process can be tested for its future life cycle, performance, or maintenance without extensive costs (Ayres, 2011).

Naturalized computational development has allowed the scaling of the notion of a digital mirror to city-scale and establish the idea of a national network of city digital twins (Rogers, 2019). Such work is nascent, and it is this emerging type of city digital twins that forms the focus of this chapter. While there is significant momentum in policy toward the concept previously described, this chapter will explore what digital cities are in their current form to examine the challenges and barriers of realizing this ambition. City digital twins are emerging from the fusion of geographic information systems (GIS) and building information modeling (BIM) data of buildings to form digital replicas at a regional scale. The current iteration and GIS/BIM fusion has many benefits and is an important step toward the creation of a digital replica (Laat and van Berlo, 2011).[a] However, given the many issues leveled at the heterogeneous portfolio of global projects termed "smart cities" since 1992 (Gibson et al., 1992), it is appropriate that a cautious approach and thorough critique of the digital twin concept be conducted, particularly in terms of planning, placemaking, and urban design, to fundamentally explore a core question: the relationship between place and digital technology.

With such large-scale ICT projects underway in several countries, the design implications and social relations with digital twins deserve unpacking. This investigation requires a discussion of criticism of city digital twins under the "smart city" umbrella to identify broad issues and pitfalls. The type of data used to construct the city digital twin is also of critical importance. Sensors may be embedded within the physical environment, and the type of information and its ethical implications raise design issues for places. Likewise the informatics gathered from embedded sensors in urban environments can provide insights into the social relations of places. The city digital twin has resulted from a technological convergence, which results from a "stacking" of technologies (Bratton, 2016). With this "stacking," it is important to consider whether this city replica is desirable or achievable and what effects will the city digital twin have on future urban spaces (Fig. 1).

Sensing and data selection

City digital twins are part of the development of smart cities, the latter having been subject to various criticisms that warrant further scrutiny here. Where smart cities have received opprobrium is in their generalized, monolithic top-down approach to cities, which focuses on technological development and progress over social concerns. For our discussion, this approach is problematic in

a. See GeoBIM, Ken Arroyo Ohori, Thomas Krijnen, Abdoulaye Diakité, Hugo Ledouxand, Jantien Stoter, *GeoBIMProjectFinalReport* (2018). TU Deflt.

FIG. 1 Virtual Singapore, © 2019 Government of Singapore. Dassault Systèmes 3DEXPERIEN-City. Virtual Singapore is a dynamic three-dimensional (3D) city model and collaborative data platform, including the 3D maps of Singapore.

two principal ways. Firstly, the emphasis on complexity in narratives surrounding smart cities typically refers to the challenges of handling big data rather than the actual intricacies of urban life. These intricacies include the movement and interactions of people who are less predictable than assumed. Secondly, having its origins in management software, the premise of a solution-orientated and adopted language of systems, processes, and outputs suggests that existing places can become more efficient through a flattening of layered realities that can be replicated elsewhere. This ignores complex knowledge from more than a century of urban planning history and analysis. In addition, we know that cities are not neutral containers that would be "fixed" with the right software platforms but are heterogeneous in their historical and spatial characteristics. Despite these significant issues, there has been an explosion in the investment of smart city development as an urban panacea for the early 21st century (Dunn and Cureton, 2020). This ongoing quest for totality, control, and management of urban systems in visions for smart cities resides on the seductive basis that real places operate in the way that complex algorithms simulate.

As Bruce Sterling has identified, there are much wider debates on smart city performance to be undertaken (Sterling, 2018). Margarita Angelidou through a literature review of major contributors to the research field cites five main criticisms of smart cities: firstly their conceptual and methodological ambiguity; secondly ICT-centric and corporate-driven utopian visions; thirdly the overlooking of citizen and stakeholders, splintered urbanism, and unequal representation; fourthly, privacy and security concerns; and fifthly lack of long-term vision for sustainable urban development adapted to local needs (Angelidou, 2017, p. 79).

Digital twins as part of the "smart city" movement are not exempt from these five concerns. With applications in healthcare, transportation, geography, and energy sectors, the digital twin market is expected to grow from USD 3.8bn in 2019 to USD 35.8bn by 2025 at a compound annual growth rate (CAGR) of 37.8% (Research and Markets, 2019). From this purview, city digital twins certainly appear to be a future trajectory for governments and global corporations. However, the idea that such digitization can create a digital twin of a city or urban area in such a scale is questionable, and focus is required on the useful elements for near-future urban design and placemaking.

Rather than accepting the entirety of the concept as part of a commercial driver for digital services, something that Luca Mora has shown and that arguably befell smart city definitions, we need to understand the agency in shaping smart places. The smart city research base remains fragmented on what a smart city *is*, from holistic, human-centered to divided and predominately corporate ICT technological vision (Mora et al., 2017, pp. 19–21). Current city digital twin data selection has focused on transport planning and digital planning services and energy sectors. While the smart city and IoT have brought an abundance of data, digital twins seek to join isolated packets for efficiency gains. Data selection and fusion is arguably important in these areas due to advanced complexity and more comprehensive records. Behavioral factors of citizens and their interactions with public space are a much more challenging area of collection, both ethical terms and in the unpredictability of the public. As Rikke Gram-Hansen writes,

> *Users of virtual and public spaces have their own logic, different from the designers'. If we cannot design or program cities or software to function the way we want, then how should we plan for the unplannable?*
>
> Gram-Hansen in Geng (2017, p. 36)

These unplannable elements represent the biggest void in city digital twin ambitions. Simulations such as Space Syntax, a software platform based on theories of spatial behaviors and movement conceived by Bill Hillier and Julienne Hanson, and Oasys pedestrian, a structural and geotechnical simulation software by Arup, make inroads to reduce the unknowable. However, the fluidity of human behavior in urban space and predictive futures remains a challenge.

Urban data, analytics, and real-time mapping

The embedding of sensors within places provides real-time information and networks, but this only addresses part of a city's real-time complexity (Fig. 2). The use of digital replicas of processes, products, or services forming a model, fed by sensors, varies across many sectors. For example, a digital wind farm provides real-time data and networks multiple sensors with artificial intelligence (AI) and data analytics to increase efficiency or varied sensors. Sensors can also be used to monitor a building, its energy consumption, and mobility within it

FIG. 2 SWA, Duffy Square, New York (designed by MNLA, Snohetta, completed 2017). The color ramp high traffic features in green and low traffic in light blue.*(From "Field Guide to Life in Urban Plazas," 2019, pp. 94–95.)*

using machine learning to aid maintenance and increase building performance. Sensors may also record pattern behavior for design enhancement. An example is SWA Group's publication *Field Guide to Life in Urban Plazas* (2019), based on William Whyte's influential study *The Social Life of Small Urban Spaces* (Whyte, 1980) that transformed New York's open space and parks and sought to produce design guidelines and frameworks for open space. Using Whyte's original techniques, as well as utilizing machine learning, the study created user heat maps and pedestrian counts. One of the behaviors meant that:

> *Stationary people tended to go to elevated areas that overlooked the plaza space. The youngest gravitated towards higher perches (3′+), and the middle-aged were drawn to slightly raised spaces (<3′). Protected perches with canopies and a sense of backing were most popular.*
>
> Schlickman and Domlesky (2019, p. 43)

The use of an elevated ten-foot-long pole with a mounting for a GoPro camera allowed machine learning algorithms to produce an overlay heat map in which the various user patterns were shown and derived design tactics were formed. The resulting design tactic from the study suggests constructing terraces and stacked theatrical seating. Imagining (or envisaging) this example within a digital twin context would mean that this single sensor would be amplified into thousands by being fused and combined within a network of sensors. These sensors could calculate facial recognition and behavior, transportation flow, and climate data among many other things to form a complex machinic city. Thus the designer both shapes public space and learns through sensor feedback on the impact of such decisions. This case highlights an approach to the shaping of space using "smartness."

The exciting possibility for reading and reporting on designed space could be accomplished by the entirety of a city-scale cyber-sensor-social complex system. City information models (CIMs) may be the next iteration of smart cities, but smart cities according to a CISCO report have predominately faced numerous challenges at the pilot stage, such as funding, adequate technology platforms, and lack of consumer demand for devices such as smart meters and building control systems (Clarke, 2013). These factors reveal negative perceptions of emerging technology, linked to privacy concerns. For example, Google Replica uses deidentified mobile data to provide simulations of population movements, including transport choices, and purpose of travel as an urban management tool. The smart systems' utopian driver is based on gaining increased insights into behavioral patterns but can mean at the same time opting into urban data collection without understanding the merits and exploits of the information collected. A clear example of this has been the growing tensions regarding data privacy for Google's parent company Alphabet and Sidewalk Labs. Their now abandoned Quayside waterfront smart development in Toronto, essentially embedding a digital twin that would also use Replica, had become a live arena for debates regarding data ethics and surveillance capitalism. Such initiatives exemplify the issues of bridging the gap between the digital and the real urban situation, that is, unachievable "frictionless futures in which the city is easily disassembled and reassembled into a coherent whole for operational optimization without question" (Dunn and Cureton, 2019, p. 19). Data are a commodity, yet we are developing cities' capacity to gather it in volumes unheard of, and the rights, trade, and governance of that commodity will become critical in implementing city digital twins. In addition to this, the possibilities are highly resourced and intensive. Making digital twins (CIMs) a key driver of smart cities locks these process and systems into the future direction of the city.

An example is Amaravati, the new capital of the Indian state of Andhra Pradesh. The city was set to be the first born with a digital twin. This ambition has been scrapped due to a range of political, social, and environmental issues (WEF, 2018). The initial 3D prototype of the city was built using Cityzenith's Smart World Pro software. However, for this model, a "real" representative where further data sets are overlaid requires constant updates and large teams for assembly and maintenance. For example, Bentley Systems views the development of real-time 3D reality assets for digital twins as paramount, highlighting the process noting that:

> *Overlapping photographs and (as needed) supplemental laser scans, largely from drones and ground-level imagery, are processed to generate spatially-classified and engineering-ready reality meshes at any desired level of accuracy – within which each digital component can be automatically recognised and/or geospatially referenced. The reality mesh can provide an efficient and immersive visual*

"twin" to intuitively navigate for finding, viewing, and querying the associated information within, or related to, the asset's digital engineering models.

BIM+ (2018)

This requires error detection, artificial intelligence, and cloud processing alongside supplementary scans for a representational base to be hosted of a city. This equates to terabytes or petabytes of data storage. This problem in city model assembly is primarily why the GIS and BIM fusion represents a real success in reducing the generic description of places with detailed semantic city models. CIMs could have an important role within a broader fourth industrial revolution prediction (Schwab, 2017) and as an important information tool for widespread climatic challenges, though fundamental data capture of the populace needs a clearer set of defined outcomes for sociospatial relationships. The challenges of delivering upon these smart city ambitions are perhaps reflected in their turbulent development. In the case of the Quayside project in Toronto by Sidewalk Labs, the economic uncertainty in the real estate market following the outbreak of Covid-19 has meant it will no longer be developed. Amaravati meanwhile has had a similar fate following the withdrawal of major investors in 2019 and the lack of political will due to change of government at the state level.

Technological stacking and fusing

Virtual Singapore is an example of a digital twin, a semantic 3D digital model, called 3DEXPERIENCity. A collaborative platform for testbedding concepts rolled out in 2018 (Dassault Systems, 2018). The project, a $73m applied digital twin concept representation of the physical city, is reliant on three core factors: data, modeling, and visualization. Virtual Singapore is both a reality and a semantic model, meaning it has classified parts of its physical fabric including its materials and infrastructural elements. The model merges static and dynamic data to incorporate climate, population movements, and demographics on a collaborative platform. It is increasingly the source for consultation and the vehicle through which to enact and govern. However, the capabilities of the platform are resource intensive to establish and maintain. Singapore already has been an early adopter and vehicle of high-volume urban sensors and high levels of government support, making this project more sustainable.

Making digital twins a key driver of a smart city may turn these elements into critical infrastructural systems through a legacy of technological lock-in. These systems could even be unaffordable for other tiers and scales of cities. In other words, initiatives such as digital twin adoption and city-level digital ecosystems may be nascent manifestations of the future for megacities as they try to address complex urban issues but may also reinforce the power geometry in those places. Also the testbed for city digital twins to date has principally been in major cities. Scaling them down to smaller urban centers is unlikely to be a flaw-free exercise. For example, these initiatives are intrinsically linked to economic growth, but at the same time, we need to consider that these systems have a relationship with wider big data, digital divides, and inequality (Lutz, 2019), and therefore modes of measurement of these initiatives and outcomes are critical (Caird and Hallett, 2019).

What are the implications of digital twins for placemaking? The individuality and specificity of place, which could be defined as a specific and bounded geographical, social, and cultural area, require digital fidelity, and the question is whether we wish to replicate to this degree. The virtual model does not account for social dynamics, the unpredictable human behaviors of everyday life, but may capture infrastructural flows and dynamics. The agency of city digital twins, at least in its current form, enables predictive urban analytics for physical states of urban space more than the cities' human actants. This is crucial both in mobility terms, and for social sciences and urban design, the city digital twin is not a panacea for its critical climatic and future challenges.

Urban computation allowed the city to be virtualized in terms of the development of urban 3D models. This appeared novel at the time but was essentially a process of digitization that supported typically static models, representing a city that had already been constructed. But the dynamics of this process have changed drastically. Through the development of powerful software engines, urban management decisions are being undertaken, projecting future form, and thus the digital has moved into a mode of dominance that changes our relationships between robotics and the built environment such as Komatsu and Skycatch (2018) using drones and automated vehicles in construction or Melike Altınışık Architects and the Robot Science Museum construction (2019) (Cureton, 2020). In this inversed situation the city model has moved from an abstract generalized description to a virtual replica that helps predict future phases, plans, and maintenance, for example, LiDAR data running in a game engine and VR that allows for bridge inspection among many other roles (Omer et al., 2019).

Governments rush to define and establish frameworks, set policy, and invest capital in the development of digital twins. The rationale behind these decisions surrounding digital twin establishment goes far beyond the planning, organization, and management of urban assets, as evidenced in the growing interest in national digital twins. In *Radical Technologies: The Design of Everyday Life*, Adam Greenfield argues that a technological stack is creating more fused and interoperable elements (Greenfield, 2018, p. 273). We are so used to different dimensions of our lives being connected and networked that we seldom seek to question this development. This *fait accompli* is particularly prominent in cities, where the suggestion that urban development may be anything other than "smart" is increasingly rendered absurd as urban futures appear to be dominated by the concept (Dunn, 2018). To investigate this, we must unpack this history of city digital twins to understand their agency, drive for creation, and how they "stack."

Digital realities, realities, and finding futures

The city digital twin is primarily about process and information management (Figs. 3 and 4). However, in computer history, the concept can be traced back to two interrelated elements: a drive for "realism" and the development of geographic representation that "stack." This realism driver stemmed from computer graphics in engineering, automotive, and aerospace sectors such as Sketchpad

FIG. 3 Zoan, Virtual Helsinki, 2019. "Virtual tourism is a globally interesting and growing theme. Consumers are increasingly aware of the climate impacts of tourism and are keen to make responsible decisions. Virtual tourism also supports Helsinki's own goal of being a pioneer in sustainable tourism while reinforcing our reputation as a city that utilises the latest digital innovations" (Laura Aalto, CEO of Helsinki Marketing).

FIG. 4 City of Helsinki, 3D+ Helsinki, 2019. Open data. There are two next-generation 3D city models of Helsinki available: a semantic city information model and a visually high-quality reality mesh model.

by Ivan Sutherland and the work of William Fetter to simplify graphic design procedures. It developed later in architecture through AutoCAD and gaming such as Micropolis (later SimCity) and Wolfenstein 3D, among many others. The realism driver in digital representations of cities is best evidenced by Zoan's development of Virtual Helsinki, which is an immersive VR of city space, to which users may interact and socialize, as well as shop. This virtual citizenship for sustainable tourism requires a replica of public space and architectural heritage, and this reality capture is the fundamental vehicle for "believability" and "authenticity" in the space in which people interact. Zoan's work in part develops from the City Helsinki 3D+ project, which is based on a semantic model at LOD2 and a reality capture model of 42,810 images, both elements being released as open data (Kutzner and Kolbe 2018). The origins for the 3D city stemmed from the 1980s and became standardized through a data model and standard called CityGML and was constructed using ground and aerial-based laser scanning, drone surveys, and spatial registries. The latest development establishes a digital twin pilot in the Kalasatama area of Helsinki being viewable via the OpenCities Planner mobile platform, which allowed users to respond to questions and surveys based on particular locales. Wind simulation and solar analysis of building orientation were just a few of the virtual simulations of the new development. Unmanned aerial vehicle (UAV) surveys would also "top up" construction phases and provide a higher resolution of a survey than traditional methods (KIRA—digi., 2019). As Stephen Cousins comments on the Helsinki city digital information model,

> *Designers also benefit from open source access to an up-to-date city model in which they can interrogate early designs and create mock-ups within the environment and, in the process, get a better understanding of how their building will fit into the environment.*
>
> Cousins (2017, p. 105)

The reality models produced as part of urban digital innovation are intrinsically linked to visualization and geographic representation, and information in the form of the development of GIS contributes to the idea of a city digital twin. The geographer Roger Tomlinson, working as a photo interpreter in the 1960s, started developing a methodology to overlay spatial data using computers, creating the first computerized system the Canada Geographic Information System (CGIS) in 1971. The Laboratory for Computer Graphics at Harvard, led by architect Howard Fisher, created SYMAP (Synergistic Mapping, 1965/6), a vector-based mapping system, applicable to urban and environmental planning, to which Philip Lewis, Ian Mcharg, and Carl Steinitz all participated. GIS scales of operation and the establishment of 3D mapping more recently (through reduced costs of data capture and development of high-res sensors) could be combined or provide context models for BIM processes as standards and frameworks became established. Thus realistic and "accurate" models of space and place have come to the fore, combined with layering approaches from GIS. These systems stack information to provide an apparent cohesive picture of that environment alongside developments in building information modeling. This technological fusion is arguably the current basis of discussion of city-level digital twins.

Jorge Luis Borges (1946) wrote *On Exactitude in Science* where cartographic science was so developed and a map of the world was produced at 1:1 scale and eventually discarded (also developed by Jean Baudrillard and Umberto Eco), which raises similar questions on city simulation: do we need an exact copy of our world? The quest for accuracy and higher levels of detail does not prove the validity of a map or model. As Michael Batty comments,

> *such systems are models of one kind in that they represent the city in digital rather than material form and may be very close to the basic physical equivalents that make up the city. But they rarely include any of the processes that determine how the city works in terms of its social and economic functions.*
>
> Batty (2018a, pp. 817–818)

Batty sets out a problem that a city digital twin requires a simulation of a future tense on the object or process that requires decoupling from the physical. The paradox is that it needs to separate itself to project its future; the term city digital twin is something of a misnomer. Batty instead uses the term *model* over digital twin as this decoupling of a virtual replica is needed for testing. Further to this, Batty states, "digital twins are not identical twins and the notion of an exact mirror is an idealization that will never be achieved" (ibid, p. 819). Therefore digital twins may simulate future scenarios, and a model is one base parameter for a probable future. But how are these scenarios stretching the boundaries using modeling for fictitious, alternative, or radical proposals? As Batty again states, "One of the reasons for doing any of this is to invent new futures which may well never turn out to be 'real things' " (Batty, 2019, p. 599). The tension between the virtual model and its degree of realness is critical here.

Digital twins of cities are purported to allow feedback mechanisms on processes for planning and maintenance, yet how they inform design decisions transgresses back into the physical city in that it may require tactical urbanism, urban acupuncture, pop-up transformation, and temporary interventions. The danger is what seems like data-informed, and therefore neutral, urban planning decisions are undertaken based on compiled data sets, even perhaps rigorously processed, and ground-truthed, but perhaps exacerbate further inequality or disruption in what data are explicitly left out in terms of its sensor framework.

Martin Tomko and Stephan Winter further expose the problematic nature of the term, comparing the city digital twin paradox of a virtual replica function as more of a cyber-physical-social process. They propose that the digital-physical properties are more like an organism that can react, predict, and act. A more suitable term for the digital twin to Tomko and Winter is more akin to a metaphor of a brain organism, one that is fundamentally attached to reality—it is twinned (Tomko and Winter, 2018):

> *A cyber–physical–social eco-system relates to an organism with a brain: the digital counterpart is coupled with the physical realm (cyber–physical) by a nerve system of sensors, actuators, and (information-processing) communication lines enabling agency, and the digital representation interacts with people (cyber–social) through the physical environment including physical communication devices such as screens, enabling also the investigation of alternative physical systems (Ibid, p. 398).*

We are not yet at a point where the city digital twin can be fully delivered but can detect a drive toward this ambition. City-level information models (CIMs) are the first step toward digital twins and have clear benefits in an increasing fidelity and deeper levels of planning and design context. They provide a unique visual communication, apparent mirrors of the world to which wide audiences can make sense of. Singleton et al. suggest that there are three underlying modeling efforts: description, prediction, and explanation (Singleton et al., 2017, p. 100). A description is a mode of summarizing data and its relationships, while a prediction is a future projection of attributes and explanation tests/hypotheses. A CIM certainly has some of the analytical capability in GIS computation, likewise BIM through building performance and environmental calculations, but it cannot accurately predict that such a design and planning intervention is ultimately an improvement. This in itself is problematic, nor can current crowd and human behavior simulation accurately predict a "frictionless future" in which highly advanced simulations still repeat the fluidity of digital space rather than the complications, tensions, and discordance evident in the reality of urban life (Dunn and Cureton, 2019, 2020,). The abundance of "real" data arguably is not a direct correlate for improved design, nor can it remove the unpredictable or "black swan" events of near, middle, and far futures. There is a utopian assumption here that the unknown future is reduced in the built environment by a purely sensed and virtually modeled basis. Modeling closer to reality does not *always* make better space, and the designer's role is more critical here.

Even if we were to deliver on the full concept of a city digital twin, to refer to Borges, and *The Library of Babel* fable, we would have a vast library of knowledge (data), which is all equitable and therefore impossible to decide what appropriate algorithmic and ML/AI action to take from it. This is critical if the city digital twin is a "mirror" or brain organism as postulated earlier, but there are urgent debates who in society is this for (Borges, 1946, pp. 78–86).

CIMs for smart cities

City digital twin ambitions in their current form are city information models (CIMs), which result from various capture technologies and a GIS and BIM fusion (Fig. 5). The continuing discourse on smart cities and their purpose, effect,

FIG. 5 Sibu_town_centre,_Sibu_Sarawak,_Malaysia, 2019. This CIM has been created by the Sarawak Multimedia Authority. It was constructed through photogrammetric surveys merging UAV and mobile LiDAR point clouds.

and scope beyond corporate ICT visions as well as the frictionless futures they present need to be understood in the development of city digital twins. While the concept is far more contained than a "smart" city, nevertheless, a complete digital replica and "mirror" of a city are certainly not possible, as time and complexity will also limit the concept. Therefore it would be more prudent to refer to the limitations of city digital twins rather than overpromote unlimited geographic software applications and solution as panaceas for cities of many scales and regions. The creation of semantic and reality capture models can be a costly exercise, and the release of open data from these activities raises questions regarding public engagement with the resulting platforms over traditional forms of public consultation regarding development. CIMs also raise questions about centrally funded open data sources and commercial exploitation. Even if CIM adoption is widespread and further systems are deployed, designers using GIS or BIM platforms need to engage in the specifications for files to create closer interoperability, explicitly defining data required (GeoBIM, p. 11). There are also data privacy concerns such as facial recognition, computer vision and machine learning, the use of smartphones for mobility studies, and transport planning and infrastructure as part of city digital twin. City digital twins are not palliative solutions for meditating between urban designed space and digital technology, more of an addition of virtual "realism" for development in AEC, and there are clear benefits here. CIMs help provide levels of detail that will significantly aid urban planning in terms of geographic realism. The promise of knowing our longer or alternative future, to test holistic scenarios outside our model of cities without physical transformation, is unachievable and will remain evasive for the science of cities. Until we can comprehend and commensurate the difference between the capacity of technologically driven desire and the capability of digital twins, the future city remains an enticing mirage.

References

Angelidou, M., 2017. Smart city planning and development shortcomings. TeMA J. Land Use Mobil. Environ. 10. https://doi.org/10.6092/1970-9870/4032.

Ayres, P. (Ed.), 2011. Persistent Modelling: Extending the Role of Architectural Representation. Routledge, Abingdon, Oxon England, New York, NY.

Batty, M., 2018a. Digital twins. Environ. Plan. B Urban Anal. City Sci. 45, 817–820. https://doi.org/10.1177/2399808318796416.

Batty, M., 2019. A map is not the territory, or is it? Environ. Plan. B Urban Anal. City Sci. 46, 599–602. https://doi.org/10.1177/2399808319850652.

BIM+, 2018. Bentley Ups the Ante on Digital Twin Technology. http://www.bimplus.co.uk/news/bentley-ups-ante-digital-twin-technology/. (Accessed 26 August 2019).

Borges, J.L., 1946. On exactitude in science. In: Collection Fictions. Penguin, London. Trans. Hurley, A. 1999.

Bratton, B., 2016. The Stack: On Software and Sovereignty. The MIT Press, Cambridge, MA.

Caird, S.P., Hallett, S.H., 2019. Towards evaluation design for smart city development. J. Urban Des. 24 (2), 188–209. https://doi.org/10.1080/13574809.2018.1469402.

Clarke, R.Y., 2013. Smart Cities and the Internet of Everything: The Foundation for Delivering Next-Generation Citizen Services. White Paper, IDC, CISCO, VA.

Cousins, S., 2017. 3D mapping Helsinki: how mega digital models can help city planners. Constr. Res. Innov. 8, 102–106. https://doi.org/10.1080/20450249.2017.1396747.

Cureton, P., 2020. Drone Futures: UAS in Landscape and Urban Design. Routledge, New York, London.

Dassault Systèmes, 2018. 3DEXPERIENCity. https://www.3ds.com/stories/how-can-technology-shape-the-future/imagining-more-sustainable-city/. Accessed 3 March 2020.

Dunn, N., 2018. Urban imaginaries: palimpsest of the future. In: Christoph Linder, C., Meissner, M. (Eds.), The Routledge Companion to Urban Imaginaries. Routledge, New York, London.

Dunn, N., Cureton, P., 2019. Frictionless futures: the vision of smartness and the occlusion of alternatives. In: Figueiredo, S.M., Krishnamurthy, S., Schröder, T. (Eds.), Architecture and the Smart City. Routledge, New York, London.

Dunn, N., Cureton, P., 2020. Future Cities: A Visual Guide. Bloomsbury, London.

Gelernter, D., 1993. Mirror Worlds or The Day Software Puts the Universe in a Shoebox.. How it Will Happen and What It Will Mean. Oxford University Press, New York; Oxford.

Geng, H., 2017. Internet of Things and Data Analytics Handbook. John Wiley, Hoboken, NJ.

Gibson, D.V., Kozmetsky, G., Smilor, R.W., 1992. The Technopolis Phenomenon: Smart Cities, Fast Systems, Global Networks. Rowman & Littlefield, Lanham, Md.

Greenfield, A., 2018. Radical Technologies: The Design of Everyday Life. Verso Books, London, New York.

Grieves, M., Vickers, J., 2017. Digital twin: mitigating unpredictable, undesirable emergent behavior in complex systems. In: Kahlen, F.-J., Flumerfelt, S., Alves, A. (Eds.), Transdisciplinary Perspectives on Complex Systems: New Findings and Approaches. Springer International Publishing, Cham, pp. 85–113, https://doi.org/10.1007/978-3-319-38756-7_4.

KIRA—digi, 2019. The Kalasatama Digital Twins Project Digital Twins Project. Ministry of the Environment, Helsinki.

Kutzner, T., Kolbe, T.H., 2018. Citygml 3.0: sneak preview. In: Kersten, T.P., Gülch, E., Schiewe, J., Kolbe, T.H., Stilla, U. (Eds.), PFGK18 – Photogrammetrie – Fernerkundung – Geoinformatik – Kartographie, 37. Jahrestagung in München 2018, Vol. 27 of Publikationen der Deutschen Gesellschaft für Photogrammetrie, Fernerkundung und Geoinformation (DGPF) e.V. Runder Tisch GIS e.V, München, pp. 835–839.

Laat, R.D., van Berlo, L., 2011. Integration of BIM and GIS: the development of the CityGML GeoBIM extension. In: Kolbe, T.H., König, G., Nagel, C. (Eds.), Advances in 3D Geo-Information Sciences, Lecture Notes in Geoinformation and Cartography. Springer, Berlin, pp. 211–225, https://doi.org/10.1007/978-3-642-12670-3_13.

Lutz, C., 2019. Digital inequalities in the age of artificial intelligence and big data. Emerging Technologies: Perspectives from Behavioural Scientists 2019, 141–148. https://doi.org/10.1002/hbe2.140.

Mora, L., Bolici, R., Deakin, M., 2017. The first two decades of smart-city research: a bibliometric analysis. J. Urban Technol. 24, 3–27. https://doi.org/10.1080/10630732.2017.1285123.

Omer, M., Margetts, L., Mosleh, M.H., Hewitt, S., Parwaiz, M., 2019. Use of gaming technology to bring bridge inspection to the office. Struct. Infrastruct. Eng. 15, 1292–1307. https://doi.org/10.1080/15732479.2019.1615962.

Research and Markets, 2019. Digital Twin Market by Technology, Type (Product, Process, and System), Industry (Aerospace & Defense, Automotive & Transportation, Home & Commercial, Healthcare, Energy & Utilities, Oil & Gas), and Geography – Global Forecast to 2025. https://www.researchandmarkets.com/reports/4807965/digital-twin-market-by-technology-type-product. (Accessed 19 September 2019).

Rogers, D., 2019. Twin tracks: the drive to create a smart digital model of the UK. Construct. Res. Innov. 10, 49–52. https://doi.org/10.1080/20450249.2019.1621591.

Schlickman, E., Domlesky, A., 2019. Field Guide to Life in Urban Plazas. SWA Group. https://www.swagroup.com/idea/plaza-life-revisited/. (Accessed 9 January 2019).

Schwab, K., 2017. The Fourth Industrial Revolution. Penguin, London.

Singleton, A., Spielman, S., Folch, D., 2017. Urban Analytics. SAGE Publications Ltd, Los Angeles.

Sterling, B., 2018. In: Traldi, D.L. (Ed.), [Not So] Smart Cities. DCASA, pp. 59–60. April.

Tomko, M., Winter, S., 2018. Beyond digital twins – a commentary. Environ. Plan. B Urban Anal. City Sci. 46, 395–399. https://doi.org/10.1177/2399808318816992.

WEF, 2018. India is building a high-tech sustainable city from scratch. World Econ. Forum. https://www.weforum.org/agenda/2018/10/india-is-building-a-green-high-tech-city-amaravati/. (Accessed 6 October 2019).

Whyte, W.H., 1980. The Social Life of Small Urban Spaces. Project for Public Spaces, New York.

Author biography

Paul Cureton is Senior Lecturer in design at ImaginationLancaster, Lancaster University, United Kingdom, and member of the Data Science Institute (DSI). His primary research interests include future cities, geodesign, GIS, UAVs, mapping, modeling, and digital fabrication. His recent publications include the monographs *Strategies for Landscape Representation: Digital and Analogue Techniques* (Routledge, 2016) and *Drone Futures: UAS for Landscape & Urban Design* (CRC Press, 2020). Paul is also working on digital twin concepts through the development of Lancaster City Information Model (LCIM) with Garsdale Design, supported by ESRI, Bluesky Ltd., and CyberCity 3D.

Nick Dunn is Executive Director of ImaginationLancaster, the design research lab at Lancaster University, United Kingdom, where he is Professor of Urban Design. He is Senior Fellow at the Institute for Social Futures, examining the insights that the arts and humanities can bring to the ways we think, envision, and analyze the futures of people, places, and planet. He has published numerous books and journal articles and is lead author on various commissioned reports. Nick's expertise on cities, futures, and health have led to curated exhibitions, keynotes, and plenary presentations around the world. He is coauthor, with Paul Cureton, of *Future Cities: A Visual Guide* (Bloomsbury, 2020).

Chapter 15

The impact of peer-to-peer accommodation on place authenticity: A placemaking perspective

Marcus Foth, Ana Bilandzic, and Mirko Guaralda
QUT Design Lab, Queensland University of Technology, Brisbane, QLD, Australia

Chapter outline

Introduction

Mass tourism is generally associated with the idea of people visiting places away from home for leisure and the commercial implications of this activity. Having its roots in the *Grand Tour*, the origins of tourism focused on the human experience of discovering new places; exploring history, cultures, cuisines, sights, and customs; and making new friends (Towner, 1985). Before tourism evolved into the multibillion-dollar industry it is today, traveling often relied on private individuals offering boarding rooms to travelers, allowing the visitors to experience a taste of authentic local life and providing also a support network while visiting an unknown locale (Brodsky-Porges, 1981). Thinking of tourism in this way highlights that place and place experiences constitute the "raw material" this industry valorizes and commercializes—assets to be cherished and protected for the sustainability of not just the industry but also its destinations for future generations to enjoy.

The effect of mass tourism on cities and their culture and character has long been debated. Back in the 1950s, Astengo was already denouncing how hotels built in the 1800s in Assisi, Italy, were basically spoiling the image of the city and arriving in some instances to advocate their removal to reinstate a more original urban character for this Italian town (Piccinato, 2010). The role

Shaping Smart for Better Cities. https://doi.org/10.1016/B978-0-12-818636-7.00015-9

of tourism as a strategic industry to support local communities is uncontested; the way to support sustainable tourism, in environmental, societal, and cultural terms, is debated in the literature (Fennell and Cooper, 2020; Scheyvens, 1999). How to preserve local place character while fostering economic growth through tourism, for example, by preserving traditional marketplaces, is also debated and contested (Gospodini, 2001; Zandieh and Seifpour, 2020).

The concept of place has been at the center of the urban discourse at least from the 1800s, when rapid urban development challenged the character of cities (Mumford, 1991). A more systematic study of place developed from the 1960s, in response to the modernist city and the general loss of traditional character in many urban centers (Arefi, 2004). In the last 20 years, place has been explored again as the physical and the digital converge to create hybrid new experiences. Since the mid-1990s the relationship between cities and information and communication technology has become increasingly profound and, as a result, has been the focus of study for scholars from social, spatial, and technical disciplines. At a time when mainstream interest was focused on the promise of the "information superhighway" (Hearn et al., 1999) and what Cairncross (1997) called the "death of distance," it was the late William J. Mitchell, then dean of the MIT School of Architecture and Planning and author of the 1995 book *City of Bits: Space, Place, and the Infobahn* (Mitchell, 1995), who suggested that place continues to matter. Rather than a decline in the significance of place through services such as telecommuting, distance education, and e-commerce, the physical and tangible layers of the city started to mix with the digital layers of the Internet and online communications (Gordon and de Souza e Silva, 2011; Graham and Marvin, 1996). Fast forward to today, utopian technology narratives elevate the idea of the "smart city" into mainstream interest and attention. Yet, there are concerns with a technocratic vision that essentially equates a city with a computer system or a commercial corporation (Foth et al., 2016; Graham et al., 2019; Mattern, 2017).

Some of the technology affordances of ubiquitous computing and smart cities have given rise to peer-to-peer platforms, platform economics, and the so-called sharing economy (Frenken et al., 2015; Frenken and Schor, 2017; Schor, 2016). While the term risks being confused with actual altruistic acts of sharing akin to the gift economy, cooperatives, or solidarity economies, the poster children of these new business models such as Uber and Airbnb are anything but charities. In stark contrast to the original idea of "Couchsurfing," Airbnb has grown to a company valued at an estimated $38 billion (Trefis Team, 2018) and raised $3 billion since its start in 2008, which makes it the second most valuable start-up after Uber in the United States (Thomas, 2015). In comparison, Uber is valued at $68 billion and received $12.9 billion in investments since its launch in 2009 (Johnson, 2017). Their quintessential business model has prompted eye-catching news headlines with Uber being heralded the largest taxi company that owns no cars and Airbnb the largest hotel chain that owns no hotel rooms.

While the mass-tourism industry has traditionally been composed of businesses and service providers strongly invested in real estate and city assets such as hotels, taxis, and tour operators, the peer-to-peer economy has changed both the global tourism marketplace and many cities (Gospodini, 2001; Jacqui, 2018; Salvia et al., 2019). While one could argue that it has had a positive impact on key aspects of the tourism industry (supply, growth, reach, audience, and competition) and tourist experiences (demand, service innovation, quality, flexibility, and diversity), it also comes with challenges, especially for cities and local neighborhood communities (Fig. 1). For example, regulatory frameworks at national, state, and especially local levels pertaining to taxes, tourism levies, permits, health and safety, insurance, etc. are still catching up with the fast-paced technological advancements of peer-to-peer platforms and smart cities. The current situation appears to enable platform operators to evade regulatory control. This in turn often fuels concerns and even outrage within the established tourism industry. A heavy reliance on peer-to-peer accommodation also takes its toll during disasters and pandemics such as COVID-19:

> *"Now that tourism has stopped and the Airbnbs are empty," he says, "they have discovered they have no neighbours. There is no neighbourhood. There is no city. If you subtract the tourists, there is nothing."*
>
> (Wainwright, 2020)

Progressive state and municipal jurisdictions have started to debate regulatory reform, and it is commendable that some have recognized the arrival of peer-to-peer platform providers as an opportunity not to merely find ways to enforce old regulation on new arrivals but to consider new approaches. For example, in 2018/2019, the Queensland Government convened a "Peer-to-Peer Economy and Short-term Letting Industry Reference Group," which comprised

FIG. 1 Apartment building complex resorting to public messages at its entrance illustrating the precariousness due to a lack of clear regulatory frameworks. *(Source: Author's photo.)*

representatives from the hotel and tourism sector, from peer-to-peer accommodation providers, from state and local government, and from academia (Queensland Government, 2018). This group quickly concluded that it is opportune and timely to use this moment in time to enact regulatory reform for the entire tourism industry in the state instead of finding new ways to enforce old regulation. This entails a broad audit of regulatory and policy entanglements comprising considerations of neighborhood amenity, disruption to the local sense of place and community, resident safety (especially, fire safety), land use planning, data collection and analysis, housing affordability and pressures on rental stock and long-term leases, joint destination marketing, and economic concerns such as the goal of leveling the playing field.

In addition to addressing regulatory concerns, we would argue that designing smart to improve cities also requires balancing technological disruption and innovation with protecting neighborhood amenity and the local sense of place and community. The habitability of existing neighborhoods in terms of character preservations and service provision is one of the recurrent concerns in the literature on peer-to-peer accommodation and the sharing economy. In response, this chapter presents a critical assessment of the impact that peer-to-peer accommodation networks have on place experiences and placemaking (Foth, 2017; Foth et al., 2015; Tomitsch et al., 2015). We focus on three themes: (a) the urban ecosystem, (b) place authenticity and localhood, and (c) a city's diversity advantage. As a way to illustrate our argument and to ground it in an empirical case, we introduce the city of Split in Croatia as a case of a local community traditionally associated with tourism and highly impacted by mass tourism's current dynamics. Our aim is to expand transdisciplinary knowledge across smart cities and urban studies in the context of tourism and hospitality. Beyond policy and regulatory reforms, this analysis also informs and puts forward emerging recommendations for placemaking strategies and tactics as a way to protect and safeguard place authenticity.

Tourism and the sharing economy

The main selling point of sharing accommodation platforms is, explicitly or implicitly, access to specific locations and their authenticity. Tourists want to experience a specific locale and enjoy the experience of place (Taboada et al., 2008). As a concept, place has been debated by philosophers, sociologists, ethnographers, and designers for over a century. The general agreement is that in experiencing a locale, we rely on tangible and intangible factors (Augoyard, 2007; Chitrakar et al., 2017; Zukin, 2009, 2010). Our sensory experience of the physical environment is layered with meanings through social practices and cultural symbols. Meaning is often indicated as the core of a sense of place; we develop meaning through experience, and we structure it through memories. In urban studies, tourism is often discussed as a challenge to established places, creating competition between locals and tourists (Chien and Ritchie, 2018).

In some instances the city of the tourists is isolated from the city of the locals. In other instances, they overlap, and in some cases, they create conflict, as recent cases in Barcelona, New Orleans, or Venice have proven. The rise of peer-to-peer accommodation has also generated new gentrification dynamics in these cities, such as *touristification* or *studentification* (Sequera and Nofre, 2018; Smith and Holt, 2007). The urban center, in fact, has become the battleground between long-standing residents and new economy investors. Permanent residents often cannot compete with investors who buy properties to be temporarily let to tourists, or other temporary residents such as students, through peer-to-peer platforms (Hughes, 2018). Consequently, in cities with a high incidence of tourism-based occupancy, the habitability, originality, and authentic character of historic centers have been challenged. The character of a locale is grounded in the intangible "sense of place" often referred to as *genius loci* (Norberg-Schulz, 1979). This is the combination of different factors, which include physical features, such as urban form and layout, architectural styles, and recurrent use of color palettes, and social practices, for example, daily rituals, ways to gather and interact, food, traditional clothing, and accessories. The commercialization of spaces or traditional social practices has a direct impact on the sense of place of a locale altering traditional meanings. Mass tourism is often denounced for commercializing cultural values and threatening the real essence of a place and of a community (Chien and Ritchie, 2018). Ultimately, these processes can challenge the authenticity of a location and the habitability of a city for its citizens.

To understand the impact of peer-to-peer sharing providers on place authenticity, it is useful to understand their business model, too. Reinhold and Dolnicar (2017) show how Airbnb creates value for guests, hosts, and third parties through its services. Their analysis reveals that Airbnb's design decisions are interdependent among cost-efficient and convenient processes, safety and risk mitigation, and service design geared for growth and member lock-in. Airbnb quickly became a controversial platform. Its matchmaking does not exclusively operate on urban and community commons, but focuses more on the business interest of the company (Sharp, 2018), also called "sharing for profit" (Kannisto, 2017, p. 206). Therefore the term "commercial sharing platforms" (Sharp, 2018, p. 2) describes holistically this type of sharing economy systems. Cities sometimes see these platforms as illegal and exploitative, as they are able to transfer wealth somewhere else at the expense of local communities (Labaeye and Conway, 2018). As a result, governments and the tourism industry face the issue of responding to such tensions, balancing the impact on society caused by digital disruption (Dolnicar, 2019; Fang et al., 2015).

Amsterdam was the first municipality to regulate Airbnb (Böcker and Meelen, 2017). Cities such as Paris and Lisbon signed an agreement with Airbnb supporting responsible home sharing (Ede, 2018). Australia's state governments in Victoria, New South Wales, and Tasmania reacted with bills on short-term (rental) accommodations (Duncan and Christensen, 2018; Grimmer et al., 2019).

The intention was to address issues affecting long-term residents who would suffer from short-term occupants and their noisy and disruptive parties. "[I]t remains to be seen whether these legislative solutions are in fact enacted and become law" (Duncan and Christensen, 2018, p. 3), more so their efficacy in solving the issues. Futuristic thinking supports a strategy that achieves scale through a global platform but builds solidarity through democratically controlled and locally owned services (Gorenflo, 2018).

Further, Frenken and Schor (2017) criticize sharing economy platforms such as Airbnb for only focusing on their direct benefits and not considering indirect and complex consequences. They distinguish between (a) hosts renting homes during their business trips or holidays away from home and (b) hosts owning a property solely for renting to tourists. Key issues include arguably unfair competition for hotels and B&Bs that comply with local regulations, which is also found by Hajibaba and Dolnicar (2017), as well as the increase of peer-to-peer discrimination (Cheng and Zhang, 2019; Edelman and Luca, 2014). New York City has found itself in an affordable housing crisis and published data on Airbnb's complicity (Stringer, 2018). The city's report attributes approximately a 9.2% increase of rental rates to Airbnb, but some New Yorkers suffered from a rental increase as high as 20% in parts of Manhattan and Brooklyn (Stringer, 2018).

In contrast to commercial sharing platforms, the Sharing Cities Network "has developed a narrative of the sharing economy as a transformational global movement founded on inclusive sharing and support for the urban commons to address social justice, equity and sustainability" (Sharp, 2018, p. 2). The book *Sharing Cities* provides examples of the actual (noncommercialized) sharing economy connecting individuals, communities, enterprises, and governments (Shareable, 2018). The Sharing Cities Network (shareable.net) aims for democratizing local economies through ecological sustainability, strengthening the commons and support for social justice in issues concerning technology, waste, food, mobility, energy, housing, and many more (Sharp, 2018). Another example of a well-known sharing accommodation platform is Couchsurfing, which started its operations as an actual nonprofit enterprise (Yoonjoung Heo, 2016).

To evaluate the impact of sharing accommodation platforms on place, we offer a critical appraisal based on three themes: (a) the urban ecosystem, (b) place authenticity and localhood, and(c) a city's diversity advantage. To ground the discussion in a specific case study, we use the city of Split, Croatia, as an example of a local community highly dependent on tourism.

Split, Croatia

Split is the second largest city in Croatia; located on the Adriatic Sea, it is a UNESCO listed settlement, due to the presence of Roman Emperor Diocletian's palace. The historic old town center, characterized by white stone buildings creating a maze of tiny streets that occasionally meet in squares, is only accessible

FIG. 2 *Top*: Picture was taken from the western end of the promenade; *bottom*: taken from the eastern end of the promenade. *(Source: Author's photos.)*

to pedestrians. It provides benches to rest in public areas such as squares, parks, and promenades that are the core of its citizens' social life (Fig. 2). Cafés, news agents, hotels, hairdressers, restaurants, and smaller shops occupy the building's ground level, while the levels above are residential (Fig. 3). The western peninsula is occupied by Split's national park and open to the public for walks, sports, and swimming in small bays along the coast. Outside the old town the outstretch of the city is mostly characterized by residential houses and high rises that have access to cafés, smaller parks, playgrounds, and small soccer grounds within walking distance. Split follows a modernist approach, with residential suburbs clustered around civic facilities and connected by arterial roads.

According to the latest census[a] as of 2011, Split's population was just above 178,000. Yet, in 2017, 720,300 tourists visited Split, staying for a total of over 2 million nights (Ministarstvo turizma Republike Hrvatske, 2018). If the number of tourists were split equally throughout the year, Split would register a 33% increase in population. However, tourism in Croatia and Split is most intense between May and September, boosted by some main economic and cultural events, such as *Croatia Boat Show*, *Split Music Festival*, the theater festival

a. www.split.hr/Default.aspx?sec=875.

FIG. 3 *People's Square* or Pjaca as called by locals is surrounded with restaurants, cafés, bars, and shops. It is a place where locals meet and chat over drinks and pass through to other squares or the promenade. *(Source: Author's photo.)*

Splitsko ljeto, and the electro music festival *Ultra*. Split was also chosen as set for the television drama *Game of Thrones*, which has also contributed to its rise in popularity. After the country's civil war, Split was mainly a transit point for tourists, who used the city as a gateway to surrounding islands, such as Hvar. Since 2000 however, Split has increasingly attracted tourists for longer stays. Croatia, especially its coastline, has a long tradition of providing private accommodation to tourists (Vlahov and Vuletić, 2016). Due to social and economic issues in Croatia, private accommodation renting has been one of citizens' most preferred additional source of income (Vlahov and Vuletić, 2016).

Tourism and the urban ecosystem

Cities are complex systems, and to thrive, they need a balance in terms of functions, activities, and social groups (Adeniyi et al., 2018). When a city invests heavily in tourism as an industry to drive economic growth, two main issues generally arise: (a) scarcity of housing options (Bainbridge and Armitage, 2017) with affordable, long-term tenancy leases available to local residents and

(b) disappearance of shops and services catering to the specific needs of local residents, for example, hardware stores. Ultimately the habitability of a city for its own residents is challenged. While traditional accommodation options for travelers, such as hotels, mainly impact on some specific areas of the city that evolve as a tourism core, peer-to-peer accommodation platforms make tourist presence more pervasive and embedded within the urban fabric. This results in a reduction of the actual resident population and/or increase in the transient population, especially in the case of unhosted accommodation.[b] As a result, patterns of economic demand in the urban ecosystem shift from local needs to tourist needs, and local retailers may find more profitable catering for the more lucrative needs of tourists. The dynamics of this population swap can be complex and driven by specific contextual factors; a general pattern observed in several tourist locations shows how progressively dwellings can be converted to temporary accommodations, pushing local residents out of central areas. Rent increases and a lack of amenities and services to cater for locals are both aspects of this progressive commercialization of city centres. This chain reaction is often exasperated by tourism-induced premium pricing that introduces artificial surcharges on everyday items and services. It is also accelerated by property investors who operate quasihotels relying on peer-to-peer accommodation platforms while accumulating properties in the same building or in the same area for this purpose.

In Croatia, as in other countries, a tourist tax is in place to balance the funding of services to tourists and locals, but some owners might decide to evade taxation and maximize profit by not declaring their guests. According to the Croatian tax administration office, 50% of hosts did not share their tax file number or shared a wrong number with online accommodation platforms or falsely recorded income that was less than the taxable threshold (Koretić, 2018). Siniša Topalović, director of Horwath Hotel, Tourism & Leisure Consulting in Croatia, calls for better regulation to stop this gray economy practice. Croatia's first attempt to tackle this issue was a law introduced in July 2018 requiring online accommodation platforms such as booking.com and Airbnb to require landlords to share their tax file number with the platform (Koretić, 2018). In Topalović's opinion, an analysis of Airbnb's impact on Croatian tourist destinations is needed to then develop suitable regulations responding to the peer-to-peer accommodation sector (Topalović in Biberović, 2016). Although this might be a long process, he argues that cities are likely to cooperate, because on the one hand they value their reputation as a destination and are interested in sustaining economic growth from tourism development, but they must also secure the long-term livelihood and viability of their neighborhoods for their citizenry (Topalović in Biberović, 2016). Topalović summarized that Airbnb

b. Hosted accommodation on peer-to-peer platforms are those where the host stays with the guest, for example, a spare bedroom rented out as part of a local's home. Unhosted accommodation represents entire houses, flats, and apartments available to guests for short-term letting.

provides new and relatively safe and cheap accommodation opportunities when compared with conventional hotels, but they need to be put into the same competitive playfield with existing registered accommodation options. Further, he highlights the downside of Airbnb's service that leads to locals moving out of the city center, or the increase of rental prices for long-term tenancies, or even an entire decrease in the availability of long-term lease tenancies in tourist resorts in Croatia (Biberović, 2016).

The rise of Split as a tourist destination combined with the popularity of peer-to-peer accommodation platforms has been impacting on the scenery and livelihood of the historic old town. Amenities and spaces originally meant for locals have now been converted to tourist activities. Some examples include:

- An old cinema turned into a nightclub.
- A one-dollar shop [called *kineski dućan* by Croatians], converted into an upmarket restaurant.
- An electronics shop becomes a bistro during the day and used as a disco at nights and weekends.
- Foreign citizens with Croatian parents moved to Split and now own restaurants, coworking spaces, or fashion retail shops, or organize pub crawls targeting wealthy tourists and a younger "hipster" audience—yet neither market segment represents the average local adult or youth in Split.

These changes can inject new life in the local economy, but they also affect the daily practices of locals. Pub crawls are being offered targeting backpackers during peak season, trying to show them the city and its nightlife in just a few hours. But this activity does not belong to the actual local culture. The local youth would rather bring their own drinks to the promenade and a pier known as *Matejuška*, where they would sit, chat, play the guitar, or sing. This small part of the promenade is also the meeting point of retired and older people who like to sing old folk songs a *cappella*. The tourist industry's commodification of "true local" products, services, and experiences to guests might raise money and boost the local economy. Yet, it comes at a costly price to the integrity of the local urban ecosystem in that it only creates a short-term injection of income averting from the authentic and genuine activities of locals who opt to take their business elsewhere. This type of mass tourism greatly influences the intangible essence of place.

Another dynamic that can be observed is the aforementioned return of Croatian migrants' children (CroExpress, 2019), who operate retail shops, mainly restaurants, and fashion shops, so as to take advantage of the tourist boom. These operators cater for wealthy tourists and a younger "hipster" audience; they might hold Croatian citizenship, yet will not have significant ties with the country or Split in particular. Marina Perić Kaselj, who is the director of the Institute for Migration and Society, has the impression that returning Croatians or their children will pass on their optimism to locals for a change in working culture and lifestyle to create a new trend that would hopefully take off even more in the future (CroExpress, 2019). These developments and hope for positive change may not yet take into consideration alienation of local traditions and social patterns.

Many tourists choose to rent peer-to-peer accommodation precisely because they seek the authentic experience of living like a local and living with the locals. Yet, if locals are being displaced as a result of this very process, the increasing tourist influx threatens the authentic place and city experience for all. This brings us to the next theme in our analysis.

Place authenticity and localhood

Issues in preserving a local sense of place—the *genius loci* as outlined earlier—are common when facing pivotal economic and social changes. The complexities of a local population overnumbered by tourists, thus challenging the capacity of local amenities and infrastructures, are a problem as old as the first Christian jubilee in 1300. Cities hosting fairs or markets in time have evolved to manage the temporary imbalance between visitors and locals (Mumford, 1991). Yet, what we experience today is unprecedented in intensity and magnitude. Today, cruise ships can flood a small island or city, such as Ibiza in Spain, Venice in Italy, Mykonos in Greece, Warnemünde in Germany, and Split, with thousands of foreigners multiple times per month. Contemporary mass tourism challenges the very essence of authenticity of a locale provided by the everyday practices of its local residents. While the income from tourist spending may boost the local economy, it comes at a cost, and finding the right balance remains a controversial challenge for local authorities and place designers (Arefi, 2004; Chitrakar et al., 2017; Vlahov and Vuletić, 2016).

The City of Copenhagen (2017) was one of the first to embrace an explicit tourist marketing strategy to mitigate this imbalance, introducing the term "localhood." Imagery and prose were created to attract tourists to Copenhagen so they can experience true Danish hospitality and genuine relationships with locals.

> *Localhood is a long-term vision that supports the inclusive co-creation of our future destination. A future destination where human relations are the focal point. Where locals and visitors not only co-exist, but interact around shared experiences of localhood. Where our global competitiveness is underpinned by our very own localhood. And where tourism growth is co-created responsibly across industries and geographies, between new and existing stakeholders, with localhood as our shared identity and common starting point.*
>
> (City of Copenhagen, 2017)

However, localhood is different from "feeling at home" (Zhu et al., 2019). It is more about "living like a local" and thus subject to the precariousness of place authenticity that stems from the dynamic nature of the relationship between local residents and their environment. The commodification of local traditions, food, artifacts, clothes, or tradition makes them accessible to everybody, devoiding them of the essential meaning of being local, exclusive of a specific population, demographic, or group. If disturbed or permeated by tourists, especially in large numbers, the risk may be that locals will behave differently, cease the activity, move the activity elsewhere, or even respond with hostility. The economic process of offering place authenticity as a resource to be "consumed" by tourists renders it akin to other economic phenomena such as the prisoner's dilemma or the tragedy of the commons (Hardin, 1968; Ostrom, 1990). If authenticity is being consumed—both digitally and physically—and thus reduced, it requires strategies to counterbalance this by reinvesting time and effort to produce and maintain place authenticity.

In terms of pillaging authenticity, there are two phenomena heavily challenging the sense of place of some locales. In terms of tangible form, replicas of buildings or environments question the uniqueness of a place (Guaralda, 2017). Examples of this approach include the Venetian hotels, attempting to reproduce the vibe of Venice in Las Vegas in the Nevada desert or on Macau Island. In terms of the intangible authenticity, the rise of the so-called James Bond effect threatens traditional meanings of places (Reijnders, 2010). Local history, traditions, and values are overridden by the fictitious one presented in successful media franchises. This is often accompanied by a rapid peak of short-term benefits and economic growth, often followed by a steep decline as tourist interest ebbs away. It further highlights the discrepancy between the "film-authentic" yet—due to cinematographic suspension of disbelief—imagined authenticity of fantastic places such as *Rivendell* or *King's Landing* and the actual places in the real world that provided the stage set for the movie. Iconic stage sets for movies such as *Lord of the Rings* in New Zealand and *Game of Thrones* in both Dubrovnik and Split in Croatia have been attracting tourists interested in being in the alternate reality provided by movies more than experiencing the traditional place that was there before the movie set. Apartments have been built excessively in tourist resorts along Croatia's coastal locations such as Split. While this led to an increase in tourism capacity, it also decreased place authenticity (Vlahov and Vuletić, 2016). Moreover, as availability of accommodation increased, prices decreased (Vlahov and Vuletić, 2016). So far, local authorities did not manage to employ policy interventions and smart regulation to avoid these negative consequences on the built form and protect place authenticity and localhood (Vlahov and Vuletić, 2016).

Airbnb introduced a new service called *Experience* on its platform that is aimed at safeguarding some key aspects of place authenticity. It allows accommodation owners to offer unique experiences to their guests, for example, vineyard tours, local cooking workshops, or discovering hidden places that are different from classic tour guides and adventure trips. While it is the guests' choice to book the *Experience*, it is an opportunity to bring tourists and locals closer together (Perić, 2018). The hosts get access to a new source of additional income, and the guests can connect with a local person (Perić, 2018). Although this brings locals and tourists together, maintaining the authenticity of these services remains a challenge. As for the earlier example of tourist pub crawls in Split, a business solution with commercial profit in mind could be neither a traditional nor an authentic local experience.

Kaštelan (2015) reports the opinion of a Spanish tourist in Dubrovnik, Croatia, to highlight the issue of losing localhood that concerns many locals. The Spanish tourist shared his hopes that the responsible authorities and providers will find a solution to regulate and balance the high numbers and turnover of tourists in a way that the local people still feel comfortable (Kaštelan, 2015). For him, it was a real pleasure to see children still playing in the streets, the older people sitting and talking, and having local shops. He thinks this is fantastic, and Dubrovnik should invest more effort to keep this spirit, as other destinations that he visited fall into the trap of high profits that turned the place into completely empty scenography for the masses devoid of place authenticity.

On the threats to city authenticity and localhood, Barbarić (2017) indicates that since tourism got hold in Split, it has become hard to find a building, shed, or attic that has not been refurbished into an apartment in the historic old town. She also claims that the number of private accommodation options has increased in the suburbs and the outskirts of the city. The boom in private supply is reflected in the city's 24,000 registered tourist beds, most of which are private accommodation (Barbarić, 2017). This leads further to the increasingly impossible task for local people to find a dwelling, especially during summer, because of short-term letting, reserved for, and only affordable by, tourists (Barbarić, 2017). This invasion led to a loss of authenticity. Tourists might find themselves less often in conversation with locals while instead finding products tailored to them rather than experiencing the *old Split*.

Local students are victims of this phenomenon, too. Although the university semester and exams end only in June/July, some landlords expect students to move out at the beginning of May, when the summer season starts. For many students, this means having to commute between the hinterland and the city during the exam term, something that makes their life harder and alters the social mix of the city. Therefore tourists experience a *nonauthentic Split*. The perverse extremes of this trend toward converting more and more housing into peer-to-peer accommodation came to the fore 3 years ago when two sisters advertised a tent for two on their balcony for about AUD 20 per night (Barbarić, 2017).

One of the key criticisms that peer-to-peer accommodation providers face is the allegation that they inject temporary visitors into oftentimes well-established neighborhood communities. Yet, there are also instances where the integration of the needs of locals and tourists has been successful, for example, when tourists are recognized as offering local communities a diversity advantage—our third and final theme.

Diversity advantage

Cities and the sociocultural diversity they harbor and attract have been described as the engines of innovation and economic growth (Bairoch, 1991), and diversity has been seen as a key factor of a city's success (Jacobs, 1969). Innovation thrives on diversity (Duranton and Puga, 2000; Jacobs, 1969; Ottaviano and Peri, 2006; Wood and Landry, 2008). Global cities, such as London, Paris, New York, and Tokyo, are contributing to world economic growth and innovation largely because of the cultural diversity of their populations (Sassen, 2011). Wood and Landry (2008) propose the notion of a city's "diversity advantage." They argue that "policy is often dominated by mitigating the perceived negative effects of diversity, and little thought is given to how a diversity dividend or increased innovative capacity might be achieved" (Wood and Landry, 2008, p. 369).

Tourists come from all sorts of walks of life, yet conventional mass-tourism industry market segmentation often ignores the sociocultural diversity of tourists themselves and reduces them to income groups according to spending capacity—from the backpacker to the business class traveler and high roller. For instance, the fact that a young backpacker may be in a gap year after having just graduated from a top university in their home country is often overlooked. Tourists constitute a great opportunity and an as yet untapped potential to add sociocultural diversity to destinations' local communities. However, this nascent potential will remain dormant without adequate strategies—both placemaking strategies and digital design solutions on peer-to-peer accommodation platforms—to take advantage of the diversity of visitors and their expertise and sociocultural background for enriching neighborhoods and local communities. Such new strategies require lateral thinking across social, technical, and spatial strata to imagine new interfaces—in both place and platform—for locals to meet tourists and vice versa (Bilandzic et al., 2013; Casadevall et al., 2018).

How could the concept of placemaking for diversity advantage apply in Split? Since 2013 the Europe edition of the *Ultra* music festival has entertained electronic music fans in Split. The festival started off as a two-day event in the first 2 years but has now extended to 3 days. The opinions of locals about it are divided and create a heated public debate. Complaints target not only the behavior of binge-drinking, drug-taking, and rubbish-leaving tourists but also the noise produced by the festival. However, many locals benefit from it through peer-to-peer accommodation platforms such as Airbnb. The festival brought a new hip and young tourist group to Split. It brought new opportunities and life to the city. Prior to the 2018 festival, organizers reported that ticket sales were strong and that there will be an even larger foreign audience than in previous years (Jadrijević Tomas, 2018). Although the festival might have controversial aspects, the diversity of people coming to the city during the festival could be beneficial to various stakeholders. Besides the obvious economic benefits for its organizers, DJs, local restaurants, and local hosts, the new target group of tourists might benefit the city's vibes, culture, and lifestyle. The aforementioned

Experience feature on Airbnb could assist in this regard and connect festival guests with locals on the basis of shared interests and shared encounters (Willis et al., 2009). This could be further complemented with hybrid (digital and physical) placemaking strategies such as curating situated experiences and programming local gatherings and in-person meet-ups for precisely this purpose.

While Split is a well-known summer destination, the city council makes an effort to attract tourists during winter, too. In December and beginning of January, winter markets are held on the promenade and two squares in the historic old town. They sell Christmas home accessories, food such as sausages or fried sweets, and mulled wine and beer and offer programs for kids and adults to keep attendees entertained with music from live DJs. They are events popular with the locals and find increasing attention from domestic and foreign tourists (Adiona Travel, 2016). If tourists keep coming to Split during winter or even throughout the year, responsible authorities should address the problem of overtourism and find solutions to protect the local long-term renters. Furthermore, they might want to address through policies and regulation exorbitant rental prices on Airbnb, for example, increases of up to 500% during the *Ultra* music festival compared with other nights in the peak season. However, rather than seeing tourists and the supply and demand for accommodation as the only issues, tourists also represent an opportunity that can help Split to thrive as a smart place for innovation (Bilandzic et al., 2018). Tourists bring diversity to Split: different cultures, languages, knowledge, professional backgrounds, and experiences. If Split's fledgling local start-up and innovation ecosystem finds a way to tap into the tourism community, it may provide a source of creative thinking, inspiration, and collaboration to boost the city's innovation capacity.

What are some other ways that Split can unlock and benefit from the diversity advantage that tourists bring? We have started to explore the notions of "workations" (coworking holidays) and coliving (Mahlberg, 2016; Razavi, 2015). The former has already been taken up by travel agents as a new tourism category; the latter is emerging as a new architectural paradigm. The number of innovation hubs or coworking spaces increases worldwide (Bouncken and Reuschl, 2016; Kojo and Nenonen, 2014). These spaces are popular workspaces for freelancers or start-ups in their early days. Split is now also home to a few of them, for example, Saltwater Split, Amosfera Coworking, Work-in-Progress (WIP), and CoCreative. These spaces mainly attract tourist freelancers and innovators who are on workation, that is, a combination of work and vacation. They struggle attracting and keeping local people in Split. WIP is funded by Remote Year and organizes team building events for its freelancers. It also encourages the tourist freelancers to connect with the local community. For instance, one group came up with the idea to rescue stray cats and find them homes with local families. Building connections between locals and tourists can be a win-win for both parties. On the one hand, it enriches tourists' experience. On the other hand, locals get the opportunity to learn from tourists and recognize the potential for local development, or they may be curious to discover the world of innovation spaces and become entrepreneurial.[c] That may be a step forward to combat Croatia's high unemployment rate of 22.7% (HINA, 2018).

Exchange has always been at the core of urban life; developing new activities and fostering new synergies can be a key to soften the pressure on traditional values, which are increasingly commercialized, and open new original paradigms of urban life.

Conclusions

In this chapter, we have analyzed peer-to-peer accommodation platforms and how they impact across the three broad themes of (a) the urban ecosystem, (b) place authenticity and localhood, and (c) a city's diversity advantage. The analysis shows how these platforms change place experiences for both locals and tourists using the case study of Split. Similar impacts have been reported in Byron Bay, Barcelona, New Orleans, and Venice. We found that a heavy reliance on peer-to-peer accommodation platforms can jeopardize a city's resilience during times of crisis (Wainwright, 2020) and potentially even compromise the local government's sovereignty (Couture and Toupin, 2019; Haché, 2014; Lynch, 2020), because of the data and place governance influence taken by a gig economy corporation such as Airbnb for its own profit. Although people are highly mobile and travel around the world more than ever, tourism is an industry that can quickly terminate. For instance, few people thought of spending their vacation in Croatia during its war of independence, and the latest spread of COVID-19 paused leisure travel globally. Tourism can benefit local economies, but its impact might be short term or terminate suddenly. Therefore we argue that designing *smart for better cities* should neither be left nor led by tech corporations alone without local leadership, participation, knowledge, and experience. By way of concluding this discussion, we want to offer some possible suggestions in three broad areas: policy, platform, and people.

In terms of policies and regulations, the debacle around the world in trying to restrain Uber has shown how today we need new tools and approaches to deal with the new sharing economy and its impact on the habitability of our cities. Instead of seeing peer-to-peer accommodation platforms in competition with traditional accommodation providers, what is needed is to rethink the sector in a more holistic way. Rather than considering policy to be enforced through punitive measures, regulatory instruments can learn from platform dynamics and trial incentives as a way to tap into collective intelligence and volunteered geographic information for crowdsourcing of data (Lausen et al., 2016; Mitchell and Highfield, 2017; Schuler et al., 2015). A better understanding of the urban ecosystem can inform better planning for locals and visitors. Additionally, while a holistic perspective to policy development and regulatory reform has merits, it is

c. For example, StartInno in Australia, www.startinno.com.

not advisable to embrace a one-size-fits-all approach. Geographic and industry-specific differences and requirements have to be taken into consideration.

Second, in addition to policy responses, tourism stakeholders should be proactive and come together to jointly pioneer platform innovation driven by tourism demand and to promote and enhance localhood and place authenticity. Platform-native regulatory responses could become alternatives to government-led policy interventions to foster and safeguard local identity. For example, a new website for tourists is ecobnb.com that advocates sustainable accommodation. Accommodations will be listed only if they meet at least 5 out of 10 environmental criteria: organic or local food; green building; 100% renewable energy; energy saving lights; solar thermal panels for hot water; ecological cleaning products; more than 80% waste recycling; car-free accessibility; recovery and reuse of rainwater; and water flow reducers. Similarly, Airbnb Plus provides new search categories for people looking for high-quality and standard services. It has already been introduced to major cities like Toronto, Barcelona, and Rome and will soon be in use in even more cities (including Split and Dubrovnik in Croatia). It allows people to personalize their search for accommodation with more specific characteristics, for example, architectural style, availability of a balcony, or even the "type" of their host (e.g. a musician and sport lover). While this may cause (new) issues of competition between accommodation hosts, this feature could be extended to allow tourist demand to drive local neighborhood improvements or offer platform benefits to locals, too.

> *Creating new value and improving the quality of private accommodation is one of the key challenges facing not only the makers of tourism policy but also a number of other interested stakeholders involved in the functioning of the segment of private accommodation.*
>
> (Vlahov and Vuletić, 2016)

These examples highlight the potential to operate within the opportunity space of platform design to achieve not only commercial goals but also government and community objectives such as reinvesting into the local community and meeting sustainability expectations.

Third, tourists bring social capital that remains dormant until it is activated. This is less about social engineering and more about human horticulture (Gilchrist, 2000). While for most contemporary urban environments it is probably illusionary trying to pursue romantic visions of place-based village communities (Foth et al., 2008), cities can provide spaces for serendipitous and shared encounters (Casadevall et al., 2018; Rehm et al., 2018; Thiel et al., 2015; Willis et al., 2009; Zuckerman, 2011). For instance, the "Skunkworks Finder" (Bilandzic et al., 2018; Casadevall et al., 2018)—if added to peer-to-peer accommodation platforms—could assist tourists, especially business visitors, in discovering smart places and connecting with the local innovation ecosystem.

The local start-up community might also benefit from new ideas and knowledge exchange with business travelers.

Furthermore, locals have started to recognize that peer-to-peer accommodation platforms can offer significant benefits when it comes to offering emergency accommodation for refugees, the homeless, and victims of domestic violence or severe weather events.[d] In 2016 the Victoria State Government in Australia signed a partnership with Airbnb to bolster Victoria's emergency response. In the event of an emergency or natural disaster, Airbnb will connect local residents who have been displaced—as well as emergency management personnel and volunteers who are coming in to assist—with local Airbnb hosts who are opening their homes free of charge. Under the partnership, Emergency Management Victoria and Airbnb are working together to identify Airbnb hosts who will commit to opening their doors in the event of an emergency, provide emergency preparedness educational materials and facilitate access to community education programs for Airbnb hosts, and use Airbnb technology to notify hosts and guests about significant emergencies.

Peer-to-peer accommodations are not a new phenomenon in our urban life, for example, they bring back tourism to its roots of a cultural experience aimed to have a taste of a different culture and society, living with the locals as temporary locals. The main issue with the current process is its magnitude and its extent, which find city managers and communities unprepared to manage the unprecedented scale of mass tourism facilitated by online platforms. This is also a reaction to structural issues with provision of dwellings in our cities, which are often not suitable to support contemporary life. The convergence of deep changes in our economy, in tourism, and in the building sectors require changes in the way we relate to places, cities, the environment, and society (Foth and Guaralda, 2017). After all, design can create digital spaces and physical places that permeate difference and foster diversity; for a local at home is a tourist somewhere else.

References

Adeniyi, A., Guaralda, M., Dias de Carvalho, R., 2018. The contextualization of divergent outlooks in a greenfield master-planned community: a pathway towards reflexivity. Int. Plan. Stud. 23 (1), 81–100. https://doi.org/10.1080/13563475.2017.1344541.

Adiona Travel, 2016. Advent u Splitu Kućice na Rivi—Adiona Travel. Adiona Travel. October 18 http://adionatravel.hr/advent-u-splitu-kucice-na-rivi/.

Arefi, M., 2004. The pedagogy of the American city: revisiting the concepts of place, non-place, and placelessness. Urban Des. Int. 9, 103–117. https://doi.org/10.1057/palgrave.udi.9000121.

Augoyard, J.-F., 2007. Step by Step: Everyday Walks in a French Urban Housing Project. University of Minnesota Press.

---

d. Airbnb has been starting to tackle some aspects of these issues head on with their *Open Homes* initiative: airbnb.com.au/openhomes. See also: fortune.com/2017/06/07/airbnb-refugees-housing/.

Bainbridge, A., Armitage, R., 2017, June 18. Airbnb's Impact on Byron Bay: 'I'm a Lady Hobo, Couch Surfing in Paradise'. ABC News. https://www.abc.net.au/news/2017-06-18/airbnb-leaving-byron-bay-locals-homeless/8612784.

Bairoch, P., 1991. Cities and Economic Development: From the Dawn of History to the Present. University of Chicago Press.

Barbarić, D., 2017. Cure sa Sućidar Hilla kao splitski Del Boy i Rodney: možemo se mi smijati što iznajmljuju šator na balkonu, ali i osnivači megapopularnog Airbnb-a krenuli su od zračnog madraca. Slobodna Dalmacija. June 26 https://www.slobodnadalmacija.hr/dalmacija/split/clanak/id/493493/cure-sa-sucidar-hilla-kao-splitski-del-boy-i-rodney-mozemo-se-mi-smijati-sto-iznajmljuju-sator-na-balkonu-ali-i-osnivaci-megapopularnog-airbnb-a-krenuli-su-od-zracnog-madraca.

Biberović, M., 2016. Airbnb je pozitivna disruptivna promjena u hrvatskom turizmu - ako se regulira! Netokracija. May 25 http://www.netokracija.com/airbnb-sinisa-topalovic-horwath-htl-omgcommerce-118953.

Bilandzic, M., Schroeter, R., Foth, M., 2013. Gelatine: making coworking places gel for better collaboration and social learning. In: Proceedings of the 25th Australian Computer-Human Interaction Conference (OzCHI 2013), pp. 427–436, https://doi.org/10.1145/2541016.2541027.

Bilandzic, A., Casadevall, D., Foth, M., Hearn, G., 2018. Social and spatial precursors to innovation: the diversity advantage of the creative fringe. J. Community Inform. 14 (1), 160–182. http://www.ci-journal.net/index.php/ciej/article/view/1455.

Böcker, L., Meelen, T., 2017. Sharing for people, planet or profit? Analysing motivations for intended sharing economy participation. Environ. Innov. Soc. Trans. 23, 28–39. https://doi.org/10.1016/j.eist.2016.09.004.

Bouncken, R., Reuschl, A., 2016. Coworking-spaces: how a phenomenon of the sharing economy builds a novel trend for the workplace and for entrepreneurship. Rev. Manag. Sci. 12 (1), 317–334. https://doi.org/10.1007/s11846-016-0215-y.

Brodsky-Porges, E., 1981. The grand tour travel as an educational device 1600–1800. Ann. Tour. Res. 8 (2), 171–186. https://doi.org/10.1016/0160-7383(81)90081-5.

Cairncross, F., 1997. The Death of Distance: How the Communications Revolution Will Change Our Lives. Harvard Business Review Press.

Casadevall, D., Foth, M., Bilandzic, A., 2018. Skunkworks finder: Unlocking the diversity advantage of urban innovation ecosystems. In: Australian Conference on Computer-Human Interaction (OZCHI 2018), Melbourne, VIC., https://doi.org/10.1145/3292147.3292169.

Cheng, M., Zhang, G., 2019. When Western hosts meet eastern guests: Airbnb hosts' experience with Chinese outbound tourists. Ann. Tour. Res. 75, 288–303. https://doi.org/10.1016/j.annals.2019.02.006.

Chien, P.M., Ritchie, B.W., 2018. Understanding intergroup conflicts in tourism. Ann. Tour. Res. 72, 177–179. https://doi.org/10.1016/j.annals.2018.03.004.

Chitrakar, R.M., Baker, D.C., Guaralda, M., 2017. Emerging challenges in the management of contemporary public spaces in urban neighbourhoods. Int. J. Archit. Res. 11 (1), 29–43.

City of Copenhagen, 2017. The End of Tourism as We Know It. Wonderful Copenhagen Website. January 30 http://localhood.wonderfulcopenhagen.dk.

Couture, S., Toupin, S., 2019. What does the notion of "sovereignty" mean when referring to the digital? New Media Soc. 21 (10), 2305–2322. https://doi.org/10.1177/1461444819865984.

CroExpress, 2019, December 10. Jedno od ključnih pitanja; Koliko se istinski vraća pripadnika dijaspore u Hrvatsku i kakvi su učinci njihovog povratka u Domovinu? https://www.croexpress.eu/hr-iseljenistvo/21127/jedno-od-kljucnih-pitanja-koliko-se-istinski-vraca-pripadnika-dijaspore-u-hrvatsku-i-kakvi-su-ucinci-njihovog-povratka-u-domovinu/.

Dolnicar, S., 2019. A review of research into paid online peer-to-peer accommodation: launching the annals of tourism research curated collection on peer-to-peer accommodation. Ann. Tour. Res. 75, 248–264. https://doi.org/10.1016/j.annals.2019.02.003.

Duncan, W.D., Christensen, S.A., 2018. A tale of two cities: attempts to regulate Airbnb occupants in strata schemes in Sydney and Melbourne. Aust. Prop. Law Bull. 33 (5), 73–76.

Duranton, G., Puga, D., 2000. Diversity and specialisation in cities: why, where and when does it matter? Urban Stud. 37 (3), 533–555. https://doi.org/10.1080/0042098002104.

Ede, S., 2018. Equitable regulations of short term rentals. In: Shareable (Ed.), Sharing Cities—Activating the Urban Commons. Lightning Source, pp. 52–53.

Edelman, B.G., Luca, M., 2014. Digital discrimination: the case of Airbnb.com. In: Business School NOM Unit Working Paper., https://doi.org/10.2139/ssrn.2377353.

Fang, B., Ye, Q., Law, R., 2015. Effect of sharing economy on tourism industry employment. Ann. Tour. Res. 57, 264–267. https://doi.org/10.1016/j.annals.2015.11.018.

Fennell, D., Cooper, C., 2020. Sustainable Tourism: Principles, Contexts and Practices. Channel View Publications.

Foth, M., 2017. Lessons from urban guerrilla placemaking for Smart City commons. In: Proceedings of the 8th International Conference on Communities and Technologies, pp. 32–35, https://doi.org/10.1145/3083671.3083707.

Foth, M., Guaralda, M., 2017. We Should Create Cities for Slowing Down. The Conversation. April 30 http://theconversation.com/we-should-create-cities-for-slowing-down-75689.

Foth, M., Choi, J.H.-J., Bilandzic, M., Satchell, C., 2008. Collective and network sociality in an urban village. In: Proceedings of the 12th International Conference on Entertainment and Media in the Ubiquitous Era, pp. 179–183, https://doi.org/10.1145/1457199.1457239.

Foth, M., Brynskov, M., Ojala, T., 2015. Citizen's Right to the Digital City: Urban Interfaces, Activism, and Placemaking. Springer, https://doi.org/10.1007/978-981-287-919-6.

Foth, M., Hudson-Smith, A., Gifford, D., 2016. Smart cities, social capital, and citizens at play: a critique and a way forward. In: Olleros, F.X., Zhegu, M. (Eds.), Research Handbook on Digital Transformations. Edward Elgar Publishing, pp. 203–221, https://doi.org/10.4337/9781784717766.00017.

Frenken, K., Schor, J., 2017. Putting the sharing economy into perspective. Environ. Innov. Soc. Trans. 23, 3–10. https://doi.org/10.1016/j.eist.2017.01.003.

Frenken, K., Meelen, T., Arets, M., van de Glind, P., 2015. Smarter Regulation for the Sharing Economy. The Guardian. May 20 http://www.theguardian.com/science/political-science/2015/may/20/smarter-regulation-for-the-sharing-economy.

Gilchrist, A., 2000. The well-connected community: networking to the "edge of chaos". Community Dev. J. 35 (3), 264–275. https://doi.org/10.1093/cdj/35.3.264.

Gordon, E., de Souza e Silva, A., 2011. Net Locality: Why Location Matters in a Networked World. Wiley-Blackwell.

Gorenflo, N., 2018. Introduction. In: Shareable (Ed.), Sharing Cities—Activating the Urban Commons. Lightning Source, pp. 20–37.

Gospodini, A., 2001. Urban design, urban space morphology, urban tourism: an emerging new paradigm concerning their relationship. Eur. Plan. Stud. 9 (7), 925–934. https://doi.org/10.1080/09654310120079841.

Graham, S., Marvin, S., 1996. Telecommunications and the City: Electronic Spaces, Urban Places. Routledge.

Graham, M., Kitchin, R., Mattern, S., Shaw, J., 2019. How to Run a City Like Amazon, and Other Fables. Meatspace Press.

Grimmer, L., Vorobjovas-Pinta, O., Massey, M., 2019. Regulating, then deregulating Airbnb: the unique case of Tasmania (Australia). Ann. Tour. Res. 75, 304–307. https://doi.org/10.1016/j.annals.2019.01.012.

Guaralda, M., 2017. Australian urbanism and the meaning of public space. Placemaking and urban design in south-East Queensland. In: Bologna, R. (Ed.), New Cities and Migration. DIDAPress, pp. 169–184.

Haché, A., 2014. Technological sovereignty. In: Coredem (Ed.), For Free Information and Open Internet, pp. 165–171. https://www.coredem.info/IMG/pdf/pass11_an-2.pdf#page=164.

Hajibaba, H., Dolnicar, S., 2017. Substitutable by peer-to-peer accommodation networks? Ann. Tour. Res. 66, 185–188. https://doi.org/10.1016/j.annals.2017.05.013.

Hardin, G., 1968. The tragedy of the commons. Science 162 (3859), 1243–1248. https://doi.org/10.1126/science.162.3859.1243.

Hearn, G., Mandeville, T., Anthony, D., 1999. The Communication Superhighway: Social and Economic Change in the Digital Age. Allen & Unwin.

HINA, 2018. Eurostat: Hrvatska i dalje među zemljama s najvišim stopama nezaposlenosti. Index. August 1 https://www.index.hr/vijesti/clanak/eurostat-hrvatska-i-dalje-medju-zemljama-s-najvisim-stopama-nezaposlenosti/2014853.aspx.

Hughes, N., 2018. "Tourists go home": anti-tourism industry protest in Barcelona. Soc. Mov. Stud. 17 (4), 471–477. https://doi.org/10.1080/14742837.2018.1468244.

Jacobs, J., 1969. The Economy of Cities. Random House.

Jacqui, A., 2018. Domesticity on-demand: the architectural and urban implications of Airbnb in Melbourne, Australia. Urban Sci. 2 (3). https://doi.org/10.3390/urbansci2030088.

Jadrijević Tomas, S., 2018. Izbacili su podstanare i očekivali laku zaradu, ali su doživjeli debakl: stotine apartmana i vila zjapi prazno za Ultru! Slobodna Dalmacija. May 21 https://www.slobodnadalmacija.hr/scena/mozaik/clanak/id/547447/izbacili-su-podstanare-i-ocekivali-laku-zaradu-ali-su-dozivjeli-debakl-stotine-apartmana-i-vila-zjapi-prazno-za-ultru.

Johnson, M., 2017. Will Uber Be the Hottest IPO of 2017? NASDAQ. NASDAQ. April 10 https://www.nasdaq.com/article/will-uber-be-the-hottest-ipo-of-2017-cm772240.

Kannisto, P., 2017. Sharing for profit: a new business model? Ann. Tour. Res. 66, 206–208. https://doi.org/10.1016/j.annals.2017.07.003.

Kaštelan, I., 2015. Rusa oduševio hrvatski pršut i rakije, Španjolca vino, a Nizozemac se u Dubrovniku osjećao kao u zabavnom parku. Slobodna Dalmacija. August 15 https://www.slobodnadalmacija.hr/stil/putovanja/clanak/id/283493/rusa-odusevio-hrvatski-prsut-i-rakije-spanjolca-vino-a-nizozemac-se-u-dubrovniku-osjecao-kao-u-zabavnom-parku.

Kojo, I., Nenonen, S., 2014. Evolution of co-working places: drivers and possibilities. Intell. Build. Int., 1–12. https://doi.org/10.1080/17508975.2014.987640.

Koretić, D., 2018. Kreće se u hitnu izmjenu zakona zbog onih koji apartmane iznajmljuju na internetu. Jutarnji. June 13 https://www.jutarnji.hr/vijesti/hrvatska/krece-se-u-hitnu-izmjenu-zakona-zbog-onih-koji-apartmane-iznajmljuju-na-internetu-nova-obveza-koja-na-snagu-stupa-u-srpnju-trebala-bi-ih-dovesti-u-red/7469112/.

Labaeye, A., Conway, R.T., 2018. Technology. In: Shareable (Ed.), Sharing Cities—Activating the Urban Commons. Lightning Source, pp. 206–207.

Lausen, L., Rittenbruch, M., Mitchell, P., Horton, E., Foth, M., 2016. CrowdRisk: exploring crowdsourcing of risk information. In: Proceedings of the 28th Australian Conference on Computer-Human Interaction, pp. 165–169, https://doi.org/10.1145/3010915.3010953.

Lynch, C.R., 2020. Contesting digital futures: urban politics, alternative economies, and the movement for technological sovereignty in Barcelona. Antipode 52 (3), 660–680. https://doi.org/10.1111/anti.12522.

Mahlberg, T., 2016. Co-Living is Demolishing the Line Between Work and Life. The Conversation. July 6 http://theconversation.com/co-living-is-demolishing-the-line-between-work-and-life-60614.

Mattern, S., 2017. A city is not a computer. Places J. https://doi.org/10.22269/170207.

Ministarstvo turizma Republike Hrvatske, 2018. Turizam u Brojkama., p. 2017. https://www.htz.hr/sites/default/files/2018-08/HTZ%20TUB%20HR_%202017%20FINAL.pdf.

Mitchell, W.J., 1995. City of Bits: Space, Place, and the Infobahn. MIT Press.

Mitchell, P., Highfield, T., 2017. Mediated geographies of everyday life: navigating the ambient, augmented and algorithmic geographies of geomedia. Ctrl-Z: New Media Philos. 7, 1–5.

Mumford, L., 1991. The City in History: Its Origins, its Transformations, and its Prospects. Penguin Books in Association With Martin Secker & Warburg.

Norberg-Schulz, C., 1979. Genius Loci: Towards a Phenomenology and Architecture. Rizzoli.

Ostrom, E., 1990. Governing the Commons: The Evolution of Institutions for Collective Action. Cambridge University Press.

Ottaviano, G.I.P., Peri, G., 2006. The economic value of cultural diversity: evidence from US cities. J. Econ. Geogr. 6, 9–44. https://doi.org/10.1093/jeg/lbi002.

Perić, L., 2018, March 28. Novost s Airbnb-a koja će se svidjeti iznajmljivačima: ova opcija oduševit će goste, a vama donijeti dodatnu zaradu. Slobodna Dalmacija. https://www.slobodnadalmacija.hr/novosti/biznis/clanak/id/538163/novost-s-airbnb-a-koja-ce-se-svidjeti-iznajmljivacima-ova-opcija-odusevit-ce-goste-a-vama-donijeti-dodatnu-zaradu.

Piccinato, G., 2010. Centenary paper: A brief history of Italian town planning after 1945. Town Plan. Rev. 81 (3), 237–259. https://doi.org/10.3828/tpr.2010.1.

Queensland Government, 2018. "Code of Conduct" the Key in State's Response to Short-Term Letting—the Queensland Cabinet and Ministerial Directory. The Queensland Cabinet and Ministerial Directory; Queensland Government. July 9 http://www.qld.qld.gov.au/.

Razavi, L., 2015. We're All Going on a Co-Working Holiday—Workaholics Welcome. The Guardian. January 28 https://www.theguardian.com/travel/2015/jan/28/co-working-holidays-new-travel-trend.

Rehm, S., Foth, M., Mitchell, P., 2018. DoGood: examining gamification, civic engagement, and collective intelligence. AI Soc. 33 (1), 27–37. https://doi.org/10.1007/s00146-017-0711-x.

Reijnders, S., 2010. On the trail of 007: media pilgrimages into the world of James Bond. Area 42 (3), 369–377. https://doi.org/10.1111/j.1475-4762.2009.00930.x.

Reinhold, S., Dolnicar, S., 2017. How Airbnb creates value. In: Dolnicar, S. (Ed.), Peer-to-Peer Accommodation Networks. Goodfellow Publishers, pp. 39–53, https://doi.org/10.23912/9781911396512-3602.

Salvia, G., Morello, E., Arcidiacono, A., 2019. Sharing cities shaping cities. Urban Sci. 3 (23), 5. https://doi.org/10.3390/urbansci3010023.

Sassen, S., 2011. Cities in a World Economy. Sage.

Scheyvens, R., 1999. Ecotourism and the empowerment of local communities. Tour. Manag. 20 (2), 245–249. https://doi.org/10.1016/S0261-5177(98)00069-7.

Schor, J., 2016. Debating the sharing economy. J. Self-Gov. Manag. Econ. 4 (3), 7–22. https://doi.org/10.22381/JSME4320161.

Schuler, D., De Cindio, F., De Liddo, A., 2015. Encouraging collective intelligence for the common good: how do we integrate the disparate pieces? In: Proceedings of the 7th International Conference on Communities and Technologies, pp. 157–159, https://doi.org/10.1145/2768545.2768562.

Sequera, J., Nofre, J., 2018. Shaken, not stirred: new debates on touristification and the limits of gentrification. City 22 (5–6), 843–855. https://doi.org/10.1080/13604813.2018.1548819.

Shareable (Ed.), 2018. Sharing Cities: Activating the Urban Commons. Shareable. https://www.shareable.net/sharing-cities/.

Sharp, D., 2018. Sharing cities for urban transformation: narrative, policy and practice. Urban Policy Res., 1–14. https://doi.org/10.1080/08111146.2017.1421533.

Smith, D.P., Holt, L., 2007. Studentification and "apprentice" Gentrifiers within Britain's provincial towns and cities: extending the meaning of gentrification. Environ. Plan. A. 39 (1), 142–161. https://doi.org/10.1068/a38476.

Stringer, S.M., 2018. The Impact of Airbnb on NYC Rents. Office of the New York City Comptroller. https://comptroller.nyc.gov/reports/the-impact-of-airbnb-on-nyc-rents/.

Taboada, M.B., Haworth, R.J., Spence, R., 2008. The Social Role of Design on Collaborative Destination Branding: Creating a New Journey, a New Story for the Waterfall Way, NSW Australia. DRS2008 Design Research Society Biennial, Sheffield, UK.

Thiel, S.-K., Foth, M., Schroeter, R., 2015. Ad hoc communities on the road: serendipitous social encounters to enhance tourist experiences. In: 27th Australian Computer-Human Interaction Conference (OzCHI 2015), pp. 643–652, https://doi.org/10.1145/2838739.2838768.

Thomas, L., 2015. Airbnb Just Closed a $1 Billion Round and Became Profitable in 2016. CNBC. September 9 https://www.cnbc.com/2017/03/09/airbnb-closes-1-billion-round-31-billion-valuation-profitable.html.

Tomitsch, M., McArthur, I., Hank Haeusler, M., Foth, M., 2015. The role of digital screens in urban life: new opportunities for placemaking. In: Foth, M., Brynskov, M., Ojala, T. (Eds.), Citizen's Right to the Digital City. Springer Singapore, pp. 37–54, https://doi.org/10.1007/978-981-287-919-6_3.

Towner, J., 1985. The grand tour: a key phase in the history of tourism. Ann. Tour. Res. 12 (3), 297–333. https://doi.org/10.1016/0160-7383(85)90002-7.

Trefis Team, 2018. As a rare profitable unicorn, airbnb appears to be worth at least $38 billion. Forbes Magaz. May 11 https://www.forbes.com/sites/greatspeculations/2018/05/11/as-a-rare-profitable-unicorn-airbnb-appears-to-be-worth-at-least-38-billion/.

Vlahov, A., Vuletić, D., 2016. The role of stakeholders in creating new values of private sector accommodation in Croatia. In: Tipurić, D., Kovač, I. (Eds.), Proceedings of the 4th International OFEL Conference on Governance, Management and Entrepreneurship. CIRU, pp. 741–751. https://www.bib.irb.hr/939632.

Wainwright, O., 2020. Smart lifts, lonely workers, no towers or tourists: architecture after coronavirus. Guardian. April 13 https://www.theguardian.com/artanddesign/2020/apr/13/smart-lifts-lonely-workers-no-towers-architecture-after-covid-19-coronavirus.

Willis, K.S., Roussos, G., Chorianopoulos, K., Struppek, M., 2009. Shared Encounters. Springer, https://doi.org/10.1007/978-1-84882-727-1.

Wood, P., Landry, C., 2008. The Intercultural City: Planning for Diversity Advantage. Earthscan.

Yoonjoung Heo, C., 2016. Sharing economy and prospects in tourism research. Ann. Tour. Res. 58, 166–170. https://doi.org/10.1016/j.annals.2016.02.002.

Zandieh, M., Seifpour, Z., 2020. Preserving traditional marketplaces as places of intangible heritage for tourism. J. Herit. Tour. 15 (1), 111–121. https://doi.org/10.1080/1743873X.2019.1604714.

Zhu, Y., Cheng, M., Wang, J., Ma, L., Jiang, R., 2019. The construction of home feeling by Airbnb guests in the sharing economy: a semantics perspective. Ann. Tour. Res. 75, 308–321. https://doi.org/10.1016/j.annals.2018.12.013.

Zuckerman, E., 2011. Desperately seeking serendipity. In: Proceedings of the ACM SIGCHI Conference on Human Factors in Computing Systems (CHI '11)., https://doi.org/10.1145/1978942.2167183.

Zukin, S., 2009. Changing landscapes of power: opulence and the urge for authenticity. Int. J. Urban Reg. Res. 33 (2), 543–553. https://doi.org/10.1111/j.1468-2427.2009.00867.x.

Zukin, S., 2010. Naked City: The Death and Life of Authentic Urban Places. Oxford University Press.

Author biographies

Marcus Foth (PhD, FACS) is Professor of Urban Informatics in the QUT Design Lab at Queensland University of Technology, Australia. He is also an Honorary Professor in the School of Communication and Culture at Aarhus University, Denmark. His transdisciplinary research focusses on interaction design, smart cities, community engagement, media architecture, and sustainability. Professor Foth has co/authored over 200 peer-reviewed publications. He founded the Urban Informatics Research Lab in 2006 and the QUT Design Lab in 2016. He tweets @sunday9pm.

Ana Bilandzic is a PhD Candidate with the Urban Informatics Research Group in the QUT Design Lab, Brisbane, Australia. Her research is on social and spatial precursors to innovation in Casual Creative Environments, for example, coworking spaces, maker spaces, and hacker spaces. She explores the spread of these spaces in Brisbane and strategies of users to find and navigate them. Further, her research unpacks a better understanding of innovation precursors in spaces that focus on social innovation.

Mirko Guaralda (PhD, MHEd, DArch) is an academic at Queensland University of Technology (QUT). His work focuses on people-place interaction, enquiring into the complex issues of urban density, place quality, and community engagement. He was a researcher at the US Study Center of Sydney in 2012; in 2017, he has been visiting professor at the Thammasat University of Bangkok, Thailand. Since 2018, he has also been engaged in research and teaching at the Jiangxi University of Science and Technology (China).

Chapter 16

Smart and informal? Self-organization and everyday

Jaime Hernández-García and Iliana Hernández-García
Department of Aesthetics, School of Architecture and Design, Pontificia Universidad Javeriana, Bogotá, Colombia

Chapter outline

Introduction

Smart and *informal* seem to emerge as two opposite concepts and practices within the discussion of urban growth (Sassen, 2013). However, they may possess more similarities than differences as the development of informal settlements follows self-organization and everyday patterns, akin to those found in smart technologies (i.e., computing systems). In this sense, researchers can use computational logics to analyze how informal settlers produce and transform an urban place, and conversely, researchers can investigate conventional interpretations of *smart* using *informal* logics. The objective of this chapter lies in considering the opportunity of reconceptualizing the smart through referencing the informal, using two main concepts. The first is self-organization (Heylighen and Gershenson, 2003), drawn by looking at the computational logics aligned with smart. The second is everyday practices (Hernández-Garcia, 2013), applying the important concept of informality as seen in informal settlements. This chapter aims to discuss how these two concepts can work indistinctively, for both smart and informal urbanism. In this sense, self-organization and the everyday allow individuals to reflect about smart in relation to the informal and their link to computational systems (Rauws, 2016). Hence, this chapter offers a reconceptualization of the smart through informality particularly that characterizing informal settlements; it also explores how smart technologies can help to measure and provide a different understanding of everyday practices and

Shaping Smart for Better Cities. https://doi.org/10.1016/B978-0-12-818636-7.00012-3

self-organizational processes that govern informally developed areas. Hence, by reconceptualizing the smart, one is also reconceptualizing the informal.

Why smart? Smart technologies applied to cities are changing the world (Kitchin and Lauriault, 2014). The full package of technologies, apps, and devices now reconfiguring life in cities includes the use of smart phones, GPS, electromobility, and biotechnology. Furthermore, with the proliferation of smart technologies, the human urban experience is taking on new meaning (Zaera-Polo and Anderson, 2017). Smart cities can be viewed as composed of complex networks, capturing and interweaving knowledge, actions, and selection of choices (Rauws, 2016). These networks can offer alternative ways of understanding the city and thinking about it, as they deal with scenarios that are undetermined and uncertain, and how they are activated through many spontaneous decisions and actions that people take in urban areas (Hernández et al., 2014). These unplanned decisions and actions actually comprise a form of knowledge that allows a different way of thinking about and understanding cities. In this sense, smart technologies can transcend and almost deny traditional planning practices and be used, not to plan cities, but to measure and understand what is not planned.

Why informal? The term "informal economy" was initially used in the 1960s to describe economic activities falling outside the modern industrial sector, primarily within the developing world (Lewis, 1955). Soon after, *informality* was coined to address activities and practices in different fields lying outside the mainstream, including those deemed unregulated, unplanned, and often illegal (Barros and Sobreira, 2002), including informal housing and informal settlements. However, similar to smart technologies now changing the world, informal settlements have already done so as they dominate the landscape of urban and housing production in the cities of Latin America (Kellett, 2002). Conolly (2013) argues that informal settlements constitute the paradigm of the built environment in the region, making cities distinguishable from elsewhere. Such settlements comprise not only a large portion but also an active and dynamic part of cities in physical, social, and cultural terms. These settlements also facilitate alternative ways of thinking and performing social, cultural, economic, urban, and spatial practices (Alsayyad and Roy, 2004; Roy, 2009; Fiori and Brandao, 2010). Characterized as primarily developed through residents' self-help practices, the settlements garner little participation from public or private bodies (Brillembourg and Klumpner, 2010; Hernández-Garcia, 2013).

Methodologically, this chapter brings together smart urbanism and informal settlements from a theoretical point of view to discuss the ways they could be more similar than how they are commonly understood. A "technomethodology" was used to link ethnomethodology (common in human behavior studies, particularly in informal settlements) with the interpretative work of computational systems (Dourish and Button, 1998). Within this framework, this chapter's main contribution lies in the theoretical discussion of *smart from the informal* or rather from the recognition of informal. Consequently the discussion focuses on the relationship between smart and informal, framing it through the ideas of self-organization and the everyday. The chapter concludes with a suggestion to reconceptualize smart cities as self-organized systems fuelled by everyday practices; by doing so, we are also reconceptualizing informal settlements as undetermined places, constantly transformed by uncertainty.

Smart and informal

Smart cities can be seen as contributing to the quality of life in urban agglomerations by offering sustainable alternatives to managing them through the use of technologies. When residents can access these technologies, they offer a myriad of benefits that can improve well-being in cities: connectivity, interaction in real time to solve problems, the use of clean technologies and mitigation of environmental pollution, and the use of large amounts of data to improve the city's efficiency.

On the other hand, informality plays an important role too. Informal settlements comprise a large portion of Latin American cities, constituting not only a spatial practice but also a social response to urban challenges. Fiori and Brandao (2010, p. 188) argued, "Urban informality is inexorably interwoven with the city as a whole—at all scales and levels—and has to be seen as another way of being in the city and constructing it." These settlements are also seen as innovative and creative: "Today we recognize the innovative genius of the urban poor in taking advantage of the specific cultural opportunities to survive and improve their living conditions" (Alsayyad, 1993, p. 5). While today's analysts associate informality with poverty, marginality, and deprivation, they also acknowledge its contribution to the overall sociospatial development of Latin American cities. Yet, improving the quality of life for the people living in those areas will require accomplishing much more.

By envisaging the combination of the contributions of smart cities with those of informality, we can initially attract some smart-related criticism. It can be argued that smart technologies offer cities interventionist strategies by means of control, power, consumerism, and the imposition of ways of living that communities do not always desire (Krivy, 2018). In particular, urban leaders promoting smart urbanism often give little consideration to people's desires, expectations, and potential. Sometimes, it is the technology, rather than people, to become the objective; this tendency shows even more clearly in informal settlements, when technology is offered as a life-changing possibility (as when corporations cite an economic interest in producing these technologies) without considering the specificities and more urgent priorities of these areas. In this sense, while the market aims to impose the same or similar consumerist strategies in different cities around the world, in the case of informal settlements, these come at higher social costs (Feldman and Girolimo, 2018) and achieve little success. In general the intersection of smart cities and informal settlements has been insufficiently studied, and most people today understand them as nearly opposite entities. Yet, we would argue that if the locale is smart, it can also be informal and vice versa. They are not mutually exclusive. *Smart* may represent more advanced technology and greater formality, with the *informal* signifying less advanced technology, local processes, and greater emphasis on livelihoods. Technology is generally provided by corporations, the informal by people themselves (Chambers and Evans, 2020).

Smart and informal are closer than expected. For example, smart technologies can prove useful in the study of place production in informal settlements, specifically through the use of mobile devices such as phones, tablets, or computers for registering or measuring motilities, trajectories, public transport, or the use of public space (Kitchin et al., 2017). Smart technologies can prove useful to generate information by means of computing systems that can measure in real-time people's everyday choices and decisions. These processes are self-organized, open, and indefinite. They are also not planned or anticipated by anyone, representing the result of people's direct actions and interactions in real time. One can link them with the production of informal space and social dynamics, within a complexity framework, making the creativity of everyday sociospatial practices visible.

Furthermore, smart technologies and informal settlements may hold more similarities than differences, particularly in terms of how self-organization provides a link between the two; self-organization is both a characteristic of informal settlements and, simultaneously, a condition of technologies following computational systems. The making of place in informal settlements is more iterative and entails self-organized and everyday actions and interactions, where design decisions are made continuously and do not follow anticipated formulas; yet this sociospatial production is akin to how complex systems work (Hernández et al., 2014). In this sense, smart technologies can make it possible to explore the poetics of the spatial transformations area residents make, as well as heighten understanding of how people produce their spaces involving creative mechanisms. In doing so, we can link together two apparent opposite practices: the smart and the informal. In this sense, we are seeking to understand ICTs not as something *additional* or *strange* to informal settlements, but a very relevant and pertinent part of them. In other words, smart technologies can help us understand better how these settlements work—as places that are constantly changing, very adaptable to social and spatial circumstances, and increasingly complex (Rauws, 2016; Hernández et al., 2014). Aside from this, high technologies are becoming more and more enmeshed in the development itself of informal settlements, with the advent of an array of new devices, to facilitate self-organization, self-spatial production, and networks. For example, this can be seen in the increasing use of smart phones and social networks: in Vanguardia (2013), we see how a resident of one of the favelas in Rio de Janeiro, Brazil, upon hearing that the two cable cars serving the high slope *favela* where he lives are broken, captures this on his mobile phones and sends them to the Governor of Rio de Janeiro via Twitter.

In this sense, we suggest realigning the definition of "smart" to also see it as a vehicle that can help understanding place, particularly when this is developed informally. One can view smart technologies as analytical lenses that aid in reading events and social practices, specifically through the use of computing logics and robust information databases. Exploring new ways of interacting between people and place is now possible (i.e., with mobile devices that not only do help with instant communication between people but also can point out issues of the space and contribute to solve/transform it, as in the earlier example); these explorations are mainly characterized by the residents' interactive approaches, making it possible to gain new understandings of people's contributions. By reassessing how informal settlers produce their space in the city, we are also reconceptualizing smart technologies.

With the earlier framework the following sections examine the relationship between smart and informal, framing this through the conditions of *self-organization* and the *everyday*. These urban development–related concepts are linked to each other in their capacity to explain how we can now view everyday sociospatial life in informal settlements as self-organized processes entailing complex computational logics. On the one hand, *self-organization* remains unplanned, unanticipated, and not deterministically based on cause and effect, being notably connected with uncertainty. On the other hand, *everyday* refers to small creative novelties that suddenly show up as a result of people's actions and interactions between themselves and with the physical space. Moreover, when these conditions are working together, the everyday generates a type of spatial production that is self-organized with an emerging social order. A more detailed discussion of this will follow, aiming at contributing to a different understanding of urban informality linked with smart technology logics. By doing so, we hope to lead readers to rethink the role of "smart."

Self-organization

Self-organization is grounded in both evolutionary theory and computing (Gilbert, 2000). A system or group of systems may exhibit increasing complexity, depending on their degree of freedom. They may gain the ability to generate valid possibilities as complexities rise (Hernández, 2016). The crossroads of these fields can shed light on our understanding of how complexity grows in self-organized systems in relation to freedom. Self-organization, complexity, and freedom are connected or intertwined as the following descriptions illustrate.

Considered as hybrid expressions of smart technologies, self-organization systems open up a wide range of possibilities to understand and study city development (Dourish, 1995). These systems highlight the relationship between smart and uncertainty, computing logics and nonlinear processing, transmission and organization of information (Epstein, 2006). In this sense, self-organization speaks about the impossibility of predicting or planning within the smart technologies' domain. Perhaps surprisingly, self-organization connects with the idea of sudden appearance of progressive but uncertain development. This perspective allows us to interpret smart systems as self-organized ones that include a high level of freedom and highly divergent from formally planned processes. These traits greatly reveal how cities work, particularly informal settlements; in this sense the study of self-organization (Kauffman, 2003) emerges as key for identifying how people make decisions concerning their space.

Self-organization can prove helpful for understanding the production of informal space based on smart technologies. Following the computational logics behind smart technologies, time increases its importance as the factor leading evolution, which always happens forward, following time's arrow (Prigogine and Stengers, 1997). This concept emphasizes the impossibility of turning back to the past by undoing the facts, as decisions become actions merged with time, not anticipated or planned by anyone. This view does not regard time as negative or entropic but a key factor for the production of everyday life and creativity. Self-organization combines nonlinear systems as Prigogine and Stengers (1997) proposed, in the same sense that the fourth principle of thermodynamics does not point to negative entropy in which everything tends toward destruction as time flows. The passing of time is seen as a positive and creative enabler. In other words, time is intensive, not linear, and with each interaction (time + action) something new can be created. In this sense, time is a vital factor for creation and possibility. Therefore a form of creative entropy exists, enabled through smart technologies. Interestingly, when applied to informal settlements, this concept helps to explain how space in these areas is produced from the very same action of inhabiting the areas, based on options and adaptations (Holland, 2004), or in short, on random decisions and actions. For example, we can consider the time allocated to transform an open space of an informal settlement into a park (i.e., with urban furniture, playgrounds, and green and paved areas) by means of settlers' decisions and actions that are randomly made based on implicit options and adaptations. This produces not only the park but also social fabric, empowerment, and resilience. The time used for this is not an entropic waste; it is time merging with actions and creating something new and positive. In this light, we argue how self-organizations systems, akin to smart technologies, help to understand sociospatial production in *barrios*.

As residents' actions affect the way space is transformed, their behavior also is relevant in understanding how self-organization works. People act in different ways and then make decisions involving space, with different timing involved. While actions take place in the present time, they may hold implications for the future and even correlations with the past. In this sense, we can view those actions as possible horizons that depend on individual or collective decisions. The experiences in the present time transform the interpretation of the past and the ideas of the future. In other words, time and actions mediated (and boosted) by behaviors speak about self-organization processes for the creation of space in informal settlements akin to the same characteristics arguably found in computational logics behind smart technologies. This includes for both informal settlements and smart technologies the capacity of adaptation and multiple ways to face contingencies. Therefore people's behavior adds another element to self-organization in terms of time linked to their actions and decisions, as well as to the uncertainty but robustness this implies.

As presented earlier, *barrios* are "... examples of complex subsystems within a complex urban system. Their morphological characteristics combined with their development process are traditionally understood as chaotic and unorganised" (Barros and Sobreira, 2002). Yet, communities in informal areas are able to address problems without hierarchy or any particular formal procedure; they tackle their difficulties by their own means—specifically through bottom-up social initiatives. Thus plausible collective behavior may spontaneously emerge. Jacobs (1993) has studied this kind of behavior as something important in the transformation of urban neighborhoods. In this sense, this concept reinforces the idea of how *barrios* are, in reality, self-organized systems that may hold different and alternatives futures. Although they have no plans designed for them, these areas can work as interconnected networks that exhibit everyday life. Johnson (2001) presented the relationship between self-organized systems and urban places that do not follow set laws, but instead arrange themselves by showing differences in alternatives and interests.

Highly populated areas, like informal areas, display high levels of complexity. Computational logics can help addressing these complex systems by providing data in real time to explore self-organization forms people use to make decisions (Dourish, 1999). The levels of complexity are understood in terms of each system's freedom and robustness. In other words, one can argue that the sociospatial production of space in informal settlements, similar to a self-organized system, exhibits freedom—yet not anticipation nor planning. The result (the sociospatial production) proves closer to people's needs and expectations. In this regard, we can understand situations of complexity through the freedom of self-organized systems (Maldonado, 2016).

With this framework, we can argue that self-organization as an important condition of computational systems allows us to think of the *smart* in a different way. That is, we can consider *smart* as more than an implementation and use of technologies but as an emerging relationship with no external influences or hierarchical orders. In this sense, smart could represent an alternative urbanism, a way to understand communities in coliving and cooperative modes, inhabiting the shared urban spaces that self-organization has produced.

The everyday

The sudden, uncontrollable, and irreversible lead to the production of the "everyday." This is how alternative possibilities merge with uncontrolled and unplanned experiences (Kosko, 2000). In this sense, features in common to everyday and smart technologies include the random and the autonomous that produce daily life and can be linked to computational systems (Barros and Sobreira, 2002).

The "everyday" describes a process that simultaneously creates and is created that emerges in autonomous and random ways, with no particular authors nor planning strategies. The concept simply refers to people living and transforming their space on a daily basis (Dourish and Button, 1998). The everyday just happens or occurs without a teleological vision or particular design by planners or architects. Informal space production is considered an exercise of diversity and uncertainty, randomness, and autonomy, fuelled by everyday actions and interactions (Beza and Hernández-García, 2018). Everyday activities developed in informal settlements give shape and life to these places, with public and private spaces resulting from these practices. Yet not only is the physical space transformed; social interactions and cultural expressiveness merging with the space also change and take particular shapes and meanings. For example, streets and street corners get new meanings, not only as places to meet but also as reference points for community, as gathering places for young people to rap, as places to have public and open-air picnics, or to give community and political information and speeches, and so on (Beza and Hernández-García, 2018). In addition, the space changes accordingly: temporarily with platforms, chairs, or marquees or more permanently possibly with the closure of a street to make it pedestrianized, with flowerpots, with permanent stands, etc.

Jacobs (1993) pointed out the relationship between uncertainty and everyday life in cities, understanding urban life as a distant phenomenon from centralized planning. In doing so, urban sociospatiality is developed, randomly and uncertainly. In this sense, Certeau (1993) analyzed aspects of the social construction of space concerning the improvised everyday use of the city, in direct contrast to ideas and theories of planners and urban developers. He argued that the city is built from the everyday journeys and from the great and small actions of the ordinary users. For Certeau (1984, p. XIV), the everyday is about "the ways to act [...] through which the actor re-appropriates the organised space with socio-cultural production techniques." And everyday practices may "bring to light clandestine forms of creativity [and self-organisation] of the group or of the dispersed individual, tactical or substitute, trapped within the nets of discipline" (1984, pp. XIV–XV).

In this sense the everyday arises as an important characteristic of informal settlements, like a place of multiple possibilities. The residents' daily and random activities in the *barrio* involve creativity. High levels of creativity are needed to develop urban and housing facilities with little help from government and experts. This concept is explained in terms of the everyday as found in *barrios*, in relationship with space and time, with diverse dimensions and participants. The everyday appears suddenly and surprisingly, in relation to processes of transformation of social life and place. One can understand these processes and outcomes through computers' simulation with artificial systems based on evolutionary algorithms—leading to answers about how the everyday emerges in societies and under what conditions and times. This computing process leads to a specific use of smart technologies in cities. It can also help to understand how the everyday is related to the act of dwelling and living.

Self-organization, as explained by Slaughter (1996), reveals the importance of the everyday. This social construction of possibilities may become a social construction of smart futures, expressed in the production of space. Its spontaneity, difference, and lack of order may generate alternatives for urban design. "The perspective of daily life is the point of view from the bottom which makes reality visible ..." (The Research Group for the New Everyday Life, quoted in Madanipour, 1996, p. 73). In this sense, one can understand the production of space in *barrios* as smart social practices emerging from everyday actions and interactions. In this sense, Slaughter described the space produced by self-organization and everyday: "The notion of the emergence can be approached concretely by observing its use in everyday life (walking, driving, etc.)" (Slaughter, 1996, p. 755). What is more, the small actions and everyday pursuits of social life can produce and transform the living space.

With this framework in place, we suggest another possibility to think of the connection between the smart and the everyday as interactions of everyday networks mediated by technological possibilities. In other words, from each interaction of everyday activities in informal settlements emerge multiple horizons and possibilities for a shared production of space and urban life. In this sense, we can understand the smart and particularly smart urban agglomerations as smart networks, in which interactions between networks occur at every minute when we walk, play, work, live, and dream.

Reconceptualizing smart and informal

The focus of this chapter aimed to provide a theoretical discussion and lay the groundwork for a new view and different understanding of urban smart technologies, introducing the concept that their normally assumed link to the formal and planned can give way to the informal and unplanned as well. The arguments sought to avoid the usual connection with technological production as applied to cities, while exploring other forms of technology as a tool to measure and interpret the everyday in informal settlements like self-organized systems. This focus intended to avoid associating the smart with only the controlled, top-down managed, and rational aspects of urban development. Instead the chapter's goal lay in offering possible associations of smart technologies with other forms of sociospatial production and the social production of knowledge deriving from the people, their dwelling strategies, and the production of informal urban space.

Informal settlements are seen as innovative and smart from the point of view of complexity. In terms of architecture and urbanism, these environments are subject to different interpretations. Among them lies the idea that the informal implies not an absence of technology but a failure to recognize it. One can see the relationship between smart technologies and the production of place in informal settlements, not only in the use of public transport, public space, or mobile devices such as phones, tablets, or computers for registering or measuring trajectories. Rather, one can also understand information and communication technologies as open logics to generate information and trace everyday practices in the processes of self-organized transformation and improvement of urban space.

Although these settlements possess visible potential and offer patterns of self-social and spatial organization, they have to be discovered and valued. This need for new enlightenment and understanding proves a characteristic of complexity and now of smart, due to its hybrid nature. The understanding of smart can undergo transformation through a new view of informal, which reconsiders the qualities of these self-organized environments. These settlements reflect the interpretations of the everyday actions of the people in and with their living spaces. Innovation originating through the everyday leads to free choices. The built environment in informal settlements and their self-organized social and spatial processes are echoed by the use of smart technologies expressed in computing logics. This can be seen, for example, in the use of mobile phones in informal settlements that can help to self-organize people to face their social and spatial problems by getting together, exchange information, meet their needs, and so on, in two words: building networks. These are of course free choices that people make, based on everyday needs, struggles, and expectations. In this sense, smart technologies can help to understand these self-organized processes, measure, and simulate them. Therefore a wide, rich selection and presentation of choices and languages that smart technologies may measure confirm the need to reconceptualize smart and informal. They convey a range of possible and open meanings in which smart became informal, literally.

Within the aforementioned panorama, we argue that smart cities are, in essence, networks of knowledge, actions, and selection of choices—in the same sense as are informal settlements. Therefore one can define the smart by looking at the common features of complex networks, in terms of technologies and the people's everyday practices. These practices are understood as smart and informal if we think of them as self-organized processes or even better as self-organized adaptive systems. Smart technologies and smart cities then become forms of knowledge and technology transitions to more of a way of thinking than a tool.

With an understanding of smart urbanism increasingly seen more in terms of people-based practice and knowledge, it becomes clearer how this can relate to self-organization and the everyday as practices that configure urban life. In turn, these practices can imply that the intelligence of knowledge networks is ephemeral and ordinary at the same time. Such practices involving smart and informal show the value of "ordinary people" and "ordinary actions" to develop the social and the spatial. Smart can contribute to the definition of the informal, and the informal can help understand the smart. With this in mind, we may find it possible to confirm the assertions of some authors (Kellett, 2002; Harvey, 1996; among others), who argue that the built environment must represent the emergence of self-organized social groups and the emergence of multiple and diverse possibilities based on the everyday.

Crawford (1999, p. 28) defines the production of space as that which presents an "absence of an [anticipated] identity itself," implying the permanent change and new construction of meaning of people-space interactions. The notions of space and people are transformed and permanently redefined to encompass the transitory and changing activities they include and perform, particularly within the case of informal settlements. Likewise, the term *everyday*, as more than "fixated in time and space, undergoes a permanent transformation in time since users reinterpret and reorganise it on a permanent basis" (Crawford, 1995, p. 5). We argue that this kind of everyday comes from self-organized systems as an expression of smart urbanism.

We recommend that other scholars and practitioners begin to look at smart and the informal as collective and adaptive self-organized systems fuelled by everyday practices, where the social emerges as everyday creativity. In the end, *smart* and *informal* describe the life that occurs essentially unplanned, undetermined, and constantly transformed by uncertainty and freedom.

References

Alsayyad, N., 1993. Informal housing in a comparative perspective. On squatting, culture, and development in a Latin American and a Middle Eastern context. Rev. Urban Reg. Dev. Stud. 5, 4–15.

Alsayyad, N., Roy, A. (Eds.), 2004. Urban Informality: Transnational Perspectives From the Middle East, Latin America, and South Asia. Lexington Books, Oxford.

Barros, J., Sobreira, F., 2002. City of Slums: self-organisation across scales. UCL Working Papers, Series, 55, pp. 1–10. In: Paper presented at the International Conference on Complex Systems (ICCS2002), Nashua, NH, USA, June 9–14, pp. 1–12.

Beza, B., Hernández-García, J., 2018. From placemaking to sustainability citizenship: an evolution in the understanding of community realised public spaces in Bogotá's informal settlements. J. Place Manage. Dev. 11 (2), 192–207

Brillembourg, A., Klumpner, H., 2010. Roles of engagement: Caracas and the informal city. In: Hernandez, F., Kellett, P., Allen, L. (Eds.), Rethinking the Informal City: Critical Perspectives From Latin America. Berghahn Books, Oxford, pp. 119–136.

Certeau, M., 1984. The Practice of Everyday Life. University of California Press, Berkeley.

Certeau, M., 1993. Walking in the city. In: During, S. (Ed.), The Cultural Studies Reader. Routledge, Londres, pp. 126–133.

Chambers, J., Evans, J., 2020. Informal urbanism and the Internet of Things: reliability, trust and the reconfiguration of infrastructure. Urban Stud. https://doi.org/10.1177/0042098019890798.

Conolly, P., 2013. La ciudad y el hábitat popular: Paradigma latinoamericano en Teorías sobre la ciudad en América Latina. Universidad Autónoma Metropolitana, México.

Crawford, M., 1995. Contesting the public realm: struggles over public space in Los Angeles. J. Archit. Educ. 49 (1), 4–9.

Crawford, M., 1999. Blurring the boundaries: public space and private life. In: Leighton, J., Crawford, M., Kaliski, J. (Eds.), Everyday Urbanism. Monacelli Press, New York, pp. 22–35.

Dourish, P., 1995. Accounting for system behaviour: representation, reflection and resourceful action. In: Procedures of Computers in Context 95, Aarhus, Denmark, August.

Dourish, P., 1999. Embodied interaction: exploring the foundations of a new approach to human-computer interaction. Transactions on Human-Computer Interaction. Draft paper.

Dourish, P., Button, G., 1998. On technomethodology: foundational relationships between ethnomethodology and system design. Hum.-Comput. Interact. 13 (4), 395–432.
Epstein, J.M., 2006. Generative Social Science. Studies in Agent Based Computational Modelling. Princeton University Press, New Jersey.
Feldman, P., Girolimo, U., 2018. Smart City: Nueva cara del empresarialismo urbano? Ciudades 120 (Octubre-Diciembre), 25–33.
Fiori, J., Brandao, Z., 2010. Spatial strategies and urban policy: urbanism and poverty reduction in the Favelas of Rio de Janeiro. In: Hernandez, F., Kellett, P., Allen, L. (Eds.), Rethinking the Informal City: Critical Perspectives From Latin America. Berghahn Books, Oxford, pp. 181–206.
Gilbert, N., 2000. The Simulation of Social Processes. Cemagref, Clermont Ferrand.
Harvey, D., 1996. Social process and spatial form. In: Jewson, N., Macgregor, S. (Eds.), Transforming Cities: Contested Governance and New Spatial Divisions. Routledge, London, pp. 19–27.
Hernández, I., 2016. Evolución abierta y sin límite: encontrando lo improbable y lo inesperado. In: Hernandez, I. (Ed.), Estética de los mundos posibles. Pontificia Universidad Javeriana, Bogotá, pp. 51–68.
Hernández, I., Niño, R., Hernández-Garcia, J., 2014. Estética, sistemas complejos adaptativos y ciudad. In: Estética y sistemas abiertos. Pontificia Universidad Javeriana, Bogotá, pp. 17–56.
Hernández-Garcia, J., 2013. Public Space in Informal Settlements, The Barrios of Bogotá. Cambridge Scholars Publishing, Newcastle Upon Tyne.
Heylighen, F., Gershenson, C., 2003. The meaning of self-organization in computing. In: IEEE Intelligent Systems, Section Trends & Controversies – Self-Organization and Information Systems, May/June 2003.
Holland, J., 2004. El orden oculto: de cómo la adaptación crea la complejidad. Fondo de Cultura Económica, México.
Jacobs, J., 1993. The city unbound: qualitative approaches to the city. Urban Stud. 30 (4–5), 827–848.
Johnson, S., 2001. Sistemas emergentes: O qué tienen en común hormigas, neuronas, ciudades y software. Turner, Madrid.
Kauffman, S., 2003. Investigaciones: complejidad, autoorganización y nuevas leyes para una biología general. Tusquets, New York.
Kellett, P., 2002. The construction of home in the informal city. J. Roman. Stud. 2 (3), 17–31.
Kitchin, R., Lauriault, T.P., 2014. Towards critical data studies: charting and unpacking data assemblages and their work. In: The Programmable City. National University of Ireland, Maynooth.
Kitchin, C., Evans, H., Mac, D., 2017. Smart cities, urban technocrats, epistemic communities and advocacy conditions. In: The Programmable City. Workshop: A New Technocracy, March 20–21. University of Amsterdam, Amsterdam.
Kosko, B., 2000. El futuro borroso o el cielo en un chip. Crítica, Barcelona.
Krivy, M., 2018. Towards a critique of cybernetic urbanism: the Smart city and the society of control. Plan. Theory 17 (1), 8–30.
Lewis, W., 1955. The Theory of Economic Growth. Allen and Unwin, London.
Madanipour, A., 1996. Design of Urban Space: An Inquiry Into a Social-Spatial Process. Wiley, Chichester.
Maldonado, C., 2016. Complejidad de la experiencia humana. In: García, I.H. (Ed.), Estética de los mundos posibles. Pontificia Universidad Javeriana, Bogotá, pp. 245–258.
Prigogine, I., Stengers, I., 1997. La nueva alianza: metamorfosis de la ciencia. Alianza, Madrid.
Rauws, W., 2016. Civic intiatives in urban development: self-governance versus self-organisation in planning practice. Town Plan. Rev. 87 (3), 339–361.
Roy, A., 2009. The 21st-century metropolis: new geographies of theory. Reg. Stud. 43 (6), 819–830.

Sassen, S., 2013. https://www.forbes.com/sites/techonomy/2013/11/10/open-sourcing-the-neighborhood/#e73a6924df04. (Accessed 5 May 2020).

Slaughter, R., 1996. Future studies: from individual to social capacity. Futures 28 (8), 751–762.

Vanguardia, 2013. La revolución del celular en las favelas. https://www.vanguardia.com/tecnologia/la-revolucion-del-celular-en-las-favelas-FAVL214652. (Accessed 25 June 2020).

Zaera-Polo, A., Anderson, F. (Eds.), 2017. Imminent Commons: The Expanded City. Seoul Biennale of Architecture and Urbanism. AZML.

Author Biography

Jaime Hernández-García is Professor at the Department of Aesthetics in School of Architecture and Design of the Pontificia Universidad Javeriana in Bogotá, Colombia. He is an Architect with a Master in Architecture from the University of York and a PhD from the University of Newcastle, UK. He has more than 20 years of teaching and research experience. His research interests and expertise are in informal settlements/housing, open/public space, conflict/peace studies, community participation, and local knowledge/expression. He has published widely in Spanish and English.

Iliana Hernández-García is Professor at the Department of Aesthetics in School of Architecture and Design of the Pontificia Universidad Javeriana in Bogotá, Colombia. She is an Architect with a Master in Aesthetics and a PhD in Arts and Arts Sciences from the University of Sorbonne, Paris, France. She teaches at the PhD Programme in Social Sciences and at the Master in Artificial Landscapes. Her research expertise is in virtual words, architecture/science/technology, aesthetics/arts, artificial life and built environments, and posthuman studies.

Chapter 17

Situating urban smartness: ICTs and infrastructure in Nairobi's informal areas

Prince K. Guma[a]
[a]*British Institute in Eastern Africa, Nairobi, Kenya*

Chapter outline

Introduction

Since the incoming of mobile telephony in urban Africa in the 1990s, a variety of mobile software platforms have unprecedentedly emerged. In many African cities, ICT platforms have particularly targeted informal areas, where they have led to the rise of mediated infrastructure services by private and public providers. Here, ICTs have become not only merely active players for service delivery but also key elements in the structural configuration of the urban spheres (Guma, 2019). They have come to define everyday urban life, depicting situated dynamics of urbanity and geographies of engagement in their reliance on mobile media, devices, and technologies as tools of access to basic services. While some aspects of these developments—including ICT integration for infrastructure delivery—have aroused public interest (see, e.g., Guma et al., 2019; Guma and Monstadt, 2020; Chambers and Evans, 2020), not much concerted effort has been designated to the variegated and place-based materiality of ICT-based infrastructure within contexts of urban informality in African cities. It is this contribution that this chapter pursues.

I examine the different ways through which urban infrastructure services in informal areas are being automated and how this automation is being reconstituted by its very situatedness in informal urban settings. In doing so, I build on

Shaping Smart for Better Cities. https://doi.org/10.1016/B978-0-12-818636-7.00005-6

the growing prominence of ICT in the provision of urban services in the contexts of the global South (e.g., Datta, 2015; Luque-Ayala and Marvin, 2015; Kitchin, 2015; Odendaal, 2014; Guma, 2019; Guma et al., 2019; Chambers and Evans, 2020). I contribute to Aurigi and Odendaal's (2020, p. 1) appeal for enhancing the social sustainability of smart urban spheres in the South by "framing and conceiving smart urban initiatives and schemes in a highly context-sensitive way". Accordingly, I take the cases of Mathare and Kibera, two of the largest informal areas in Nairobi, Kenya's capital. Mathare covers an area of approximately 73 ha, while Kibera covers 225–250 ha. Mathare is situated 2 miles northeast of Nairobi's central business district and Kibera only 3.1 miles southwest of Nairobi city. According to recent estimates from the Central Bureau of Statistics, while Mathare's population is estimated to be 90,000, Kibera's is up to 170,000. Unlike Kibera, Mathare is structurally a flood plain along the Nairobi River. Both settlements are marked by the presence of stark and distinct informal, low-cost shelters that are privately provided by enterprising groups both within and outside of the settlements.

I draw from two projects shaped by their integration of ICTs, sculpted as pilot projects, marketed as social and periurban community-based initiatives, and deployed with the aim of dealing with the challenges of urban service and infrastructure supply. The first project is commonly referred to as Water ATMs, designed and funded by Grundfos Lifelink, a Danish water engineering private company, and deployed in partnership with Nairobi City Water and Sewerage Company (NCWSC), the largest water utility provider in Kenya. The second project is M-Maji ("M" for mobile and "Maji" Swahili for water), a mobile-based data management platform for sharing information about water availability in informal areas, deployed by a self-organized "goodwill" collective. Beyond their design and nature as strategic business concepts by private and public companies, this chapter examines these projects as infrastructure-centric solutions to urban challenges of water planning and development in Nairobi's informal areas and has been sought to address challenges to delivery of services in informal areas, including curtail nonpayment and illegal taps in such areas. As such, I examine these projects as infrastructure ensembles that are relational and situated.

In making this contribution, I demonstrate processes through which ICT deployed by private and public actors are being appropriated by various stakeholders according to a diverse set of considerations (Jaglin, 2008). Accordingly, my approach is rooted in perspectives that offer a more nuanced and creative way of thinking about infrastructure development projects beyond technologically deterministic understandings. I argue that while artifacts are designed to shape user practices and enforce discipline upon users, they are not necessarily neutral; neither do they enter people's lives as closed-ended domains. Instead, they are exploited by the wide array of users who are always bound to "hybridize" and "informalize" them. As such, situated urban spheres are not simply laboratories and test grounds for imported and "black-boxed" technologies, but spheres in which technologies become "molded and appropriated to suit livelihoods" (Odendaal, 2014, p. 32). Thus I examine water technological infrastructures as systems that are always in flux, open to processes of tinkering, refashioning, and recalibration wherein they are constantly being "reworked"; "always relative to the different constituencies, populations, and agencies at work" (Dourish and Bell, 2011, p. 194).

Methodologically, I employ the empirical strategy featuring the principle of data triangulation and basing on three datasets from research carried out in Nairobi since March 2015. The first dataset is from a review of policy, strategy, and related documents pertaining to innovation in East Africa. The second dataset drew from interviews. I interviewed (and engaged with) different stakeholders and joined them during their day-to-day activities for a combined period of 4 weeks in specific days of the week. Additionally, I conducted a total of 18 in-home interviews, collecting a range of viewpoints from household users and consumers who had a variety of occupations such as teachers, shop and kiosk owners, and jua-kali worker. I also talked with engineers and officers from NCWSC, infrastructure development professionals and practitioners, and consultants and experts from the private sector. Interviews were mostly largely semistructured and open-ended in style. The third dataset is primarily borrowed from observations that emerged out of key structured city walks in Mathare and Kibera (Krisch, 2009). This constituted a type of observation "to find out more about a social space through a collection of impressions and perceptions gained through direct field experience" (Spatscheck and Wolf-ostermann, 2009, pp. 9–10).

Accordingly the chapter proceeds with a review of literature on urban infrastructure in the digital age, with the aim to reframe the debate by focusing its analysis on the informal nature of sociotechnical innovations. Then, I delineate the heterogeneous landscape of water supply in Nairobi's informal areas. I examine the nature of ICT-based water projects in Mathare and Kibera and how these projects have become reconfigured in these contexts of urban informality. I conclude that processes in the deployment of ICT-based infrastructure development projects in informal settlements of Nairobi radically displace the notion of the ideal type or standard of urban smartness. As such, through their contextualization and appropriations within their informal settings, they reflect the situated nature of ICTs and infrastructures in the global South.

Researching ICTs and urban infrastructure in the global South

ICTs have attracted a growing interest among scholars in the social sciences. At the beginning of the new century, different studies played an integral role of bringing to attention the incremental impact of ICT toward enhancing the speed and proficiency of organizational processes of service delivery (e.g., Oyedemi, 2004, 2009; Foster and Briceño-Garmendia, 2009). While making significant contributions for understanding the emergent ICT platforms in service provisioning in the global South, this literature mostly possesses a deterministic and evangelistic view of ICT platforms as a transformative and modernizing force. Whether celebratory or dismissive, this literature is mostly based on a model of media transmission, possessing the potential to lead to direct effects such as shaping urban development processes. It places significant prospects on the potential of the mobile phone for "leapfrogging" conventional development process (Etzo and Collender, 2010, p. 662). It is "largely concerned with developmental objectives and technology diffusion" (Odendaal, 2011; Odendaal, 2010, p. 41). It examines ICT platforms "with the dubious underlying assumption that there exists an inherent way that ICTs foster development" (Kamga and Cishahayo, 2010). As such, this work bears resemblance to "ICT for development" approaches, which tend to see technology as a modernizing force to be introduced into underdeveloped settings. As such, it aligns with first generation of literature that views ICT platforms as monolithic or placeless technologies as opposed to their relevance as tools used by diverse people, in diverse real-world locations, embedded in urban environments and shaped by different "ordinary" facets and routinized place-based practices.

Beyond this work, however, there has been a rise of contemporary studies engaged in examining new and emergent ICT-based infrastructures in African urban settings. Particularly in the water sector, research in this area has noticeably evolved with studies endeavoring to address this concern by investigating the advent and role of ICT-based payments (Krolikowski, 2014; Hope et al., 2011; Foster et al., 2012) and prepayment technologies (Loftus, 2009; von Schnitzler, 2008, 2013) in varied urban contexts. Here, scholars have examined how these technologies create automated and virtually cashless payment pathways citing business models, structural transformations, and users' everyday practices. This work has examined the potential of these technologies for precipitating profound sociospatial transformations in the service provisioning sector—shaping higher cost recovery and effective forms of load management between supply and demand (see, Guma, 2019; Guma et al., 2019). It has portrayed far-reaching expectations about the role and impact of smart technology in the supply and use of basic services in the global South (ibid.), in contributing to the attainment of smart urbanism (ibid.).

Over all, this literature has been beneficial for explicating the ways in which smart technologies become integral components for addressing multitudes of social, economic, and environmental problems. It has examined the procedural and technical shifts in service delivery enabled by ICTs (Hope et al., 2011; Foster et al., 2012), the role of ICTs in solving complex urban problems of inadequate access (Heymans et al., 2014), the extant effects in organizing and regulating urban territories (Baptista, 2015; von Schnitzler, 2008), and the resultant sociospatial temporalities (Odendaal, 2014; Guma, 2019). However, despite the significant contributions of this literature, focus on the situated nature of ICT-driven urbanisms, particularly at the community scale of the city, still remains lacking. Especially for Africa contexts, more research is needed to better understand ordinary materializations of ICT-based infrastructure deployments and appropriations within informal settlements.

Thus there still remains room for research that better interrogates not only how ICT-based infrastructures are shaped or shape different populations in diverse real-world locations but also how they are embedded in urban environments and shaped by different "ordinary" facets and routinized place-based practices (Robinson, 2006). In other words, there is need for a reframing of urban infrastructures within located geographies in the digital age. Such a reframing needs to expound on the new and emergent research agenda on urban smartness and its situatedness in a specific context. It is this contribution that incites this study. I examine the advent of ICTs deployed by urban utilities and private organizations and the local articulations of ICT-based infrastructures for urban development in situated contexts of an African city. As such, my interest lies in how the proliferation of ICTs shapes everyday infrastructures and urban lives in informal settlements. I am interested, in part, in how these technologies, for utilities, enable constant access to information, remote payment process and bill processing, and improve accounting formalities in the delivery of networked infrastructure within informal settlements, formerly unserved or underserved by the state. In so doing, I seek a better understanding of ICT-based infrastructure in informal urban areas, problematizing them not only as static and technocentered domains but also as open and dynamic systems whose configuration is reliant on their situatedness (see, e.g., Guma and Monstadt, 2020).

As such, I transcend a common focus on the idealized but unrealized visions that often tend to dominate social imaginaries and critiques of technologically mediated cities and infrastructures in the contemporary present. By examining links between technology-focused innovations and the messy realities and ambivalences of these innovations as reflected through local institutions and urban areas, I present a more nuanced picture of the discursive terrain of ICT platforms within the context of particular urban areas of the city. Accordingly, I offer a more situated approach, following Aurigi and Odendaal's (2020, p. 1) lead for a "significantly more socially sustainable" understanding ICT-based infrastructure, one that transcends a "prevalently prepackaged, product-like version of the smart city, as the 'smart in a box.'" Incorporating ICT-based infrastructures into context-sensitive placemaking and city-making, I demonstrate how the technological scripts of these infrastructures, through their heterogeneous reworkings, are not "future proof" (Edwards et al., 2009, p. 371); neither are they "universally explicable" (Baumgardt, 2018, p. 47) even when their developers and planners might appear to think of them as such. I pay better attention to the context-dependent nature of urban infrastructure development in the digital age (Datta, 2015; Aurigi and Odendaal, 2020; Guma and Monstadt, 2020) and transcend widely critiqued "stereotypically top-down, neoliberal, and repressive visions of the smart city" (Shelton and Lodato, 2019). I argue that such an attempt is important for adding nuance to extant studies in the global

South, particularly in urban Africa on the rise of the infrastructure automation within contexts of urban informality.

ICTs and infrastructure in Nairobi's informal areas

As it has for long been difficult for public and private sectors to plan and provide basic services to Nairobi's informal areas given their informal land tenure systems, a key challenge and barrier to access of reliable services has persisted. Urban authorities in Kenya have acted not only to prohibit the construction and development of permanent structures but also to restrict the provision of infrastructural services to the community. As such a large number of urban populations have for long remained either completely unserviced or faced sporadic difficulties in the midst of water rationing in the city (Wamuchiru, 2017). This state of exception constitutes a rather more generalized mode of urbanization in Nairobi. Predictably, it has attracted audacious groups of people who have "smuggled" and circumvented the county's main grid from neighboring suburbs by using underground pipes, creating syndicates of de facto self-assembled chain distributions of crisscrossing spaghetti lines synonymous with DIY engineering. This class of residents, often referred to as gangs or cartels, has made a substantial gain and profit through unregulated and untaxed redistribution in the community. Over the years, it has engaged in gray and organic infrastructural networks and has grown and developed to became a dominant force in the community, reengineering far-reaching sociotechnical influence and transformation of local processes of production while, at the same time, bypassing key county and national planning requirements and recommendations to urban life and development. Landscapes of water provision in informal areas, including Mathare and Kibera, have become complex and dynamic, with formations of operation different and outside of the realm of the central networks gradually evolving. These arrangements speak to the heterogeneous nature of southern urbanism (Robinson, 2006; Roy and Ong, 2011) where infrastructures evolve in ways that diverge from the "networked city" ideal (Coutard and Rutherford, 2015). While more prevalent in the poor, fringe, and unplanned settlements where inhabitants rely more on informal alternatives to meet water needs, they are also frequent in rich and business communities (Mutisya and Yarime, 2011). And yet, while such alternatives may be able to meet most immediate needs and demands of the urban communities, they are hardly, if ever, sustainable.

Since 2008 NCWSC—the largest water provider in Nairobi—has set out to provide formal access to informal areas through a process of institutional change, forming a specialized unit referred to as the Informal Settlements Department (ISD) solely responsible for extending water and sanitation services to low-income areas in the city (WSUP, 2018). This unit has implemented a set of structural reforms, strategies, and service delivery approaches to facilitate the realization of citywide service provision. As such the unit has devoted resources to mobilize stakeholders including private companies and NGOs with a view of extending services to low-income areas and reducing levels of nonrevenue water in these areas (WSUP, 2018). Through this unit, NCWSC has initiated different projects in partnership with private companies and other development cohorts to expand access to clean water and sanitation beyond wealthier areas (NCWSC, 2014). Among its advances the ISD has leveraged an expanded role of new technological solutions and sought ICTs in its operations both for enabling access to basic services such as water and engineering new sociotechnical developments for the benefit of urban populations (Ndugu, 2015; United Nations, 2015). Challenged by obduracies of service provisioning and urban informality, NCWSC has innovated a series of projects targeted for low-income areas, employing them with the aim to realize overarching goals of expanding its network grids and increasing revenue streams. This has largely been through collaborations and close partnerships with development agencies; financial institutions including microfinance institutions; targeted state and private agencies and initiatives including the Kenya Informal Settlements Improvement Project; and ICT companies such as mobile network operators.

However, the utilization of ICTs for service delivery in Nairobi's informal settlements still echoes Benner's (2014, p. 237) claim of how the "adoption of digital technologies remains incomplete in many traditional industries". Rightly so, Brenner states how much of what has been adopted to date provides at most incremental advances to the rather splintered infrastructure network grids in the city (ibid.). Notwithstanding, these advances are nonetheless transformative—particularly in the water sector (Ndugu, 2015). ICTs are increasingly solving complex urban problems of service delivery in formerly unserved and underserved communities. They are gradually reorganizing water access, boosting "smart" and sustainable service delivery in the city (e.g., NCC, 2014). They have been applauded as tools with the capability to overcome sociospatial disparities of urban splintering and lead to the achievement of equitable access to basic services. Ultimately, ICTs have the potential to shift the identity and image of the city's diverse urban areas and "to turn Nairobi into a modern, smart, digital city of choice to live" (Guma and Monstadt, 2020, p. 13; Sonko, 2017). Thus they have come to reflect the urge for private and public companies to innovate modalities of water delivery and access for vulnerable and informal urban areas. This urge is reflected in the city county's visions and priorities that have become earmarked in the Vision 2030 and Nairobi Metro 2030 Strategy documents for Nairobi.

Water ATMs in Mathare

Water ATMs are designed and funded by Grundfos Lifelink, a Danish water engineering private company. They have been deployed in urban informal and rural remote areas of Kenya. The very first ATMs (such as the one in Image 1) were deployed in Mathare in June 2015 under a partnership between Grundfos and NCWSC. These ATMs are connected to the centralized water supply grid

IMAGE 1 A water ATM in Mathare, 2016. *(Photo taken by author.)*

provided by NCWSC and regulated to meet the standards required by state and local regulations. The utility company handles their funding, management, repair, and awareness raising. For NCWSC, water ATMs are part of a targeted and collaborated initiative by the company that focuses on informal settlements at an institutional level. The ATMs are thus part of a wider complex of sustained and overarching effort at extending and formalizing access, and they serve as a solution to water access in informal settlements.

Water ATMs are customized for low-income contexts including informal urban areas. As such, they are a low-tech and low-cost modality that enables access to water at a nominal charge. They are self-contained and automated and are designed with the ability to store clean water and provide this water in the required quantity and at any hour convenient to the consumers and providers. Also, their dispensing machines are an architectural innovation of the normal water kiosks in Mathare (see Image 1). A typical ATM has a built-in filtration system. This kiosk comes in different variants such as stand-alone dispensing units, container booths, and delivery vans with dispensers. The machine is typically composed of an overhead water storage tank. This tank is designed for water storage (in some cases) in case of rationing to allow for 24-h supply. The kiosk is mostly mounted inside the dwellings of the users, in visible and accessible locations to allow monitoring of consumption and maintenance. The kiosk is composed of an ATM dispenser: an electronic water meter is mounted on the window of a water kiosk and is activated by a water ATM card for dispensing water. It is connected to an integrated and cloud-based water management system (WMS), a system that processes and reports all water data and credit transactions and operations. The water ATMs consist of smart or intelligent water ATM–designed and ICT-based machines that accept payment from prepaid smart cards.

Upon loading credit the water consumer uses the smart card to make their payments. They load credits onto their smart cards, either through a local water credit vendor or via different mobile payment options provided by ICT firms in Kenya. Operating like vending machines, users swipe the smart cards against a reader in the water ATM to dispense water from the tap. They place or tap their card the smart card in the machine slot to see their credit balance and the water price. Upon verifying that the requested amount does not exceed the user's balance, the reader is able to debit the account. The user then pushes the water button to start tapping water, which is metered on the display. When the process is completed, the consumer then pushes the water button again to stop the process.

In Mathare, at least four water ATMs have been deployed by NCWSC. Their everyday management is placed on the village chairperson in association with a committee constituted by residents. They are attended to by a kiosk operator, a sociologist, and public health specialist, whose basic work is simple: to act as the custodian, protect the substructure from vandalism or misuse, and speak on behalf of Grundfos Lifelink and NCWSC. Residents, especially women and younger girls, mentioned that they do not have to worry about water any more or move long distances to fetch water. They argued that now they are able to spend doing other things. For instance, most users who compared the costs and the inflated prices from informal vendors argued that the new system was much more affordable (i.e., one KES for 10 gal of water—a fee that s subsidized by NCWSC). Other users mentioned that the smart cards were mostly reported as being "transparent and reliable" as "efficient for the water service provision." Thus many were positive about their experience with the system that is easy to use and provides them the option of buying water credits through their mobile phone. Particularly, they were happier with the charges as many claimed to have been overcharged for the service by the ordinary informal vendors.

However, while the benefits of this intervention in the community cannot be denied, many issues about the technology were pointed out in the interviews. Firstly, I found during my site visits in the settlement that some residents resented the card-based option for water access. Thus I found that the cards had a weak security threshold and provided less user control when it came to authorizing payments and transactions from the card. So, this system had in part been forfeited. Instead of users owning and using their own cards, it was often the case that an agent—often from the utility—would be stationed at the water ATMs dispensers with a multiuse card that they would swipe on the machine on behalf of the customers who needed to fetch water. Upon fetching the needed water, customer paid the agent in cash or mobile money. Secondly, the water flow of the ATM was not very reliable. According to the residents, water was only released once in 4 days, and even on such days when water was expected to flow, there was less sufficient availability for residents. Thus, sometimes, residents had to buy water from other suppliers in the settlement including from

IMAGE 2 Water ATMs in Mathare, 2020. *(Photo taken by author.)*

the expensive private tankers. Thirdly, some water ATMs in the slum were at the time nonoperational. Those that were operational had become increasingly reconfigured (such as in the second photo under Image 2 above) and were now usurped by community groups in the settlement. These groups included committees, task forces, and associations, with the most common of these being water and sanitation committees and cooperatives as well as other tightly knit youth and women's groups in the settlement. In addition to offering water supply under formal arrangements, the water ATMs were also increasingly unearthing new and more hybrid and domesticated modalities. As such, in their assimilation of a diverse range of place-specific and situated authenticities, water ATMs have evolved quite contrary to expectation as can be witnessed in Image 2.

The M-Maji platform in Kibera

The M-Maji platform, a two-way mobile-based data management platform for sharing information about water availability in informal areas, was launched as a temporary solution for water access in Kibera aimed at formalizing an informal water value chain. Its application is designed to map clean water vending point of sales and enable residents to locate them through their mobile phone. Thus it allows registration of informal water vendors, submission of SMS notifications of their availability of water, daily rates, and location. Registered water vendors have the option to provide updated information of the last time the water was purified and the results of testing for quality purposes. These data

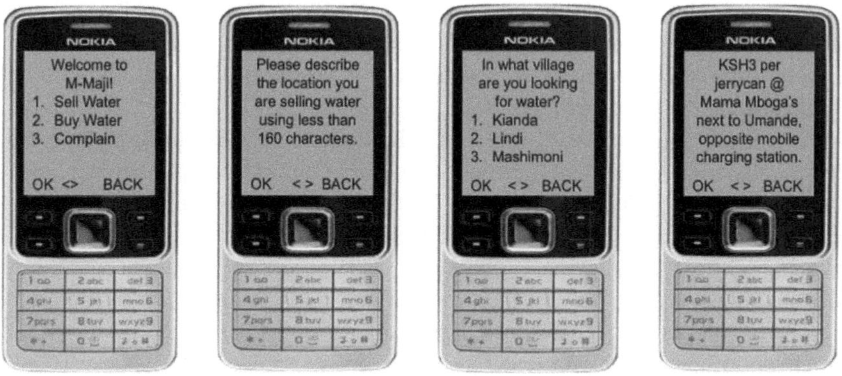

IMAGE 3 Workings of the M-Maji platform.

are stored in the M-Maji database in real time. With this information, water buyers, by initiating an Unstructured Supplementary Service Data (USSD) session with M-Maji database, are able to find the closest water vending point of sales with the location-relevant listings for the cheapest, cleanest water. Within the same USSD session, users can also update the listings by price, quality, ratings, or geographic location. Moreover the platform also can allow the M-Maji team to conduct surveys and water quality tests and to review complaints and vendor ratings to alert future customers of negative stories in terms of water quality and service levels (Cavaretta, 2013) (Image 3).

M-Maji is a mobile-for-development model. First, it employs a "human-centered design" guided by the philosophy of problem solving by designing products, processes, and experiences anchored with end users. Second, it is shaped by the urge for novelty, empowerment, and liberation of disadvantaged communities with better information about water availability, price, and quality. What they refer to as a "liberation technology" that emerges from "a real and sustained engagement with end users" (Jeon et al., 2011). Third, it improves human welfare and provides "real freedom" to poor consumers in Kibera. Fourth, M-Maji tackles the information side of the problem: its proponents recognized that increasingly mobile-based development projects in the water sector that targeted informal areas in Nairobi were focused on technological solutions and inattentive to user needs and the preferences and capacities of communities. They also recognized that communities in Kibera were increasingly struggling to locate and evaluate the quality of their water on a daily basis and that the lack of information pushed the prices of water to be higher than they normally would have been based on supply and demand.

Unlike the Water ATM project that is largely connected to and extends the centralized water supply grid by a public utility, M-Maji was conceived and deployed by a "goodwill" collective including a team of Stanford University students focused on the urge to design liberation technologies for underserved communities. It was funded through a grant from the Stanford Program on

Liberation Technologies, the Stanford Center for Philanthropy and Civil Society, the Stanford Center for Innovation in Global Health, and the Stanford Freeman Spogli Institute. It was deployed through collaboration with the Department of Health and Human Services (CDC), UnaMesa that is a public charity that works with the private sector to create technological innovations and platforms, and Wezatele, a company that provides mobility solutions in commerce, supply, distribution, and payment integration. The rationale behind the platform was to leapfrog institutional difficulties in delivering clean water by informal providers and thus seeks to alleviate problems of water access for residents and reduce information asymmetries and the time, money, and burden spent in searching for clean water in the settlement by providing a smart algorithm than word of mouth and improve the efficiency of informal water markets.

In Kibera, Umande Trust, a local community-based development organization, led the pilot implementation and marketed M-Maji to the community, collected and spot-checked accuracy of field data, and obtained user feedback. Moreover, it facilitated collective, bottom-up, and "grassroot" action and sought to create partnerships by organizing and coordinating self-organization efforts where numerous mobile community-led networks, bottom-up technosocial projects and semiformal initiatives have increasingly been adopted as a means to fulfill the need for utility services as electricity, sanitation, water supply, and refuse collection, thereby reinforcing and reclaiming old forms of community activism and citizen participation. This was sought to be based upon internally organized frameworks that constitute committees, task forces, and associations, with the most common of these being water and sanitation committees and cooperatives.

M-Maji in Kibera was launched to a lot of fanfare. Leaflets, posters, and brochures were printed and distributed, and marketing was carried out through community events, prizes (water tanks, water purification, endorsements, t-shirts, and radio advertisements on local radio (i.e., Pamoja FM)), pitches to clinics, community organizations, and women's groups. Speeches were given by politicians, and residents were increasingly registered onto the platform. The launch was described as a huge success (see Image 4), and a significant celebration took place in the settlement in anticipation of a new dawn.

During the first 6 months in operation, 7000 requests had been received and, on a daily basis, roughly 30 water vendors (out of 158 registered) advertised on the platform (Cavaretta, 2013). On many days, about 50–100 queries from users would get registered (ibid.). However, the M-Maji platform did not go far beyond the pilot stage. By the time of my fieldwork in 2015, M-Maji had lost all initial hopes of ever scaling up due to a variety of reasons. For example, the platform had limited financial capital for growth and development. This was in part due to the students' lack of ability to augment partnerships with state and private institutions including service providers and telecoms such as Safaricom, which by the time had fueled an extraordinary expansion in access to financial services. Moreover, there was no social or economic incentive for Safaricom to get involved. The targeted audience of the platform was already

IMAGE 4 The launch of M-Maji in Kibera, March 2012. *(Photos taken by Umande Trust.)*

using Safaricom anyway, and the project being a development rather than business model was not appealing for the company. The same goes for other technology companies who could have picked up the idea but did not believe in its ability to maximize profits and recuperate costs and investments. State institutions (including NCWSC) could not get themselves to recognize M-Maji as a formal actor and instead branded it as an informal operation. As it remains part of the city county's long-term strategy to get rid of informal coproviders, which the platform primarily targeted, it was increasingly hard for M-Maji to find state actors willing to partner.

But what is commendable about the M-Maji platform is that it leverages and builds upon Kibera's informality and lack of access to the central network grid. It is shaped by situated realities propelled by and through the settlements' ordinary realities and technologies such as the use of feature phones. It integrates the settlement's very own informal modalities of service provision into the application of the platform, thereby seeking to formalize the very informality of water supply and access in the settlement. This is an admirable aspect of the design and novelty of M-Maji. However, the limitations of this platform concern its inability to more critically engage with the local dynamics of place and context for purposes of realizing its goals and objectives. In other words, while M-Maji's proponents were successful in facilitating collective, bottom-up, and grassroots action at its initiation, further work ought to have been done to engage other stakeholders beyond plain reliance on goodwill actors in the community such as Umande Trust. More organization and coordination linking community-led networks, community activism, and citizen participation with tangible public and private sector organizations would have been important for complementing and sustaining the project beyond its limited duration.

Conclusion: Situating urban smartness

The infrastructure projects portrayed in this chapter illuminate the contingent nature of ICT platforms in informal areas that, while originally portrayed as autonomous, are increasingly complex, dynamic, and hybrid. For example, while formal technological designs reflect the signification of hegemonic ICT-based projects, informality has had a part to play in the development of ICT-based water infrastructures. The infrastructure projects studied have increasingly been shaped through everyday realities of urban informality. A case in point is how as architectural in situ innovations of the old or normal kiosks in Nairobi's informal areas, the Water ATMs are a symbolic shift from the practices of self-assembled syndicates of water access in informal areas. Where the informal regimes in the settlement had become synonymous with perpetual water theft and powerful gangs and cartels, ATMs were envisaged to minimize these issues for NCWSC and attract more consumers who Grundfos Lifelink and NCWSC believed required proper services. Thus the observed ICTs have evolved quite contrary to popular expectation and conventional wisdom. While Water ATMs have been more effective platform for Mathare, M-Maji can be classified as a completely failed platform, where this effectiveness-ineffectiveness and failure-success both derive from their situated realities within the context of urban informality.

While typically framed within the realm of urban spatial planning and redevelopment interventions—and are designed to promote the deployment of new service models and ICT-based platforms and innovations for a more efficient and attractive city—they have been reflective of the ways in which urban infrastructures materialize in different ways than envisioned. Obviously, this predicament poses a great challenge for city makers and planners intent at modernizing informal areas through ICT platforms and determined to "conform" to a globally sanctioned approach. It exposes the extent to which informal areas have informalized ICT platforms. This implies a debunking of conventional, distinct, and long-standing classifications. New ICT-based projects, while business and development oriented, are also social and enable communal and collective use. They are greatly shaped by the variegated nature of urban informality wherein the ingenuity of the urban residents in the settlement has led to the production of even more provisional and improvised configurations. Ultimately, they are socialized through a backdrop of ingenuity and innovation by urban residents who through their everyday process of urban survival and coping, renegotiate or interfere with their functioning and operation.

Despite numerous plans and policies to improve water access in informal areas through ICT platforms, key challenges and barriers of access persist. As noted earlier, partly due to their complex land tenure system, informal areas have seen city authorities prohibit not only to the construction and development of permanent structures but also the provision of infrastructural services to the community. As such a large number of urban populations remain either completely unserviced (by the main grid) or at least face sporadic difficulties mostly due to lack of proper piped infrastructure. The absence of basic service provision by the state in informal settlements has seen residents search other ways of survival employing such mechanisms as fetching water from neighboring places by foot, relying on social networks and infrastructures for survival, and employing animal carts or hand-push carts, vehicles, or tanker trucks or other localized means of motorbikes referred to as tuk-tuks or boda-bodas.

Beyond these, heterogeneous modalities of water supply still persist in the informal areas studied. Urban residents, for instance, also employ public standpipes, overhead water tanks and different vending options, and concentrated supply chains of vending that are dependent on strategic clusters of informal water distribution and connect from or bypass centralized grids and networks. Moreover, kiosks also remain some of the main sources of provision particularly for informal areas. Majority of the kiosks often illegally connect to the central grid and are sometimes controlled by black-market cartels. They sell water for twice or up to many times higher than the authorized tariff (i.e., by inflating prices up to 50 KES per 20 litters), thus making a substantial gain and profit. Thus the urban provisioning landscape in informal areas studied remains significantly covered by informal and self-organized entrepreneurs and coproviders—realities that automated water infrastructures in Mathare and Kibera still have to contend with.

What this means is that envisioning urban smartness in contexts of urban informality calls for the acknowledgment of the fact that such contexts are highly heterogeneous spaces that cannot entirely be served by a standard model. This reflects the extent to which local structures will always coevolve with ICT platforms in Nairobi, portraying the urban landscape of service provisioning in the digital age as one that remains shaped and counterpoised by private or informal actors and ambivalent outcomes that reflect scenarios in which the deployment of techno-utopian dreams in the form of ICT platforms falls short of expectations. These processes reflect the situated nature of urban smartness. Herein, urban smartness lies not in the imagined or standard globally sanctioned technological ideals for Southern or African cities like Nairobi, but rather in the ordinary and situated responses to such ideals. When viewed through water ATMs in Mathare and M-Maji in Kibera, urban smartness is embedded in the responses of the urban actors and residents. Thus it transcends standardization and serves as a reminder of how urban populations will always live beyond standard ideals and designs.

From the aforementioned summations an implication for urban planners and practitioners can be made concerning the need to consider the reality through which ICT platforms—particularly in contexts of urban informality—will coexist with others and be shaped in diverse ways. Such a consideration should be more effective than the belief in ideal models of technological design. Utility providers need to be more realistic and seek to achieve hybridized modalities that speak to the everyday contexts of the urban poor in these communities.

In this case utility providers need to be alert to the practice in which urban populations leverage patchwork and in situ incremental and self-organized infrastructures such as has been shown to be the cases provided in this chapter. Moreover, urban planners and practitioners could play a more significant role to facilitate a suitable environment in which new modalities could coexist with the networked ideal through appropriate technologies. Thus they need not approach ICT-based platforms as technocratic but as highly heterogeneous and context specific.

References

Aurigi, A., Odendaal, N., 2020. From "smart in the box" to "smart in the city": rethinking the socially sustainable smart city in context. J. Urban Technol., 1–16. https://doi.org/10.1080/10630732.2019.1704203.

Baptista, I., 2015. "We live on estimates": everyday practices of prepaid electricity and the urban condition in Maputo, Mozambique. Int. J. Urban Reg. Res. 39 (5), 1004–1019.

Baumgardt, L., 2018. In/visible infrastructure: thinking (along) with Martin Heidegger about infrastructural breakdowns in South Africa. Middle East-Topics and Arguments 10, 40–51.

Benner, C., 2014. New technologies and work transformations in African cities. In: The Routledge Handbook on Cities of the Global South. Routledge, Abingdon.

Cavaretta, A., 2013. Examining the Viability of Transportation Development to Reduce Urban Poverty in the Developing World Through Accessibility: The Case of Kibera-Nairobi, Kenya. Doctoral dissertation, University of Florida.

Chambers, J., Evans, J., 2020. Informal urbanism and the Internet of Things: reliability, trust and the reconfiguration of infrastructure. Urban Stud. https://doi.org/10.1177/0042098019890798.

Coutard, O., Rutherford, J. (Eds.), 2015. Beyond the Networked City: Infrastructure Reconfigurations and Urban Change in the North and South. Routledge.

Datta, A., 2015. New urban utopias of postcolonial India: "entrepreneurial urbanization" in Dholera smart city, Gujarat. Dialogues Hum. Geogr. 5 (1), 3–22.

Dourish, P., Bell, G., 2011. Divining a Digital Future: Mess and Mythology in Ubiquitous Computing. MIT Press.

Edwards, P.N., Bowker, G.C., Jackson, S.J., Williams, R., 2009. Introduction: an agenda for infrastructure studies. J. Assoc. Inform. Syst. 10 (5), 6.

Etzo, S., Collender, G., 2010. The mobile phone 'revolution' in Africa: rhetoric or reality? Afr. Aff. 109 (437), 659–668.

Foster, V., Briceño-Garmendia, C.M. (Eds.), 2009. Africa's Infrastructure: A Time for Transformation. The World Bank.

Foster, T., Hope, R., Thomas, M., Cohen, I., Krolikowski, A., Nyaga, C., 2012. Impacts and implications of mobile water payments in East Africa. Water Int. 37 (7), 788–804.

Guma, P.K., 2019. Smart urbanism? ICTs for water and electricity supply in Nairobi. Urban Stud. 56 (11), 2333–2352. https://doi.org/10.1177/0042098018813041.

Guma, P.K., Monstadt, J., 2020. Smart city making? The spread of ICT-driven plans and infrastructures in Nairobi. Urban Geogr., 1–22. https://doi.org/10.1080/02723638.2020.1715050.

Guma, P.K., Monstadt, J., Schramm, S., 2019. "Hybrid constellations of water access in the digital age: the case of Jisomee Mita in Soweto-Kayole," Nairobi. Water Altern. 12 (2), 725–743.

Heymans, C., Eales, K., Franceys, R., 2014. The Limits and Possibilities of Prepaid Water in Urban Africa: Lessons From the Field. World Bank.

Hope, R.A., Foster, T., Krolikowski, A., Cohen, I., 2011. Mobile Water Payment Innovations in Urban Africa. School of Geography and the Environment and Skoll Centre for Social Entrepreneurship at Saïd Business School, Oxford University, UK.

Jaglin, S., 2008. Differentiating networked services in Cape Town: echoes of splintering urbanism? Geoforum 39 (6), 1897–1906.

Jeon, S., Bendavid, E., Cohen, J., Hoffmann, K., Winograd, T., 2011. A mobile water project: mobile-for-development meets human-centred design. Arguing About Justice. Essays for Philippe Van Parijs, p. 245.

Kamga, O., Cishahayo, F., 2010. Information and communication technologies for African development: proportional technologies and an ethics of uses. In: Future Imperatives of Communication and Information for Development and Social Change, Bangkok.

Kitchin, R., 2015. Making sense of smart cities: addressing present shortcomings. Camb. J. Reg. Econ. Soc. 8 (1), 131–136.

Krisch, R., 2009. Sozialräumliche Methodik der Jugendarbeit. Aktivierende Zugänge und praxisleitende Verfahren, Weinheim.

Krolikowski, A., 2014. Can mobile-enabled payment methods reduce petty corruption in urban water provision? Water Altern. 7 (1), 235–255.

Loftus, A., 2009. Rethinking the political ecologies of water. Third World Q. 30 (5), 953–968.

Luque-Ayala, A., Marvin, S., 2015. Developing a critical understanding of smart urbanism? Urban Stud. 52 (12), 2105–2116.

Mutisya, E., Yarime, M., 2011. Understanding the grassroots dynamics of slums in Nairobi: the dilemma of Kibera informal settlements. Int. Trans. J. Eng. Manage. Appl. Sci. Technol. 2 (2), 197–213.

NCC, 2014. The Project on Integrated Urban Development Master Plan for the City of Nairobi in the Republic of Kenya: Final Report. Japan International Cooperation Agency (JICA), Kampala.

NCWSC, 2014. Nairobi City Water and Sewerage Company Strategic Plan 2014/15–2018/19.

Ndungu, W.L., 2015. Web Based Mapping of Water Supply Situation in Mathare. University of Nairobi School of Engineering. Degree of Masters of Science in Geographic Information Systems, Department of Geospatial and Space Technology of the University of Nairobi.

Odendaal, N., 2010. Information and Communication Technology and Urban Transformation in South African Cities. Doctoral dissertation, University of the Witwatersrand.

Odendaal, N., 2011. Splintering urbanism or split agendas? Examining the spatial distribution of technology access in relation to ICT policy in Durban, South Africa. Urban Stud. 48 (11), 2375–2397.

Odendaal, N., 2014. Space matters: the relational power of mobile technologies. Urbe. Rev. Bras. Gest. Urbana 6 (1), 31–45.

Oyedemi, T., 2004. Universal access wheel: towards achieving universal access to ICT in Africa. Afr. J. Inf. Commun. Technol. 2004 (5), 90–107.

Oyedemi, T.D., 2009. Social inequalities and the South African ICT access policy agendas. Int. J. Commun. 3, 18.

Robinson, J., 2006. Ordinary Cities: Between Modernity and Development. Routledge, London.

Roy, A., Ong, A. (Eds.), 2011. Worlding Cities: Asian Experiments and the Art of Being Global. vol. 42. John Wiley and Sons.

Shelton, T., Lodato, T., 2019. Actually existing smart citizens: expertise and (non) participation in the making of the smart city. City 23 (1), 35–52.

Sonko, M., 2017. If we continue at this pace, we will surpass our promise to turn Nairobi into a modern, smart, digital city of choice to live, work and invest in. Twitter. 8:50 a.m. Retrieved from https://twitter.com/MikeSonko/status/940217308806606848.

Spatscheck, C., Wolf-ostermann, K., 2009. Social Space Analyses and the Socio-Spatial Paradigm in Social Work. Working paper 2009-1. School of Social Work, Lund University. http://www.soch.lu.se/images/Socialhogskolan/WP2009_1.pdf. (Accessed 20 June 2019).
United Nations, 2015. Transforming Our World: The 2030 Agenda for Sustainable Development. https://www.sustainabledevelopment.un.org. A/RES/70/1.
von Schnitzler, A., 2008. Citizenship prepaid: water, calculability, and techno-politics in South Africa. J. South. Afr. Stud. 34 (4), 899–917.
von Schnitzler, A., 2013. Traveling technologies: infrastructure, ethical regimes, and the materiality of politics in South Africa. Cult. Anthropol. 28 (4), 670–693.
Wamuchiru, E.K., 2017. Rethinking the Networked City: The (Co-)Production of Heterogeneous Water Supply Infrastructure in Nairobi, Kenya. PhD Thesis, Technische Universität, Darmstadt.
WSUP (Water and Sanitation for the Urban Poor), 2018. A Journey of Institutional Change: Extending Water Services to Nairobi's Informal Settlements. WSUP, UK.

Author biography

Prince K. Guma is a research fellow and assistant country director at the British Institute in Eastern Africa, Nairobi, Kenya. His research work draws on the intersection of STS, urban studies, and postcolonial studies to examine the contingent and place-based articulations of cities and how these are mediated through the diffusion and uptake of ICT-based plans and infrastructures. His findings are hoped to provide a menu for new explorations, and to enhance our understanding of urban and digital possibilities in Africa.

Chapter 18

EMTHONJENI—Public space as smart learning networks: A case study of the violence prevention through urban upgrading methodology in Cape Town

Kathryn Ewing[a] and Michael Krause[b]
[a]*Department of Architecture, Planning and Geomatics, University of Cape Town, Cape Town, South Africa,* [b]*Violence Prevention Through Urban Upgrading NPC, Cape Town, South Africa*

Chapter outline

Introduction: Urban space and safety

South African cities, particularly Cape Town, experience rapid urban growth on the periphery. The result is often one of insecure living conditions in low income expanding informal settlements, coupled with high crime rates and poor service delivery at the margins. Past and present South African approaches to development support clean visual urban form, which forces a level of control and order, encouraging disconnected and fragmented neighborhoods, and resulting in a "closed city" (Sennett, 2007). Less attention is given to the space in-between buildings or the public space, ensuing left-over, dangerous and neglected urban

Shaping Smart for Better Cities. https://doi.org/10.1016/B978-0-12-818636-7.00013-5

spaces in these areas. It has been argued that South Africa is one of the most violent countries in the world (Brown-Luthango et al., 2017) where "bad living environments in the form of a growing number of informal settlements represent one of the most visible manifestations of poverty and inequality in South Africa" (p. 2). Residents living in informal settlements in South African cities experience extreme levels of vulnerability. The terror of crime in the public domain extends to losing control of the street, especially after dark. This is all too common for many citizens living in informal settlements where there is minimal lighting, nor structured, surfaced pathways and a pattern of interspersed dangerous spaces. Minimal access by police results in leaving public space safety to under resourced local neighborhood watch groups or local undemocratic appointed strong men. While there is a growing discourse in South Africa around participatory in-situ upgrading of informal settlements (Cirolia et al., 2016), the reality on the ground is that upgrading processes predominantly evolve or result in either mass relocations by uprooting a whole community to a green field site, or roll-over upgrading that completely negates existing forms of settlement patterns or public space networks and replaces negotiated spaces with specific engineered solutions. This exclusive development process continues to entrench exclusion where citizens have little influence on becoming part of the co-creation of the living environment. The state continues to impose structural, institutional and cultural violence onto the urban poor. As a result, the people living in low-income environments are not seen as key owners nor active participants in the co-creation of the public realm.

This chapter expands on experimental learning and incrementally co-created and co-owned examples from the Violence Prevention through Urban Upgrading (VPUU) programme. VPUU's approach is co-creation of informal urban space through discovery and social enhancement of the vernacular urban design deriving from many years of negotiated development amongst the settlement makers. The *emthonjeni* (public space) project within Monwabisi Park (MP) informal settlement is used as a case study. VPUU promotes the public realm as a strategic building block and a key design tool at the urban neighborhood scale. Upgrading of the [informal] public realm can enable catalytic socio-spatial change. Such urban space requires continuous negotiation, participation and maintenance that is "open" to local dynamics. The chapter questions how activation through learning programmes, monitoring and evaluation along with evolving community-based operation plans through basic communication and technology enable such urban space/s to flourish. The discussion further expands on the concept that intervention in the public realm in an informal settlement seeks to improve the residents' quality of life and improve levels of safety through what we call "smart learning networks." The case study disrupts the conventional meaning of "smart urbanism," by acknowledging diverse and local opportunities for learning, monitoring services and knowledge sharing using digital technology and communication in a transitional and ever-evolving neighborhood with social good in mind.

To add to the argument, the agenda is based within a highly transitional and vulnerable informal context in a South African city, Cape Town. Although the concept of informality is not investigated per se, the idea of smart urbanism within the informal context is acknowledged. The work is seen as "experimental" (Luque-Ayala and Marvin, 2015) to explore and test the contradictions, risks and opportunities of smart urbanism from experience, whilst sharing multi-layered knowledge that has the potential to be upscaled and developed in [in]formal approaches and incorporated into and challenged within municipal policy and the everyday functioning of a city. In a sense, we seek to reveal an engaged process of "smart informal urbanism" and the social politics of implementation that [re]configure urban space. The case study shows an emergence of a new kind of delivery of services or learning network via ICT for development (ICT4D) considering intelligence in a new way of smartness (Picon, 2015). The ICT4D interventions expanded in the case study utilize free and open source software to be easily accessible and are designed for local community use and engagement. This pushes the boundaries and terminology of "network" into a different imaginary.

So why do we need to look at this case study through the lens of smart urbanism, given the limitations or contestations of the concept in an informal context? What makes a neighborhood smart? We engage with this question to understand the application of smart urbanism and how this theory is designed in an informal settlement or specific to public space upgrade. The intention is to explore smartness where the concept has been positively mobilized and actively co-created, where a series of urban spaces have been upgraded that connect social and technological networks through a local partnership approach. This builds on Picon's (2015) and Odendaal's (2018) arguments of the emerging picture of urbanity that is removed from mainstream notions of smartness and smart urbanism. This further opens the question as to how can we harness the potential of the public realm in South African cities in line with the Sustainable Development Goals toward inclusive, safe and sustainable communities through an alternative understanding of smart urbanism?

Rethinking smart urbanism at the margins of informality

As argued by many stakeholders (government, community, NGOs) and theorists (Hollands, 2008; Kitchin, 2014; Luque-Ayala and Marvin, 2015; Odendaal, 2006; Picon, 2015), smart urbanism (SU) or the "smart city" seems to be guiding the new development agenda in global cities of the north and south, where the anticipation is that SU will solve urban growth, climate change and many more deep-rooted challenges in our urban societies of today. Picon (2015) opens his work on the smart city with the statement, "The smart city, driven by digital technology, is poised to replace the typical networked city of the industrial era, whose success was built on its hard infrastructure, from roads to water supply and sanitation systems, not only as a technological optimum but also as a social and political project" (p. 009). Luque-Ayala and Marvin (2015) question if smart urbanism is really "emerging at the intersection of visions for the future of urban places, new technologies and infrastructures..." "...Yet our understanding of the opportunities, challenges and implications of SU is limited" (p. 2105–2106).

How can we understand the concepts of SU whilst in juxtaposition of big business and online corporations (IBM, Google, Amazon, Facebook and so on) involved in smart building, big data and intelligent systems set against a context of poverty and inequality, interspersed with the consequences of natural disaster and political unrest? Is such "smartness" (Luque-Ayala and Marvin, 2015) relevant and what of the expectations of the new e-citizen living on the margins of society, particularly given the exclusionary nature of the South African development arena? As noted by Odendaal (2006), "Some contend that the impact of ICT is ambiguous; it unifies but also divides" (p. 32). Marvin et al. (2016) and Kitchin (2014) further note that the ideological label given to the smart city (Hollands, 2008) is vague and uncertain, combined with the faddishness of the term smart urbanism that lures corporate and/or private investment, politicians vote count and technocratic control and management.

Hollands argues that a just smart city is about allowing technology to empower people. Odendaal calls for a "smart city from the bottom up" (Marvin et al., 2016, p. 82), in which people will develop smart technologies that enable social justice. However, "In the South African context Odendaal (2006) has demonstrated the complex ways in which digital technologies have unexpected consequences that can also reinforce existing social disparities" (Luque-Ayala and Marvin, 2015, p. 2111), questioning the fit between ICT and 4D (ICT4D). Information, communication and technology for development has the potential to enable learning and the exchange of knowledge, as well as the collection, storage and interpretation of information. Most importantly, it is changing the way we identify social issues and conceptualize, design and implement solutions. It thus, could increase the speed, efficiency and efficacy of basic service delivery in informal settlements.

Husar and Ondrejicka (2019) look at the nexus between technology and social innovations, where such innovations can potentially contribute to social change. Odendaal (2018) notes the "dance between technological innovation and appropriation through social action and livelihood imperatives within the highly unequal urban spaces of the global South. This is an opportunity for urban interventions to be relevant and effective" (p. 245–246). However, as argued by Luque-Ayala and Marvin (2015), "...the landscape of SU does not follow black and white logics... in practice, community involvement in SU shows that notions of 'top-down' and 'bottom-up' do not adequately reflect the complexity of issues at play" (p. 2113). "The stories of the everyday are easy to overlook in the shadow of the glossy smart city" (Odendaal, 2018, p. 255). This is further exacerbated by the notion of "dashboard urbanism" (Odendaal, 2019, p. 165), that potentially masks serious social conditions with a representation of numbers and indicators or checklist boxes to be transparent and accountable (by whom, for whom) and what is considered progress by local municipalities.

Picon (2015, p. 013), emphasizes the importance of space in digital technologies where information is now geolocated and in real-time. Digital technologies "...enhance physical reality by enriching three-dimensional space with contextual electronic content" (Picon, 2015, p. 014). The confusing contradiction is that space is generally tangible (physical manifestation of a public park, playground, sports field, library etc.) and smart urbanism is generally intangible (a-physical being "in the sky": Wi-Fi networks, Bluetooth, satellites, remote surveillance etc.). Where do the two converge? Social mapping through geographic information systems (GIS) takes on a new form of spatial representation and analysis plus efficient assessment of informal space, creating opportunities to map, monitor and evaluate the everyday and the ordinary at multiple scales (e.g., OpenStreetMaps). The positive occupation, activation, monitoring and maintenance of public space is a critical success factor toward safer communities, but how would we understand this if we are not collecting any data from the ground?

The notion of Graham and Marvin's (2001) concept of "splintering urbanism", which expresses the interplay between networked infrastructure and the urban societies or urban condition that they support (Coutard, 2005), is brought under review. Graham and Marvin (2001) argue that "unbundling configurations of infrastructure networks continuously and subtly recombine with the production of new configurations of urban space; understanding one without the other soon becomes impossible" (p. 216). So, what if the infrastructure is not physically there; it is an informal context where the built form is temporary and service delivery minimal, where not even basic standards are met (some waste removal, some accessways, electricity, communal water points and toilets, below standard for basic service delivery) and space is reconfigured by those who live there? There are minimal poles, pipes, services that are considered legal or fully functioning to provide a service or network. Access to basic services is at stake, unlike Graham and Marvin's (2001) notion, where basic network infrastructures is the "social norm" (Coutard, 2005, p. 50). But, what does this mean for smart urbanism? There is a deliberate intention by the case study to "unbundle" the theory and seek not to splinter (Coutard, 2008) but to integrate the most vulnerable citizens. It applies a network that changes and influences how the public sector interacts with its citizens (or users) around the delivery of essential services. Essential services are currently defined by the public sector by the provision of rudimentary water and sanitation provision, some form of stormwater management system, some form of rudimentary solid waste removal and some form of Disaster Risk Management. VPUU promotes the inclusion of health standards, education, climate resilience, public lighting, safety, documentation of all residents in the form of a register to provide limited security of tenure, and some form of early warning and crisis relief systems or resilience network to respond to situations like Covid-19, floods or fires or other daily or seasonal disasters. This approach includes the activation and monitoring of social and educational programmes.. This is about networks at a very local neighborhood scale that impact broader city functioning. The approach actively and purposefully includes social and educational services into the definition of essential services, illustrated through the intervention of public spaces and buildings as locations for early childhood development (ECD) learning. The idea is universal access for the child to education as a basic service. Therefore, the smart learning network makes use of technical networks (focused on ICT4D: cell phone apps and wifi hubs), social networks (focused on learning: early childhood development) and spatial networks (infrastructure focused on the public realm: public space and public buildings/active boxes). An everyday form of smart urbanism, where public space forms smart learning networks.

Apartheid planning actively disregarded the spatial form of people spaces as dynamic elements to city structuring. In current times, there has been minimal attempt in South Africa to take up the challenge of spatial reconstruction (Watson, 2014) linked to smart urbanism which aims to address social justice and spatial equity, "…as smart cities have simply not landed in South Africa" (Odendaal, 2019, p. 166). The principles of investing in the public realm to improve the levels of safety, where streets and public spaces are socially activated, safe places designed to the human scale, monitored and assessed, are far from everyday reality in Cape Town's peripheral townships and informal settlements. Public spaces hold the potential to be points of exchange amongst citizens in a neighborhood, where they are the basic structuring elements to create a positive citizen-led, city-making and integrated approach, acting as social and digital hubs. The key is to provide just enough of a framework or network (physical and social) to understand the important elements and conflicting possibilities that will initiate and shape urban form and upgrading. This lends itself toward a new way of looking at the smart of urbanism, promoting integrated sustainable settlements trying to overcome structural, cultural and direct violence toward social improvement.

Co-creative problem solving to address the problem: VPUU

Violence Prevention through Urban Upgrading (VPUU) is a comprehensive area-based community development programme that aims toward safe and integrated sustainable communities and the improvement of quality of life for residents in low income neighborhoods in the Western Cape, South Africa. The programme contextualizes and fills with meaning the principles of Crime Prevention through Environmental Design (CPTED) (Kruger et al., 2001) to low-income neighborhoods, emphasizing negotiated development, safety promotion and socio-economic programmes over surveillance and other methods of crime prevention strategies (Violence Prevention Through Urban Upgrading, 2014). VPUU promotes the notion of prevention which is based on defining locally accepted social norms as part of a process of engagement and dialogue which then uses smart technology to support, monitor and provide evidence for those co-created social norms and standards. This is a different narrative to the surveillance and enforcement approach of predefined norms onto a community of people. Such predefined norms work on the basis the digital ground rule of 0 and 1 and have only right or wrong norms with little room for incremental, context specific adaptations and nuances. This can lead to alienation and a state of surveillance rather then using smart technology to enhance social norms for social good.

VPUU is a multidisciplinary approach that tackles the three types of violence; structural, cultural and direct violence (Krause, 2018). It uses a participatory research model, featuring the strategy of prevention, cohesion, protection and evidence research. Elements of international best practice of organizations including UN Habitat Safer Cities Programme and the World Health Organization's Life Cycle Approach are included into the overall method. It combines these best practices with the concept of asset-based development informed by the local context and community knowledge. Community members are involved in delivering and maintaining the services and interventions. Local leadership are partners in development and custodians of their local area *Community Action Plan* (CAP) and the *Public Investment Framework* (PIF)—the spatial development plan supporting the CAP. Monitoring and evaluation by local partners and communities using smart technology is integral to the process (baseline surveys, regular household surveys, indicator development, exit survey). The programme implementation is measurable and accountable to the funding partners and the community beneficiaries. VPUU acts as independent intermediary between government (local, regional, national) and communities and other stakeholders.

After visiting the programme in 2011, the late Sir Peter Hall (Urban Redevelopment Authority of Singapore, 2012) stated that, "The VPUU Programme addresses the most basic priority of urban policy – personal safety. It offers a proven solution to this problem where the issue is most acute: the informal and semi-formalised housing areas that are proliferating on the edges of major cities throughout the developing world. And it does so in a remarkable cost-efficient and effective way, through relatively simply and cost-effective urban design measures. The strategic positioning of buildings and structures that provide visual surveillance and security along open and well-lit pedestrian paths ensures safe routes to and from people's homes. In addition, voluntary community participation provides and effective means of training young people, and thereby assisting their entry into the formal economy" (p. 98).

Public space as "smart" learning networks in Monwabisi Park: A case study

The suburb of Khayelitsha was established in the 1980s as a residential township for black African working-class people. Today, it is a densely populated dormitory township situated approximately 30 km south-east of Cape Town Central Business District. Spatially, the area is characterized by predominantly low-income formal and informal housing and large tracts of underutilized land and a poorly maintained open space system that is characterized by unsafe pedestrian routes. High levels of unemployment and poverty exacerbate the problems of crime. As stated in the Commission of Inquiry into Allegations of Police Inefficiency and a Breakdown in Relations between SAPS and the Community in Khayelitsha (2014), the area experiences very high rates of contract crime. In 2018, the murder rates in Khayelitsha were 192 murders per 100,000 persons (Crime Stats SA, 2018). This creates a high perception of insecurity. Khayelitsha is a classic example of social engineering of the apartheid state—a result of the state imposing structural and cultural violence onto one group of the population. Monwabisi Park (MP) informal settlement is located on the southern edge of Khayelitsha. It was a citizen led process to occupy land adjacent to the Wolfgat Nature Area to the south of Mew Way in 1996.

VPUU started working with the community of MP and the local municipality, the City of Cape Town, as a co-creation and partnership approach in 2009. The vision for MP was to build safe and integrated communities by upgrading the existing informal settlement without disturbing existing social networks by relocating people outside of the area. It was considered one of the five pilot sites as part of the Informal Settlement Transformation Programme. The approach involved the fundamental idea of establishing a sustainable, holistic framework for informal settlement upgrading. "Holistic" refers to an all-encompassing transversal view, avoiding the 'silos' that informal settlements are all too often labeled under or portioned into specific municipal line departments. The approach entails creating a vision that is about livelihood systems, not individual houses. "Sustainable" refers to where "deep and wide roots have been grown and woven together between partners so that there are strong interdependent bonds that cannot easily be broken" (Krause, 2011, p. 098–099). Important to note is the VPUU Programme in Monwabisi Park occurred in a context where the upgrading of shelters (houses) was not the primary focus of the upgrading intervention.

In August 2009, a local stakeholder group was formed and a Baseline Survey conducted. By 2009, there was very limited infrastructure in MP, with only 358 toilets and 153 communal water standpipes. Poverty and crime were major challenges for the people living in the settlement. The perceived safety in 2009 was that most respondents felt crime was a "very serious problem". In February 2010, a Community Action Plan (CAP) was signed by the Mayor of the City of Cape Town (first of it's kind) and the local project leadership, known as the Safe Node Area Committee (SNAC). In 2013, as part of a locally enabled enumeration process, it was estimated that the population of MP was approximately 17,800 residents of which 11% were under the age of 6 years old living in 6867 households. The settlement consists of 4 Sections, A, B, C and M.

The systematic use of basic GIS was introduced early into the programme working with community members to facilitate a baseline survey and an in-depth site analysis to understand and map social conditions, basic household demographics, perceived crime levels and service delivery elements and impact (e.g. electricity, water points, paths and walkways etc.). Along with this, weekly photographs were taken using digital cameras and later cellphones by a local community group at strategic points over a period to monitor change (settlement scale). Most of the existing walkways were 'tracked' by community members, using hand-held GPS. These were then re-traced in terms of priority desire routes working with community members walking the area and additional mapping in focus groups, workshops and reference groups. Highlighted public walkways were school routes (in and around the settlement), everyday routes (to public transport) and access to water points, which were considered in terms of time (day and night) linked to safety. Aerial photography from the City of Cape Town and mapped information collected from the ground and further negotiated, formed the key spatial resource or "base map", which is ever evolving, as is the nature of informality.

By 2010, the proposal of spatial layers of intervention, overlaid with the information from the base map, baseline survey, ongoing monitoring and evaluation information through live quarterly surveys (using a cell phone app) and the community input via the CAP provided guidance to develop a design framework for public investment and community ownership: a Spatial Reconfiguration Plan (SRP) for the informal neighborhood of MP—an incremental public investment plan. The SRP identified the nature of interventions and priorities relative to strategies, social programmes, timeframes and budgets in consultation with the local municipality, an intermediary, the local leadership and relevant stakeholders and partners. Fundamental to the spatial layers and strategies of the SRP, was a phased framework designed to consider development over time and assess different alternatives in terms of community trade-offs, municipal requirements, regulations and capacity, engineering possibilities and site constraints. Small catalytic projects (Hamdi, 2004) were considered *by all* as a priority to radically improve the quality of life of the residents in MP in the short to medium-term. Out of this process and with further engagement with community and other stakeholders, emerged the emthonjeni (public space) project.

Emthonjeni—Spaces of celebration and the everyday

The MP community and the facilitators of SUN Development (implementing agent) on behalf of VPUU engaged on a visioning process as part of the development of the SRP. A major shift arose out of this process toward asset-based community development and positively occupied public spaces referred to as emthonjeni or celebration places. Emthonjeni are multifunctional places. They are dynamic and can evolve (or even disappear) as its contribution to overall community development adapts with time. The isiXhosa word emthonjeni refers to "being at the fountain"—a place by the water. Traditionally, communities gather around to fetch water, do washing, play or simply spend time catching up on latest news, gossip and chatter. Such places exist in MP around public water taps or communal standpipes, however given the negative and unsafe environment, such places were only discovered through a collaborative process of local knowledge, social mapping (informed by the base map) and design. These emthonjeni guided the primary spatial structure of the in-situ plan for incremental upgrade in MP. The pattern, distribution and scale of public space are closely interlinked with the existing location of various neighborhood and education facilities and the street network to build on the existing structure of the settlement in terms of long-term development potential, whilst also guiding short-term social and spatial intervention projects.

Emthonjeni are carefully negotiated spaces of the residents of MP. The upgrade of small scale emthonjeni is a practical solution to improve safety as they create safe places for children to play and for the provision of selected managed utilities and services. These spaces are focused internally to the neighborhood cluster and encourage social cohesion. They are located along pedestrian desire routes and are the breathing space along pathways which connect neighborhoods. Clustering around small open spaces helps to provide space for economic opportunities and assist in providing safe public spaces close to home. The public space projects are associated with a construction methodology which includes the utilization of local labor and skills development. However, over time some of these spaces have seen to lack place management and became underutilized where the social negotiation of the spaces faded or were contested. "Other" public spaces are defined according to engineering decisions by municipal infrastructure provision such as toilet blocks or water taps. The lack of the social negotiation process of these other spaces have created underutilized infrastructural waste spaces, accessed only by people collecting water from mismanaged water taps, generally on the edge.

Larger emthonjeni are located adjacent to areas that have the potential to become higher order structuring spatial elements such as sports fields adjacent to neighborhood centres or education hubs. The facilities are seen to be zones of connectivity both socially and digitally. Consequently, previously underutilized unmanaged space is put to optimal use whilst providing a spatial hierarchical order. Such places also play an important role in neighborhood watch and patrolling initiatives, which promote respectful community action, moral behavior, information exchange along with community policing against illegal activities using the larger public spaces as a base, landmark or reference point. These important spaces allow for strategic public investment and implementation at neighborhood level to test methodologies, processes and outcomes in an informal settlement. The design of these spaces is rooted in deep consultation with the stakeholders and neighbors. The use of the VPUU Urban Design Guidelines of surveillance and visibility; owned spaces; defined access and movement; image and esthetics; physical barriers; operation and maintenance (VPUU, 2014) as part of an inclusive design approach acted as part of a toolkit for urban upgrade. Key to such spaces are connectivity.

Emthonjeni—Locations for learning and socialization

In 2010, the emthonjeni early childhood development (ECD) concept was developed by SUN Development (on behalf of VPUU) through a series of participatory workshops with the local ECD forum, SNAC leadership representatives, ECD community-based organizations, City of Cape Town Social and Economic Development, Environmental Health and Water and Sanitation line departments during the planning stages of the SRP. Emthonjeni are viewed as public spaces that could provide a venue for "out-of-centre" learning for young children, based on two principles: firstly, safety for children, encouraging adult supervision and safe play; and secondly, equitable access to education and health for children. The idea was to ensure a population based approach where all children access a structured ECD programme—either a centre based (creches) or out of centre based (emthonjeni) to provide additional spaces to meet the right of every child to a structured ECD service irrespective of their living conditions. Emthonjeni are stimulated outdoor classrooms that support the development of the child. This is conducted through structured ECD programmes following the Unlimited Child curriculum to ensure that children can become school-ready. Over time VPUU has developed internal capacity to provide a service for the ECD programme in MP. About 200 children are currently registered that utilize 5 emthonjeni in A-Section. The outdoor classrooms are the base for the continuum of services and serve immediate neighborhood blocks of 50 to 100 households and are supported by an ECD Resource Centre.

ECD centres are the second stepping-stone that serve a neighborhood section. There are currently 6 ECD centres in A Section in MP. VPUU cooperates and builds capacity within the ECD forum of MP representing about 15 ECD centres in all 4 sections. A major success has been the agreement within MP to use one curriculum in all ECD centres working with the emthonjeni project.

The emthonjeni out of centre learning concept was taken further with the development of a third level of service provision at a public facility, known as a Neighborhood Centre (Active Box) as a home for all outreach programmes in MP. In 2017, the MP Neighborhood Centre in A Section opened its doors. This centre was constructed, and is currently operated and managed by VPUU. It constitutes the first formal building in MP. An ECD resource centre is located in this building as a base for the out of centre ECD service and it also provides a toy and educational library for all ECD Centres in MP. The A Section precinct includes the Active Box, two sports courts/fields, significant urban agricultural activities, a Community Register, a base of the ECD attendance monitoring, the tap and toilet monitoring programme, a wifi hotspot, and a base for the neighborhood watch. There are future development plans to add the first Grade R class room in MP as linking element between ECD and primary school. This follows a community node of clustering facilities in an informal settlement. Essential to the Active Box model is to act as a home for local ICT4D.

TABLE 1 ECD services and public space.

Public space	Scale	ECD services
Emthonjeni	Neighborhood block 50–100 households	Basic ECD activation, stepping stones following a curriculum
Creche/ECD Centre	Section area 200–300 households	Education according to a curriculum
ECD Resource Centre (located in an Active Box)	Neighborhood area 1500 to 3000 households	Programme approach, training of care givers; shared services, toy library, M&E, Quality assurance, immunization of children, nutritional programmes
Community Library	Suburb/town <100,000 residents	Added services toward a holistic life cycle approach, teacher training

The fourth level of service provision in the case of MP is a link to the community library in Harare, Khayelitsha with the Funda Udlale (Learn and Play) facility with a reach of three wards of about 100,000 residents. Table 1 provides an overview of the 4 levels of ECD service compared to the hierarchy of public spaces in MP.

Emthonjeni—Places for smart networks

The first example of smart learning networks is the registration system of the children using the emthonjeni out of centre ECD learning system. Each child receives a unique QR code accompanied by a consent letter of the child parents. The monitoring and evaluation system allows to log each session a child uses the out of centre service. Regular monthly tests enable to assess the progress of each child toward school readiness. The curriculum is structured in 2 hour modules which are registered against the child's QR code to monitor the exposure of each child to learning content. Via the ECD forum this monitoring system is gradually extended to ECD centres which enables a comparison of centre based and out of centre services and their effectiveness toward the child's development. This progression is captured on VPUU in-house developed Open Data Kit app more recently moved onto Kobo. Fieldworkers have dedicated groups of children whose attendance and exposure to and progress in learning content is measured, regular home visits to pregnant mothers are measured, week end activities, cleaning of emthonjeni spaces and parents involvement in meal preparation are currently monitored. Having the vision to offer ECD services to all children within an informal settlement requires a close interaction of a register of residents and the needs analysis and resource allocation for ECD services.

Since VPUU conducts full scale enumerations (and on-going household surveys) it is possible to define the extent of the need for ECD services in an informal settlement. The enumeration is conducted via smart phones that have an ODK/Kobo app. The enumeration is kept updated via a live community register. This community register is managed by local volunteers that receive a stipend by VPUU. Residents can receive a confirmation of residence from that local office if the locally agreed protocol has been followed to ensure that the people are residents of the informal settlement. This data is managed via a cloud data base, geolocated on GIS. The interrelation between an accurate and updated local register and the provision of social and educational services is evident. Odendaal (2019) notes the "objectification of urban life through enumeration and quantification plays an important role in targeting policy and focusing planning efforts" (p. 248). The smart system enables the needs definition for provision and utilization of public spaces: open spaces as well as facilities such as the neighborhood centre facilitating an ECD resource centre.

Currently the public sector has a very narrow definition of essential services (primarily taps and toilets), thereby VPUU developed a mobile inspection tool in the form of an app, called CitySpec/CheckIT, to allow a greater efficiency in the monitoring and repair of this provided technical infrastructure. VPUU trains local community members to report on faulty taps, toilets and street lights in their areas. The reports are then relayed to the municipality who are given repair needs including geolocations, a detailed list of spares required to repair the fault and a contact person who can take the City of Cape Town municipality repair team to the faulty item, if required. This is done at least once per week and formal fault reports are logged with the municipality. Geolocated maps are generated and hotspots of over usage in terms of number of households sharing the essential services are identified (in some instances up to 300 households use one water tap exceeding the national standard 12 times) and shared with local leadership and municipal officials. The general functionality at the beginning of the project in 2009 was at about 60%–65% of services provided in the area over a 6-month period. Currently, this service standard is improved to about 90% and has sustained at this level of functionality. Both the citizens of MP and the municipal officials where Cityspec/CheckIT is utilized, acknowledge the impact of accurate data driven and evidence-based work in keeping the functionality of public infrastructure and quality public spaces (that are used for educational programmes). In some cases, evidence had shown inappropriate design where water run off flows into adjacent houses and as a result neighbors shut off taps. This has led to redesign after negotiation with the neighbors to avoid pre-emptive vandalism of water taps.

The social inclusive value of "commons" around smart urbanism linked to social good of education and place management is core to the design and implementation of VPUU's vision.

A number of wifi hotspot have been installed in Harare, Khayelitsha and MP with 1.200 unique weekly users. Based on the success of these points, the concept of a "mash network" arose and is currently under co-creation by VPUU with a wide collaboration of private developers, NGO's and tertiary institutions. A key partner in the conceptualization of the service is Zenzeleni, an NGO promoting commons in local internet services. While Zenzeleni via the Mankosi cooperative has extensive experience in rural settings, the value of the partnership with VPUU is to start moving into urban areas with a focus on informal settlements. By January 2020 the backbone infrastructure of the VNET covering key facilities in Harare and MP is operational allowing a mash network to emerge. The proceeds will feed into sustaining the network and the public space maintenance. As such the VNET is an enabler to exponentially set free opportunities for education once the ECD centres are connected via wifi. A local server is to be installed in the MP Neighborhood Centre to act as resource base for educational materials enhancing quick access to information linked to the curriculum used in ECD on an area-based approach. This is just one of the potentials of local wifi networks. The response to crisis events such as Covid 19 will show whether smart systems can assist in limiting the negative impact on marginal communities.

Emthonjeni—The impact of socio-spatial intervention

In 2017, VPUU conducted research on how MP residents have received and responded to the emthonjeni project and walkways built in the settlement. A public open space survey, using a cell-phone app, was carried out and collected data from 77 interviews which were located approximately within a 200 m radius of an emthonjeni in all four Sections (A, B, C and M Sections). 62% of the respondents were women. The survey revealed two specific uses of the emthonjeni as both a playgroup (linked to the out of centre ECD learning programme); and social activity space (primarily a water collection and washing place, but also a community meeting space, and other programmes such as immunization and health). Important to the emthonjeni was the connection to the paved walkways, which were perceived as safer (lighting) and people could move faster through the settlement. 60% of the respondents perceived safety to have improved and emthonjeni are *very safe*. The paved walkways, however, were seen to be much safer than the informal network of paths. Detailed fieldwork was conducted to observe the public spaces over a period of 3 months.

To build on this, two focus groups were held in May and August 2017 to obtain qualitative data regarding parents, neighbors and local ECD CBO fieldworkers' responses to the emthonjeni and ECD programme. A key response was "Emthonjeni belongs to the community… it is a place where you can leave your child, in trust… unlike other open spaces, children know the emthonjeni, it's a familiar place…" From a learning perspective, "outside rules are not the same, children are open-minded and more free" and where "Emthonjeni has a positive impact on our children. Today, when we take our child, it is at the ECD inside the class and outside at the emthonjeni, and they know what to do. They know how to read, to write, recognise animals and shapes…". However, seasonal weather conditions, problematic stormwater conditions and cleanliness appear to be the biggest challenges where, "during winter there are a lot of problems with the rain. When it is raining, water is staying on the emthonjeni and the road… we have to solve conflicts and the emthonjeni needs to be cleaned" (anonymous representatives from A Section, May 2017).

In response to the maintenance and on-going safety challenges, three responses have been co-created with the citizens and the public sector. Firstly, a locally developed response called *Safety through Activation*, where a local group takes guardianship of a public space in exchange for a specific benefit that the group or the individuals of that group receive directly from place management. There are two examples present in MP: parents of children benefiting from the out of centre ECD programme volunteer to keep the 5 ECD emthonjeni clean in A Section. This is done in direct exchange for free services for the children. On a nodal level this system evolved as urban gardeners taking procession of a vacant piece of land adjacent to the neighborhood centre to benefit the group via improved nutrition and income through selling produce. VPUU provided the fence around the garden, access to water and access to the building as a base for the garden activities. In exchange the group acts as guardians for the facility and public space. The second response to maintenance is that, volunteers from within the community keep the non ECD emthonjeni clean in exchange for a personal benefit in the form of an agreed skills training course (computer, driving license etc.) to work toward personal career goals. Thirdly, tapping into public sector resources and providing the service via dedicated Expanded Public Works Programme (EPWP) staff, a government driven mass programme to bring low skilled people into public sector funded employment contracts. While access to EPWP has been established for the monitoring of the taps and toilets in 2015, the support for maintenance work only started in November 2019. The City of Cape Town has taken on the VPUU approach in an innovative manner into an area-based approach for A-Section in MP. A Management Agreement has been created that includes public space management, tap and toilet monitoring, the management of the community register, the ECD fieldworkers to look after 200 out of centre based training of children, safety volunteers, a centre manager, support for economic development, social programmes on an area basis. This is all underpinned by smart urbanism principles. Current discussions emerge around adding solid waste removal on household level to the model to reduce health risks, improve recycling, and to start a local employment system to keep the income within the neighborhood.

The emerging responses are well in line with the ideas of commons within a smart learning network and further research is suggested about the preconditions required to harness such smart urbanism and e-citizen involvement in their neighborhood and how this evolves over time.

Toward an improved quality of life in informal settlements

The case study disrupts the conventional meaning of smart urbanism, by acknowledging diverse and local opportunities for learning networks using digital technology and communication in a transitional and highly vulnerable context. Local data collection is seen as a huge opportunity to assist with more equitable service provision and assisting in overcoming the structural violence by allowing citizens access to information all toward the greater social good.

In conclusion, the VPUU emthonjeni project is a small step in a positive direction toward the transformation of negotiated space through an appropriate process of co-creation and programme intervention, and in turn has resulted in an increased level of perceived safety around specific public spaces. The intention is to provide a holistic and participative process of design, allowing for active decision-making and practical problem solving between the citizens of MP community and the public sector. This has been achieved with the support of an intermediary that provides technical expertise, a social dialogue and the capacity to mobilize resources. Monitoring and evaluation are conducted in real-time and geolocated in space; be it attendance of children in out of centre ECD programmes and the educational outcomes. As such an independent ELOM (Early Learning Outcomes Measure) assessment of a cohort of Emthonjeni children has concluded that the educational level is comparable with children in middle class neighborhoods, well above their peers in former townships. This result is taken as one element to proof that the quality of education does not depend necessarily on the physical building but on the dedication, and quality of how care takers conduct their work. The continued measuring of the functionality of essential infrastructure provided by the municipality, or the continued update of the community register were key elements to enable the buy-in by the City of Cape Town to introduce Management Agreements on neighnourhood level in informal settlements. The Active Boxes (neighborhood centres) have a key function to enable a smart learning network approach. The range of scale of emthonjeni form an important step in showing commitment with the community and stakeholders of MP not only from a spatial or physical perspective, but as a social agenda related to such spaces. Smart networks are introduced to assist in providing responses to current challenges. Limitations in terms of scale of physical infrastructure, human-centred solutions and local municipal capacity and willingness to transform and recognize an informal settlement remains a major challenge to radically improve the quality of life of the residents at a neighborhood scale in MP.

To overcome the structural violence, past and present, spatial professionals, scientists, government authorities, community leaders, economists and the private sector need to realize and recognize the process of negotiation when spontaneous settlements and vernacular minimalistic forms of a public space system emerge. Ways of exploring commons is in the process of allowing a transition for VPUU as intermediary to exit the settlement and leaving a positive legacy of a strong civil society behind.

References

Brown-Luthango, M., Reyes, E., Gubevu, M., 2017. Informal settlement upgrading and safety: experiences from Cape Town, South Africa. J. Housing Built Environ. 32, 471–493. https://doi.org/10.1007/s10901-016-9523-4.

Cirolia, L., Görgens, T., Van Donk, M., Smit, W., Drimie, S., 2016. Upgrading Informal Settlements in South Africa: A Partnership-Based Approach. UCT Press, Claremont, South Africa.

Commission of Inquiry into Allegations of Police Inefficiency and a Breakdown in Relations between SAPS and the Community in Khayelitsha, 2014. Towards a Safer Khayelitsha: The Report of the Commission of Inquiry into Allegations of Police Inefficiency and a Breakdown of Relations Between SAPS and the Community of Khayelitsha. August. Retrieved from https://www.khayelitshacommission.org.za/.

Coutard, O., 2005. Urban space and the development of networks: a discussion of the "splintering urbanism" thesis. In: Coutard, O., Hanley, R., Zimmerman, R. (Eds.), Sustaining Urban Networks: The Social Diffusion of Large Technical Systems. Routledge, London, England, pp. 48–64.

Coutard, O., 2008. Placing splintering urbanism: introduction. Geoforum 39, 1815–1820. https://doi.org/10.1016/j.geoforum.2008.10.008.

Crime Stats SA, 2018. Contact Crimes Against the Person. Retrieved from https://www.crimestatssa.com/precinct.php?id=1074.

Graham, S., Marvin, S., 2001. Splintering Urbanism: Networked Infrastructures, Technological Mobilities and the Urban Condition. Routledge, London, England.

Hamdi, N., 2004. Small Change. The Art of Practice and the Limits of Planning in Cities. Earthscan, London, England.

Hollands, R., 2008. Will the real smart city please stand up? Intelligent, progressive or entrepreneurial? City 12, 303–320. https://doi.org/10.1080/13604810802479126.

Husar, M., Ondrejicka, V., 2019. Social innovations in smart cities—case of poprad. Mobile Netw. Appl. https://doi.org/10.1007/s11036-018-01209-z.

Kitchin, R., 2014. Real-time city? Big data and smart urbanism. GeoJournal 79, 1–14. https://doi.org/10.1007/s10708-013-9516-8.

Krause, M., 2011. Violence prevention through urban upgrading. Dig. S. Afr. Archit. 16, 098–103.

Krause, M., 2018. Western Cape Government Department of Health Research Newsletter 10. Department of Health Internal Publication, Cape Town, South Africa. June 2018.

Kruger, T., Landman, K., Liebermann, S., 2001. Designing safer places—a manual for crime prevention through environmental design. In: Council for Scientific and Industrial Research. Retrieved from https://www.csir.co.za/designing-safer-places.

Luque-Ayala, A., Marvin, S., 2015. Developing a critical understanding of smart urbanism? Urban Stud. 52, 2105–2116. https://doi.org/10.1177/0042098015577319.

Marvin, S., Luque-Ayala, A., McFarlane, C., 2016. Smart Urbanism: Utopian Vision or False Dawn? Routledge, Taylor & Francis Group, London.

Odendaal, N., 2006. Towards the digital city in South Africa: issues and constraints. J. Urban Technol. 13 (3), 29–48. https://doi.org/10.1080/10630730601145997.

Odendaal, N., 2018. Smart innovation at the margins: learning from Cape Town and Kibera. In: Karvonen, A., Cugurullo, F., Caprotti, F. (Eds.), Inside Smart Cities: Place, Politics and Urban Innovation. Routledge, London, England, pp. 243–257.

Odendaal, N., 2019. Appropriating 'big data': exploring the emancipatory potential of the data strategies of civil society organisations in Cape Town, South Africa. In: Cardullo, P., di Feliciantonio, C., Kitchin, R. (Eds.), The Right to the Smart City. Emerald Group Publishing, Bingley, England, pp. 165–176.

Picon, A., 2015. Smart Cities: Theory and Criticism of a Self-Fulfilling Ideal—AD Primer. John Wiley & Sons, New York.

Sennett, R., 2007. The open city. In: Burdett, R., Sudjic, D. (Eds.), The Endless City. The Urban Age Project by the London School of Economics and Deutsche Bank's Alfred Herrhausen Society. Phaidon Press, London, England, pp. 290–297.

Urban Redevelopment Authority of Singapore, 2012. Cities in Transformation: Lee Yuan Kew World Cities Prize. Didier Millet, Csi, Singapore. Retrieved from https://www.leekuanyew-worldcityprize.com.sg/laureates/special-mentions/2012/2012-special-mentions/aht-group-and-sun-development.

Violence Prevention Through Urban Upgrading (VPUU), 2014. Violence Prevention Through Urban Upgrading. A Manual for Safety as a Public Good. [PDF Embedder]. Retrieved from http://vpuu.org.za/success-story/vpuu-manual/.

Watson, V., 2014. African urban fantasies: dreams or nightmares? Environ. Urban. 26, 215–231. https://doi.org/10.1177/0956247813513705.

Author biographies

Kathryn Ewing is passionate about cocreating safe space. Ewing's work focuses on safe cities and participative design, building on lessons from design practice for policy development and teaching and knowledge exchange in African cities. She is a Senior Lecturer and Convener of the Urban Design Programme at the University of Cape Town. Ewing has architectural, urban design, planning practice, and research experience working through a multidisciplinary approach in Southern Africa. She is one of the founding Directors and currently a non-executive Director of VPUU NPC.

Michael Krause is a placemaker who believes in negotiating solutions to shape urban environments. An Urban Designer from Bauhaus-University Weimar, Germany, he has been working since 1995 in South Africa, mainly in informal settlements including Phoenix Settlement—Gandhi's second Ashram. As CEO of Violence Prevention through Urban Upgrading NPC Michael leads a highly dedicated and transversal team of people that continuously learns with residents in low income areas how to improve the quality of life in the most difficult circumstances.

Chapter 19

Watering India's smart cities

Cat Button
School of Architecture, Planning and Landscape, Newcastle University, Newcastle upon Tyne, United Kingdom

Chapter outline

India's smart city agenda

India's cities have the potential to be world class. Smart places are central to the Government of India's imagination of modern India, as indicated by the drive to create one hundred smart cities in India. Prime Minister Narendra Modi launched the Smart Cities Mission in 2014 as a flagship program in his first term of office that would shape the future of 100 cities (Das, 2020). This plan ostensibly works toward the Sustainable Development Goals (SDG), specifically SDG 11: Sustainable cities and communities, which is in turn beginning to become synonymous with smart cities (Kaika, 2017). The documentation for the initiative puts basic infrastructure front and center in the rationale (Ministry of Urban Development, 2015; Chakrabarty, 2019), but framing them as smart issues has at the same time directed attention elsewhere.

Kaika (2017) says that smart cities might change the aims, but the *how* also needs addressing to create the radical change to our cities that we need. India's Smart Cities Mission attempts to use the ideals of various smart city narratives to address the challenges of urbanization (Ministry of Urban Development, 2015). In doing so, they are continuing trends of privatization and market-led urbanization that focuses on small affluent urban areas (Kaika, 2017; Joshi, 2016). As Das (2020) points out, the aims of the Smart Cities Missions were laudable, but a focus on the means took precedence over the outcomes leading to market-driven initiatives that favor the elite. The smart cities program is driven from the center and as such is also disempowering local governments (Tandon, 2020; Joshi, 2016, Jadhav, 2019). This leads to smart places and digital technologies being used as technical fixes for social problems (Chakrabarty, 2019).

Shaping Smart for Better Cities. https://doi.org/10.1016/B978-0-12-818636-7.00007-X

The Government of India's Ministry of Urban Development, which heads the Smart Cities Mission, avoids the difficult task of defining what a smart city is by quite rightly pointing out that there is no definitive concept and they are multifaceted entities (see also Hollands, 2008). The instructions for entering the competition give more clues as to what is seen to constitute a smart city by setting out criteria (Ministry of Urban Development, 2015). The initial competition was launched in 2015, and it has drawn a lot of attention from the media both in India and internationally. The competitive system and the overall mission have been widely critiqued by both journalists and academics such as Sampath (2016), Datta (2015a,b), and Das (2020). Water access is an issue for the poor, whereas high-tech solutions are the preserve of the rich. Here, we address how the smart cities framing can be used to encompass a more inclusive approach to designing and retrofitting places.

This chapter considers how smart city aims and processes can be reconciled by focusing on water. The chapter is based on over a decade of research on water in India and particularly drawing on discussions from September 2017 in Mumbai, Maharashtra. It should be noted that Mumbai is not part of India's Smart Cities Mission and resisted inclusion in the Smart Cities Challenge competition to become one of the 100 smart cities. This could be interpreted as a choice to spend money in other ways and/or to keep more control at local government level, rather than with national government and private companies (Das, 2020). Places can still be smart outside of official benchmarking schemes. I start by introducing Mumbai's complex relationship with water as experienced through water shortage and flooding events. I then discuss rainwater harvesting as a technology being used to address some of these concerns. I will rethink rainwater harvesting as a smart technology to address water shortage and consider how the legislation surrounding rainwater harvesting fits with the Government of India's drive to create smart cities. We should not discount technologies because they are not traditionally labeled as "smart." I then consider the effectiveness of implementing the legislation in Mumbai through the concept of "eyewash" and also bring in the issues of maintaining smart places once they are created. Finally, I draw these threads together to rethink smart places through the lens of water justice. There is little point in a smart system that regulates water supplies digitally and remotely, when there is no water supply.

Water shortage in Mumbai

The rain is lashing the apartment window. The monsoon rains have been devastating in Mumbai this year, leading to floods, transport chaos, diseases and accidental deaths. In recent years the monsoon rains have been "scanty" and water shortages have been common.

(Extract from author's notes on 22/09/2017 in Mumbai)

Water shortage and flooding often go hand in hand. When researching flooding in Vietnam, I found that farmers would rapidly start talking about drought, and when looking at water shortage in India, people often talk about flooding. "There is too much population on an island, and all is concrete. Need to let the water go down, but sea water is also rising, so water will be salty soon. If you don't wake up now, it will be too late" (resident and activist interviewed in Mumbai on 19/09/2017). At some points in the year, it seems that there is water everywhere in Mumbai as the monsoon rains pour down, but most of that rainwater runs over the tarmac and concrete straight into the sea. Mumbai was hit by severe floods in 2005 and again in 2017. Several people told me that the floods in September 2017 were even worse than the widely reported ones in 2005. Respondents complained that the roads were flooded to mid-calf level after just 10 min of rain downpour. This is due to several factors including poor maintenance and the concretization of the metropolis.

In August 2017 a well-renowned doctor's car broke down in the floodwaters just 5 min from home. Dr. Deepak Amrapurkar decided to walk the short distance, but he fell into an open manhole and drowned. The manhole cover had been deliberately removed by local residents in the (mistaken) belief that it would ease the inundation of their apartment building. This tragic example highlights the need to create smart places and systems to monitor drainage (Debroy, 2017). The municipality seems to have largely relied on the famous "Mumbai spirit" to deal with the disastrous floods. Antitamper systems could be fitted onto water and drainage systems. This could also link to maintenance of drainage and water infrastructure and be used to log when maintenance occurs. Even with improved drainage, flooding can still occur. Smart solutions could use remote monitoring to identify places of water inundation and coordinate a response. This could also draw on that "Mumbai spirit" of strong civil society by encouraging residents to use social media platforms to update on serious flooding events or use citizen science in other ways. So far, there is no evidence of using these techniques. Considering flooding goes beyond the scope of this chapter, but it is an interlinked issue to keep in mind when considering water shortage.

Water shortages are not new to Mumbai. They can even be seen as designed into the original mains water supplies and then built upon since then. The provision of mains water in Mumbai has roots in the colonial planning of the city when the existing wells and tanks were insufficient due to drought and urbanization (Zérah, 2008). In response the Water Department was created in 1860 to supply water from lakes through a centralized infrastructure system that is provided and controlled by the government (Gandy, 2008; Zérah, 2008; Anand, 2011). The colonial government did not have aspirations that mains water would be universally provided and the first system that brought water from Vehar reservoir, finished in 1860, only supplied water to wealthy households (Anand, 2011; Gandy, 2008). The water system was popular with wealthy residents, and an increase in the number of connections, combined with droughts, led to the extension of the Vehar system and creation of the Tulsi scheme by 1879 (both of which still serve South Mumbai), followed by the ambitions Tansa scheme completed in 1948 (Gandy, 2008). Even with these expansions, there were still insufficient and uneven water supplies into the city at the time of India's independence. Postindependence, there have been several more schemes bringing water from distant lakes. The most recent is the Bhatsai scheme completed in several phases from 1983 until the present day, and several more schemes to increase water supplies are in preparation (Bambale, 2012).

Although the network of supply pipes has been extended, it has not kept up with urban growth, and a shortage of water supply has been created. Mains drinking water now comes from several reservoirs or lakes outside the city but is only supplied for a few hours per day to each area of the city (BrihanMumbai Municipal Corporation, 2003; Sule, n.d.). Water is only supplied to buildings that the municipality consider legal, and water is not supplied to informal settlements constructed after 1995 (Bambale, 2012). By 2003 the municipality had acknowledged the water shortage and that the problem had accelerated faster than expected:

> *The present water demand of Mumbai is 3400 MLD*[a] *against supply of 2900 MLD. This difference was expected to grow much higher in 2021 when the projected demand would be 5400 MLD for a projected population of 15.60 millions. But the population is almost there today itself & the city is facing water shortage in some areas.*
>
> BrihanMumbai Municipal Corporation (2003, p. 1)

The Municipal Corporation of Greater Mumbai (MCGM, also known as BMC) states that in 2012 water supply levels to Mumbai were 180 liters per person per day, which is higher than supplies to London (Bambale, 2012). However, due to not all residencies being supplied, the poorly maintained and inefficient network, leakages, and thefts, the per capita water supply is much lower than that, and the dream of a 24/7 supply for all residents is a long way off (Graham et al., 2013; McKenzie and Ray, 2009). The difficulties in accessing water for the urban poor majority are a daily struggle. New legislation (in line with Sustainable Development Goal 6) that pledges to supply piped water to all citizens by 2024 has been announced by the Government of India in Modi's second term (Shrivastava, 2019; Baby and Reddy, 2019).

The MCGM still bears the sole responsibility for the lucrative water supply in the city, unlike the privatized electricity supplies (BrihanMumbai Municipal Corporation, 2003; Zérah, 2008). Ruet et al. (2002) explain that the water supply is still public because of the economic benefits as the returns far outweigh the expenditure, making water a good source of revenue for the MCGM. Although this in some ways limits the commodification of water resources, the lack of an autonomous water agency reduces the efficient management of the

a. Million liters daily.

system (Bakker, 2010). However, Ruet et al. (2002) point out that the number of divisions within the MCGM with responsibilities for part of the water infrastructure, and the large number of other institutions involved at the state level in provision of water, has created a very complicated system that has led to failures. And corruption is also a serious problem across all parts of the water supply system (Asthana, 2008).

Piped water is only delivered to each area for a few hours a day, even to rich households (Button, 2014). At the domestic scale water supply issues are frequently experienced, and this provides a lens for viewing the situation of water security in Mumbai as a whole. The variance in supply is foregrounded when there is failure in the system. "The water is finally back on! [...] There was water to flush the toilet and for one tap in the kitchen only. This means it probably comes from a tank or a borewell, definitely from a different source. It turns out the pump was broken and needed a new baring. The baring was fitted late this morning. So, I can shower before meeting friends for dinner. No rain today so it is HOT" (Extract from author's notes on 18/09/2017 in Mumbai). Many wealthy Mumbai residents cope with these shortages by creating a portfolio of water supplies (Button, 2017a). This means that there is a constant supply of running water in apartments, but it comes from a variety of sources that are carefully managed. This portfolio increasingly includes rainwater harvesting and could be the key to retrofitting more inclusive smart technologies.

Rainwater harvesting as a smart technology

"Smart" does not necessarily mean digital. Technology does not have to be digital, and smart interventions do not need to be large and expensive to be effective. Solutions need to be all-encompassing throughout cities and move away from a system of "greening by numbers and indicators" (Kaika, 2017, p. 90). Labeling a process "smart" invites the applicants and policy makers to sidestep the requirement to improve infrastructure as a small tick-box hurdle to be overcome, rather than the primary goal of the endeavor. The grounded, often dirty sometimes hidden and always expensive, basic infrastructure hardware is easy to pass over when there are shiny new possibilities on offer in a smart city vision.

Interest in the environment and "green" technologies that are central to smart places has not been high until recently in India, and this could be because the technologies do not address the main issues or are at the wrong scale:

> *India needs technologies that are cheap, small and rectified for the needs of society. Simple, small things. Every house has an "Aqua Guard" [water filter] can this be solar powered? Most people have a balcony, can they have solar lighting? Cheap and simple. Need some crisis to make everyone take notice.*
>
> MCGM official (interviewed in Mumbai 14/02/2010)

The need for appropriate technologies that address not only the requirements of the population but also the capacities and culture is required. The 20th-century emphasis on large-scale centralized energy infrastructure that was developed in response to industrialization in the Global North may not be appropriate in contemporary Mumbai where decentralized water responses are taking center stage. Thus South-South appropriate technology transfer might yield better results, with due attention given to site specificities. When the technology is appropriate and provides cost-effective solutions to the problem, such as rainwater harvesting and water shortage, a precedent is created that can be used to implement further technology interventions and legislations.

Rainwater harvesting is mandatory in Mumbai for all newly constructed domestic buildings with a plot area of over 1000 m^2 since October 2002 and over 300 m^2 since June 2007 (Municipal Corporation of Greater Mumbai, 2003, 2008). This legislation was brought in to address concerns over water shortages in Mumbai and has been retrospectively presented as an environmental policy. The environment has moved up the political agenda due to the impacts of extreme weather events and international pressures, and the MCGM has included rainwater harvesting as a prominent part of the *EcoHousing* scheme (Municipal Corporation of Greater Mumbai, 2009). It is a policy that sits between water, environment, housing, and planning legislation by changing the nature of water provision and controlling development. This means that housing, and corresponding legislation, is used as a conduit to govern water provision. Developers are not granted an occupation certificate giving permission for people to live there until a rainwater harvesting system is in place and will not be connected to the mains supply without a rainwater harvesting system in place. So far, legislation has been for new buildings only, but there is a rising trend in residents retrofitting rainwater harvesting systems to increase water supplies and also to reduce localized flooding (Button, 2017a,b).

A huge number of factors influence the design of a rainwater harvesting system, and each system must be designed for the specific site and building(s), but there are certain commonalities. By considering rainwater harvesting schematically (Fig. 1), we can see places where smart technology could enhance the functionality. This could be in the collection, storage, withdrawal, and/or usage of rainwater. It should be noted that rainwater is not to be used as potable water, but it collected for use in purposes other than drinking, cooking, and bathing.

The three routes shown in Fig. 1 that harvested rainwater takes are as follows:

A. *Water is stored in a tank, and the rainwater can be easily accessed and used.* This pathway directly replaces water that was previously supplied by water tanker. Water is made available instantly from the water stored in the tank, but it is quite limited, unless an extremely large tank is available. As this is a limited supply, it is not the favored route, but is often used in conjunction with pathway B. Water is available instantly and in the long term when the two routes are combined. They can be included separately or with route B supplies from an overflow from the tank in route A.

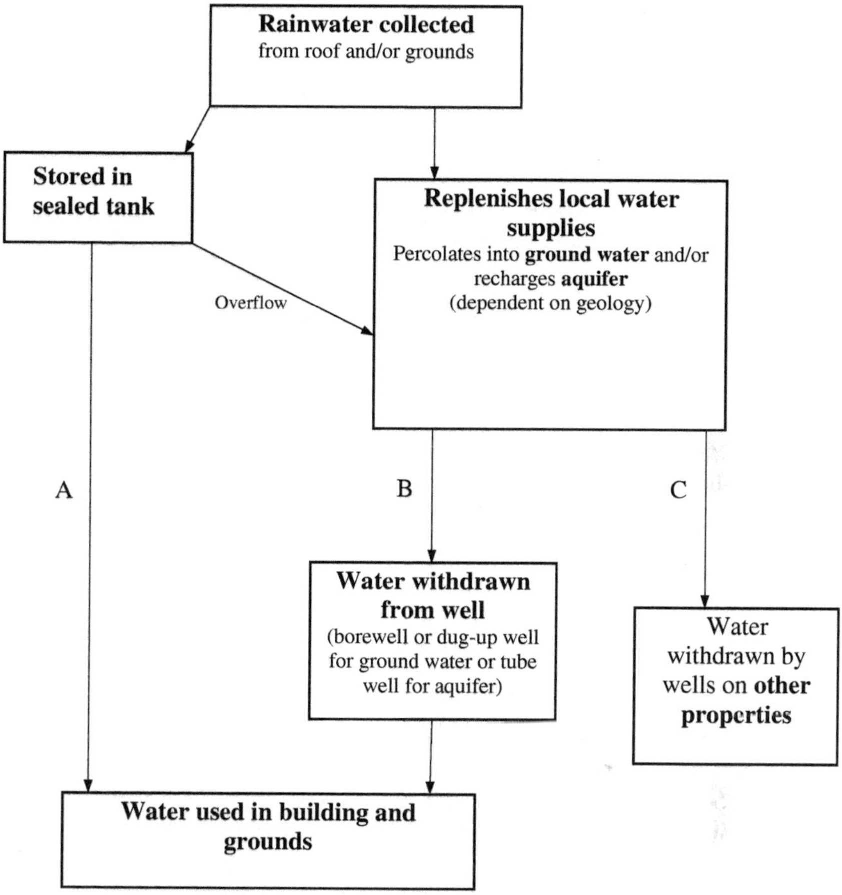

FIG. 1 Rainwater harvesting pathways schematic diagram. *(Credit: Cat Button.)*

B. *Water is used to recharge the groundwater (or sometimes an aquifer), and water is then later withdrawn using a well to be used in the building and grounds*. This is the route that I found to be favored in retrofit because it allows for a large volume of water to be added to the water table. The amount of water that is replenished and that withdrawn might differ, and there is no regulation of this. Currently the amount withdrawn seems to be less than put in, as the level in wells has risen slightly, but this may not be the case in all systems and may not continue in the future. Metering the water put in and withdrawn would be one way to monitor this, and smart metering could help regulate water supplies across a neighborhood. The collected rainwater is not bounded (as in A) and might flow away from the original property and be withdrawn elsewhere, as shown by route C. Route B is also popular because it reduces the risk of mosquitoes breeding in the water, whereas in route A chemicals would need to be added to reduce the mosquitoes.

C. *This is the route favored by developers of new-build properties and fulfills legislative requirements*. If percolation happens in gardens, then it can be used to water them as it drains away and such soakaways are popular. The water may be withdrawn by wells on other properties and so might improve the local water security. The rainwater in pathway C replenishes the local water table (or aquifer) as in route B, but it is not directly withdrawn for use in the building. There is no monitoring of how much water enters a soakaway, and this would be useful information for designing and operating smart places.

The differentiation in the water supplies that facilitates environmental innovation also limits the joining up of systems. Even when three towers are in the same compound, they have separate rainwater harvesting systems retrofitted because they are each run as a separate cooperative. Water supplies can be linked together so that we can create places with smarter water supplies. Then measure and monitor water in these complex situations to improve efficiency and make responsive infrastructure. However, the recharge and the withdrawal of water from rainwater harvesting systems are not measured. Installation of meters would be a start to making smart water systems in Mumbai.

Rainwater harvesting systems can require a lot of space. This makes them difficult to retrofit, especially in low-income homes and neighborhoods, where roofs are smaller and there is less open space. Water must be collected and then stored and/or percolated to recharge the groundwater. Rainwater harvesting could be a key technology in smart places to supply water and also to reduce localized flooding by introducing permeable surfaces, such as green spaces. A plentiful water supply then makes the greening of places possible, as shown in the previous image where the harvested water is filtered through tiered flowerbeds, both using and creating the green space (Fig. 2). The unusual system pictured in Fig. 2 makes good use of the grounds of the apartment building. The vegetation in turn improves and cools the local microclimate, making more usable outdoor places in hot and humid areas. Smart places that put water at the center become green spaces.

Creating and maintaining smart rainwater harvesting systems

I have heard from rainwater harvesting consultants and installers several times in the last decade that property developers approach them, not to install systems, but to loan them the equipment to make it look like the rainwater harvesting is in place, so that the buildings can be signed off for occupancy. There must be a smart way to check and monitor systems. "Eyewash" is a term used by rainwater harvesting consultant Gurjeet Singh Bedi to describe the situation when a (rainwater harvesting) system looks good but does nothing. This is his feeling about many companies that install rainwater harvesting systems. He says that

FIG. 2 Rainwater harvesting often both needs and creates green spaces, and so smart places become green spaces. *(Credit: Cat Button.)*

they are marketing companies really, who then call in contractors who do not do a good job. For example, one large real estate developer is now using his company for a development of 25 residential towers because they have been disappointed with work of other rainwater harvesting installers. "They now want quality" (Gurjeet Singh Bedi interviewed in Mumbai on 24/09/2017). This reflects the growing water shortages and global concerns around water security. Developers are realizing that water saving technology is a good investment, and so they should install systems that work from experienced consultants. It is an extension of the concept of greenwash, not that something is described and marketed as "green," but rather when a "green" technology is installed and does not work or is simply made to *look* as if it has been installed. These eyewashes create mirages of smart places.

There is doubt over whether rainwater harvesting legislation is being effectively enforced, as emerged in several interviews, and media reports suggest that only fifty percent of new buildings have systems in place after construction (Kulkarni, 2012). This could be symptomatic of the builders not benefiting directly from installation, and so only providing the bare minimum to be within regulations. Providing additional environmental systems would create savings for the residents, but the builder would incur the costs. The matter of checking the systems was raised with the chief engineer of MCGM who explains the process of checking rainwater harvesting systems are in place:

> *Building proposal department gives certification. This is done post construction but a licensed architect checks during the construction process.*
>
> MCGM Chief Engineer (interviewed in Mumbai 10/02/2011)

A more detailed explanation of this was not forthcoming and leaves the implementation of the rainwater harvesting legislation in doubt:

> *To be honest, it's so easy to get around it … So even though it's mandatory on paper, erm, I think not more than 20–30% of the developers are doing it.*
>
> Property Developer (interviewed in Mumbai, 16/12/2009)

> *It's not monitored. Honestly, it's not monitored and that's where I feel that the citizens are not aware of how to do, where to do, the requirement. Many a time it starts off very well, you know if functions for 3, 4 years and then it collapses.*
>
> BMC Official (interviewed in Mumbai, 20/01/2011)

This sentiment was echoed by many people in Mumbai. "Legislation is not working" (architect and educator in conversation in Mumbai on 21/09/2017). She told me that there is no groundwater recharge from systems and tanks cost *crores* (tens of millions) of rupees. Therefore developers take plans for water plants but then do nothing. She is starting to think that addressing the water shortage at the level of the individual building may not be the answer and a centralized system might be more effective.

Practitioners voiced concerns over whether the systems are installed and also whether they are then subsequently maintained. The issue of maintenance is a crucial one that arose in many interviews, with particular doubt as to whether it was being carried out at all:

> *There are certain mandatory things, as you know, that have come in also. But it's not enough just to make them mandatory. It is important to see how these things are implemented and maintained and after 10 years, are they done?*
>
> Environmental Architect (interviewed in Mumbai, 18/11/2009)

There is the question of whether there should be some obligation for follow-up monitoring of projects and whether this would be on the part of the developer, the installer, the municipality, or the residents:

> *Maintenance is part of users' responsibility. We run awareness campaigns. Very important – we see ourselves as social engineers.*
>
> MCGM Rainwater Harvesting Officer (interviewed in Mumbai, 10/02/2011)

At the time of interview, there was no plan to follow up the installation of rainwater harvesting to check maintenance, as it is considered an issue for private self-governing. Maintenance is a constant issue in India. Maintenance of systems must be folded into smart places and planned from the beginning. This must be a long-term solution as updates to technology will also be needed regularly.

Maintenance is vital to keeping the rainwater harvesting process flowing. This kind of ongoing maintenance and repair can be seen as a kind of resilience that does not just return the system back to its original state after a breakdown but allow it to change and to bounce *forward* (Shaw and Theobald, 2011). "Repair and maintenance are not incidental activities. In many ways, they are the engine room of modern economies and societies" (Graham and Thrift, 2007, p. 19). Maintenance creates instability in a system. Graham and Thrift (2007) and de Laet and Mol (2000) acknowledge the transformative potential of maintenance through incremental change or "tinkering." This maintenance can lead to innovation as systems are incrementally adapted and improved. Leaves, litter, dirt, bacteria, and mosquitoes are the focus of maintaining a rainwater harvesting system, and the levels of these could be monitored using digital technologies. This taming of rainwater and exclusion of other elements combine to provide safe clean water to be used in domestic practices (Kaika, 2005). This is particularly significant in Mumbai where maintenance often doesn't happen, but in smart places it can be vital for the process to happen.

FIG. 3 Maintenance of retrofitted rainwater harvesting system in Mumbai. (A) Clearing leaves form the system. (B) The gravel filters must be cleaned annually. *(Credit: Cat Button.)*

Maintenance is daily, annual, or sometimes on an ad hoc basis, or indeed a combination of these, depending on the specific system involved at that moment in time and space. Some rainwater harvesting systems require intensive annual maintenance; others require daily upkeep and monitoring; finally, some ad hoc maintenance is needed along the way. The systems require different types and levels of maintenance, but one commonality is that the maintenance is largely undertaken to *exclude* certain unwanted parts from the system. The following image shows one of the consultant's employees clearing leaves from the rainwater harvesting collection system. This task is undertaken during the dry season (winter), after the leaves have fallen but before the rains begin again. Before the monsoon the filters of all rainwater harvesting systems must be cleaned. The leaves, litter, and dirt are removed before the collection process starts, and these unwanted materials exert their agency and disrupt the process by blocking pipes and filters. This annual maintenance also involves the removal and cleaning or replacing of the gravel and sand in the traps and filters (Fig. 3).

The previous image shows a rainwater harvesting system being physically disassembled for cleaning and maintenance. This annual maintenance is carried out either by employees of the consultant who designed the system or by the *watchman* (janitor) in other buildings. This depends on the consultant used and the length of time since installation of the retrofit. In new builds, this kind of annual maintenance is provided by the building management after the responsibilities of the installers and construction teams has ended. Maintenance and repair may not be formally organized. The maintenance of systems is negotiated by the residents or managers of the building. The removal of rubbish and the washing of filtration screens may be necessary, and repair is also required at times of breakdown or blockage. Even systems that start as effective may become ineffective "eyewash" over time due to lack of maintenance and updating.

Water justice in smart places

Digitally enhanced smart places must not overlook the provision of water, particularly in places like Mumbai, where water supplies are far from guaranteed. As India refines its smart cities program, water security should be a key area with rainwater harvesting and graywater recycling included as smart technologies. It is a misconception that smart means digital, and what should be prioritized are solutions that respond to the specific needs of specific places. Smart places can be designed using appropriate technology, not just digital systems.

Moreover the narrow focus of India's Smart Cities Mission further increases urban inequalities. Digital technologies are expensive to install and to maintain, and they usually augment existing infrastructure, infrastructure that is already unevenly distributed. The privatization and digitizing of implementation exacerbates this. Only the wealthy can afford to live in the smart places, or projects are initiated only in wealthy areas. Smart city programs must acknowledge and address the issues of justice and power in who gets to inform, create, and access smart places. Rather than pushing certain technological fixes, there should be a pull based upon needs, and technology should respond to this. Smart placemaking must consider community needs.

Innovations in water provision can be combined with smart metering and monitoring to further improve the efficiency of service delivery to improve both the quantity and quality of the water and help demand to be predicted. Digital systems could monitor and regulate water provision, usage, and drainage, but more equitable access to clean water supplies must come be the starting point.

The designing and creation of smart places is exciting, but what happens afterward leaves a legacy for shaping future places. Mumbai has seen the problems of straightforward technology not being maintained, leading to blocked rainwater harvesting systems that are then abandoned and to poorly maintained buildings collapsing in the monsoon rains. Smart places must take into consideration the regulations and maintenance requirements, as well as the cultural attitudes and the local climate. A place isn't smart if it doesn't stay smart. So, updating and maintenance should be central to a lasting vision of smart places in India, without relying upon expensive fees from residents, as this will continue to exclude the urban poor from smart places.

Many smart city initiatives are set to become "eyewash." There is a danger that promises to make Indian cities into smarter places will lead to mere illusions of smart systems and the new utopias will turn out to be mirages. Adequate water supplies and the green spaces that improve the water supply and drainage should not be the preserve of the wealthy. The problems with rainwater harvesting systems serve as analogies for smart cities as a whole.

References

Anand, N., 2011. PRESSURE: the PoliTechnics of water supply in Mumbai. Cult. Anthropol. 26 (4), 542–564.

Asthana, A.N., 2008. Decentalisation and corruption: evidence from drinking water sector. Public Adm. Dev. 28, 181–189.

Baby, V.K., Reddy, V.R., 2019. Promising Piped Water for All by 2024: A Pipe Dream or Reality? IRC WASH Expert Blogs. 22nd July 2019. Available from: https://www.ircwash.org/blog/promising-piped-water-all-2024-pipe-dream-or-reality. (Accessed 7 June 2020).

Bakker, K., 2010. Privatizing Water: Governance Failure and the World's Urban Water Crisis. Cornell University Press, New York.

Bambale, R.B., 2012. Water Reforms—Mumbai, Maharashtra. Municipal Corporation of Greater Mumbai, Mumbai.

BrihanMumbai Municipal Corporation, 2003. Water Conservation & Rainwater Harvesting for Brihan Mumbai. BMC, Mumbai.

Button, C., 2014. Domesticating Infrastructure: Mumbai's Middle Class Housing and Rainwater Harvesting. Department of Geography, Durham University, Durham.

Button, C., 2017a. The co-production of a constant water supply in Mumbai's middle-class apartments. Urban Res. Pract. 10 (1), 102–119.

Button, C., 2017b. Domesticating water supplies through rainwater harvesting in Mumbai. Gend. Dev. 25 (2), 269–282.

Chakrabarty, A., 2019. Smart mischief: an attempt to demystify the smart cities craze in India. Environ. Urban. 31 (1), 193–208.

Das, D., 2020. In pursuit of being smart? A critical analysis of India's smart cities endeavor. Urban Geogr. 41 (1), 55–78.

Datta, A., 2015a. A 100 smart cities, a 100 utopias. Dialogues Hum. Geogr. 5 (1), 49–53.

Datta, A., 2015b. New urban utopias of postcolonial India: 'entrepreneurial urbanization'in Dholera smart city, Gujarat. Dialogues Hum. Geogr. 5 (1), 3–22.

de Laet, M., Mol, A., 2000. The Zimbabwe bush pump: mechanics of a fluid technology. Soc. Stud. Sci. 30 (2), 225–263.

Debroy, S., 2017. Body of missing Mumbai doctor found washed ashore off Worli. The Times of India. Mumbai.

Gandy, M., 2008. Landscapes of disaster: water, modernity, and urban fragmentation in Mumbai. Environ. Plan A 40, 108–130.

Graham, S., Thrift, N., 2007. Out of order: understanding repair and maintenance. Theory Cult. Soc. 24 (3), 1–25.

Graham, S., Desai, R., McFarlane, C., 2013. Water wars in Mumbai. Publ. Cult. 25 (1 69), 115–141.

Hollands, R.G., 2008. Will the real smart city please stand up? Intelligent, progressive or entrepreneurial? City 12 (3), 303–320.

Jadhav, R., 2019. What's smart about Smart Cities? The Hindu. 25 February 2019.

Joshi, B., 2016. Smart Cities Mission: flaws in a flagship programme. The Hindu. 26 June 2016.

Kaika, M., 2005. City of Flows: Modernity, Nature, and the City. Routledge, Oxon.

Kaika, M., 2017. 'Don't call me resilient again!': the New Urban Agenda as immunology … or … what happens when communities refuse to be vaccinated with 'smart cities' and indicators. Environ. Urban. 29 (1), 89–102.

Kulkarni, P., 2012. 50% of new buildings don't have rainwater harvesting system. Hindustan Times. 22nd May 2012.

McKenzie, D., Ray, I., 2009. Urban water supply in India: status, reform options and possible lessons. Water Policy 11, 442–460.

Ministry of Urban Development, 2015. Smart Cities: Mission Statement & Guidelines. Government of India.

Municipal Corporation of Greater Mumbai, 2003. Water Conservation & Rainwater Harvesting for Brihanmumbai. MCGM, Mumbai.

Municipal Corporation of Greater Mumbai, 2008. Brihanmumbai Mahanagarpulika: Rain Water Harvesting. MCGM, Mumbai.

Municipal Corporation of Greater Mumbai, 2009. Eco-housing. MCGM, Mumbai.

Ruet, J., Saravanan, V.S., Zérah, M.-H., 2002. The water & sanitation in Indian metropolitan cities: resources and management in Delhi, Calcutta, Chennai, Mumbai. CSH Occasional Paper. French Research Institutes in India.

Sampath, G., 2016. Fooled by smartness. The Hindu. 28 July 2016.

Shaw, K., Theobald, K., 2011. Resilient local government and climate change interventions in the UK. Local Environ.: Int. J. Justice Sustain. 16 (1), 1–15.

Shrivastava, R., 2019. Parched: PM Narendra Modi's new project Piped Water for All by 2024 faces nationwide water crisis. India Today. 7 June 2019, New Delhi.

Sule, S., n.d. Mumbai's Water Supply. Understanding Our Civic Issues. Bombay Community Public Trust, Mumbai.

Tandon, R., 2020. Webinar Report Impact of COVID-19 Lockdown on Urban Informals in India. PRIA, India.

Zérah, M.-H., 2008. Splintering urbanism in Mumbai: contrasting trends in a multilayered society. Geoforum 39, 1922–1932.

Author biographies

Dr. Cat Button is a Senior Lecturer (Associate Professor) in the School of Architecture, Planning and Landscape at Newcastle University, UK. Water has been Dr. Button's research focus for over a decade, particularly looking at domestic water shortages in India. She is interested in how people deal with not having enough water (and also what happens when there is too much water in the wrong place). She is increasingly drawn to *why* people make these decisions around water provision and use. She is a coinvestigator on two international and transdisciplinary UKRI GCRF research hubs. Her current research focusses on cultural, social, and spiritual significances of water.

Chapter 20

Potential and shortcomings of two design-based strategies for the engagement of city stakeholders with open data

Luca Simeone, Nicola Morelli, and Amalia De Götzen
Aalborg University, Copenhagen, Denmark

Chapter outline

Aims

Economic production and distribution processes have become interconnected at an unprecedented rate and have spread across complex networks (Castells, 1996) simultaneously operating in a multiplicity of geographic, social, and cultural markets and contexts (Julien, 2007). Within cities, as crucial nodes of these digitally enhanced networks (Willis and Aurigi, 2017), a variety of services build upon continuously flowing streams of data, which are gathered, interpreted, and processed to provide an offering that is adapted and tailored to the (oftentimes presumed) needs and wants of a variety of city stakeholders (Shepard, 2011).

On the one hand, this phenomenon touches sensitive areas such as how this massive amount of data can allow tracking individuals, groups, and even objects at an unprecedented fine-grained level (Ciuccarelli et al., 2014; Lupi et al., 2012) or how algorithms processing this huge quantity of data are increasingly

Shaping Smart for Better Cities. https://doi.org/10.1016/B978-0-12-818636-7.00001-9

regulating city dynamics and lives (O'Neil, 2016). On the other hand, big (and, especially, open) data are seen as instrumental to a better understanding of the inner functioning of organizations and cities (Ratti and Claudel, 2016) and building more inclusive services and government processes (Ratti and Townsend, 2011).

This chapter intends to focus on a specific aspect of this broad phenomenon, which is the potential of open data to feed analyses and services aimed at urban innovation and at pursuing more desirable futures (Townsend, 2013). Various cities around the world are releasing open data for anyone interested in accessing, elaborating, and exploiting these data (Kapoor et al., 2015). A core element of these processes is how to ignite and sustain the engagement of a wide variety of stakeholders with these open data. The capacity to access, analyze, and use these data might not be evenly distributed in a way to guarantee that all the city stakeholders can take an active part in the exploitation of open data and reap the benefit of it.

The chapter will examine the different design-based strategies for the engagement of city stakeholders with open data that emerged from two recent research and innovation projects (Urban Sensing and Open4Citizens). These two projects explored how open data can be used to develop new services to improve urban quality and livability while considering specific and local conditions of applications. Urban Sensing aimed at developing a software platform that could be used by urban designers, planners, and managers and other city stakeholders that wanted to analyze large amount of data (e.g., open data or data coming from social media platforms like Facebook, Twitter, FourSquare, and Flickr) and extract performance indicators for various areas of their cities (e.g., transportation, liveability, and tourism). Open4Citizens was oriented toward involving citizens into a codesign process (mostly articulated as a series of hackathons), together with IT experts, public administrations, interest groups, and start-up companies, to develop new services to improve urban quality and certain aspects of their everyday life. The aim of the project was to raise citizens' awareness of the opportunity offered by open data and create a new culture of innovation in public services.

In line with the definition proposed by the strategist Richard Rumelt—who claimed that the core of strategy should be aimed at "discovering the critical factors in a situation and designing a way of coordinating and focusing actions to deal with those factors" (Rumelt, 2011, p. 3)—this chapter is particularly concerned with comparatively examining the different key factors or leverage points (Meadows, 1999), tackled by Urban Sensing and Open4Citizens. As better discussed in the next sections, a firmer or looser anchorage to specific contexts and places is an important element of differentiation between the two projects. While Urban Sensing tried to develop a sort of context-neutral software platform, potentially ready to be quickly customized and instantiated in different urban areas and cities, the work of Open4Citizens was much more grounded into painstakingly preparing and organizing local hackathons. This chapter will examine the potential and shortcomings of these two approaches.

The chapter is organized as follows: Section "Theoretical framework" presents some theoretical concepts used to examine the cases. Section "Methods" describes the research approach. Section "Two design-based strategies for the engagement of city stakeholders with open data" presents an analytical description of the cases. Section "Discussions" discusses the results. Finally the last section concludes the chapter underlying the practical and the theoretical implications.

Theoretical framework

Open data

The growing number of government initiatives for the publication of open data is generating an important information resource, which is also incremented by a technological trend that multiplies the number of devices that are recording different aspects of human life, natural environments or urban contexts (Kalampokis et al., 2013; Simeone and Patelli, 2016). Scholarly research documented a variety of open data projects in the United States, Singapore, Mexico, Russia, the United Kingdom, and various other European countries (Davies, 2010; Turki et al., 2017), which are generally aimed at increasing the government transparency (citizens access to government data), at supporting service development by third parties and at developing a new generation of services that stimulate the economy (Janssen et al., 2012).

The present situation, though, is characterized by an asymmetry between the supply side and the demand side. The policies for opening data have been focusing on the quantitative implementation of the data repositories, rather than on the reuse of them, whereas the long-term demand side still needs to be adequately stimulated. This is mainly due to (a) the lack of a consistent framework to orchestrate and assess strategic interventions to shape an open data ecosystem (Turki et al., 2017) and (b) the absence of a consolidated practice—and a community of practice—that exchanges knowledge and experiences while working with open data (Morelli et al., 2018). These practices are nowadays mostly in the hands of people and organizations with a specific level of data literacy and competences in data science, analysis and visualization (de Götzen et al., 2018), thus undermining the possibilities to democratize the production and usage of open data.

These are complex challenges, which need to be addressed also taking into consideration the diversity of the contexts of intervention and their specific geographic, social, economic, and cultural conditions (Choi and Tausczik, 2017; Davies, 2012). Urban Sensing and Open4Citizens used two different strategies to address these challenges, each identifying specific leverage points to guide their interventions.

Leverage points

Strategy entails identifying and framing problems to be tackled, considering the available resources and capabilities (Barney, 1991, 2001; Eisenhardt and

Martin, 2000; Teece et al., 1997) and then implementing plans that narrow the perimeter of actions (Fréry, 2006) as to decrease risks and increase the chances to reach the objectives (Echevarria, 2017). A key element of strategy is the idea that efforts should be addressed toward critical points that might maximize the impact of the action (Freedman, 2013; West et al., 2014). These critical points have been termed in various ways from "center of gravity" (von Clausewitz, 1984), up to critical factors in a problematic situation (Rumelt, 2011), all the way up to "leverage points" (Meadows, 1999). From a perspective looking into system dynamics, Donella Meadows defines leverage points as "places within a complex system (a corporation, an economy, a living body, a city, an ecosystem) where a small shift in one thing can produce big changes in everything" (Meadows, 1999, p. 1).

Literature in design research has examined the notion of "strategy" from various perspectives, ranging from engineering design (Holt, 1991; Hsu, 2006, 2009) and ergonomics (Dul and Neumann, 2009), up to design management (Lockwood and Walton, 2008; Nixon, 2016; Sung et al., 2010), all the way up to collaborative design (Hyysalo and Hyysalo, 2018), architecture and urban planning (Hill, 2012; Kempenaar and van den Brink, 2018), and sustainable design and production (Niinimäki and Hassi, 2011). However, the concept of leverage points remains underexplored. This chapter intends to look into this concept and apply it to analyze two cases with a different approach toward urban innovation processes.

Two perspectives on innovation processes

Innovation processes are usually the result of a complex interaction of factors such as the different actors involved, their technical knowledge and cultures, and the contextual conditions of such interactions as articulated into innovation trajectories (Dosi, 1982; Nelson and Winter, 1982). Within such trajectories, however, the agency of innovation may originate in the action of individuals (e.g., an entrepreneur or an inventor) or emerge from the collaboration of different actors, each contributing to change through their individual problem-framing and problem-solving attitude. The first focus has been in the spotlight of the wide literature on innovation for several years, which was often inspired by Joseph Schumpeter's work (Schumpeter, 2003, 2011) and examined the action of individuals, entrepreneurs, companies (Dosi, 1982; Freeman and Perez, 1988; Kuhn, 1962), or even countries—seen as coherent innovation agents (Freeman, 1987). These actors were considered key agents behind technological change and economic innovation. The second focus is the object of more recent studies based on the idea of innovation as a product of a social process (Bijker, 1995; Bijker et al., 1987). This second perspective more closely explored the interplay between individual (innovation-oriented) action and broader socio-technical contexts, thus highlighting the mechanisms that trigger innovation.

Within design research the technological/entrepreneurial perspective mostly revolved around Roberto Verganti's studies on design-driven innovation (Verganti,

2009), which put the designer at the center of innovation processes that produced radical changes in both technological systems and in the system of meanings. The social innovation perspective, conversely, looked at innovation as a broader process and as a result of a diffuse problem-solving capability (Manzini, 2015), a process to which professional designers contribute by provoking, enabling or "infrastructuring" collaboration with broader audiences (Björgvinsson et al., 2010; Karasti, 2014). In Verganti's perspective the role of design is more explicit in the direct action of innovators, while—if one looks at broader social innovation processes—design remains a more implicit and distributed force.

Within urban contexts, innovation processes can be of varied and mixed nature, with initiatives originated by individuals, innovation agencies or entrepreneurs, and change phenomena that are happening as the result of collective and bottom-up actions. Grazia Concilio and colleagues (Concilio and Tosoni, 2019), therefore, proposed an extension of Verganti's perspective to include as possible innovation agents not only individual entrepreneurs or organizations but also wider social groups. In other words, these authors built on the idea that innovation can be produced not only by innovation experts but also by the participatory action of citizens or communities who define new meanings or new functions for the city.

The analysis of the two cases examined in this chapter will provide further considerations across different and contrasting perspectives on design-driven innovation in the urban context.

Methods

The considerations presented in this chapter stem from a case study approach (Eisenhardt, 1989; Eisenhardt and Graebner, 2007; Yin, 2009). Case studies allow identifying key insights over time (Paré, 2004), within real-life contexts (Pettigrew, 1990; Yin, 2009) and using multiple sources of evidence. The authors acted as project leaders (for the interaction design activities) in Urban Sensing and project coordinators for Open4Citizens. They have been involved in the different activities of the consortia having also the chance to gather data for the entire duration of the projects (3 years), through ethnographically inspired methods such as participant observation and semistructured conversations with key project stakeholders.

Two design-based strategies for the engagement of city stakeholders with open data

Urban sensing

Urban Sensing was an EU-funded project that explored the potential of big data and open data analysis and visualization as sources of knowledge for participatory urban design and management. The overall idea behind the project was as follows:

- to analyze various data streams (i.e., mostly open data and big amount of data originating from the citizens' and tourists' activities on Twitter, Facebook, Foursquare, and Flickr) related to three European cities: Milan, Barcelona and the Hague;
- to extract from these data some indicators on how people perceive and live the urban environment;
- to use this knowledge to feed more inclusive urban design processes (e.g., by measuring the real-time reactions of citizens to new architectural interventions).

Across a 3-year period, the team behind Urban Sensing created and tested a software platform, which do the following:

- gathers data from four social media channels (Twitter, Facebook, Foursquare, and Flickr) and various open data streams tied to key areas of the three cities examined (e.g., tourism, public transport, and housing);
- applies multiple strategies (including text mining) to analyze these data and extract indicators related to several areas of interest (such as the characterizations and perceived identities of geographic areas or the users' feeling toward local policies and urban interventions);
- visualizes the results, plotting them on a web-based map, which can be used by city designers, planners, and administrators or accessed by a broader audience interested in urban dynamics.

Fig. 1 shows a visualization elaborated within Urban Sensing (Ciuccarelli et al., 2014) and related to the area of the Polytechnic University campus in Milan, Bovisa.

In this example, the Urban Sensing software platform plots on the map the geographic locations associated to where the Twitter user tweets, also displaying the time trends (i.e., the number of contributions per day in the time span). The platform can also track tweets from a single user and, consequently, analyze how this user moves across the city. In Fig. 1 the blue dots represent the initial position of the users, while the green and red dots show the positions of the same users immediately before and after. By connecting the dots, we can clearly track users' movements over time and have an idea of the locations where, for example, users come from before getting to the Polytechnic campus and where they go after. If crossed with existing open data related to public transportation or distribution of the city population per neighborhood, the results generated by the social media analysis can provide useful information on the ways in which those citizens who regularly use social media (thus, not the whole city population) move across the city and how the public transportation does or does not adequately support this mobility patterns.

Urban Sensing aimed at investigating the potential of this and other more complex types of data and social media analyses and visualizations. The whole platform was created through a somewhat collaborative process in which various

FIG. 1 Urban Sensing: Map of the Bovisa area, Milan (August–September 2013).

local stakeholders—architects, urban planners, and city administrators—were involved, especially in the phases of user requirements and testing. However, most of the design and development process of the platform was in the hands of a small group of expert and professional designers who defined an initial number of data sets, some specific processes to analyze and visualize them and the whole user interaction model. As such the platform was an example of a tech-driven innovation, whose main agents were professional designers, programmers, and data scientists. The platform was also designed as to be fully configurable. It could be easily instantiated in different contexts by uploading local data sets and by selecting some predefined algorithms to analyze and visualize such data sets. To go beyond these predefined algorithms and to perform more fully customized data analyses and visualizations, the platform needed be modified with the help of programmers and data scientists.

The idea behind Urban Sensing was that this software platform could (1) ignite the interest of various city stakeholders including city administrators and citizens, (2) help them tinker with big and open data, and (3) make them aware of the value of these data and of the need for open data. This was considered the first step in orienting the interest of these stakeholders toward open data as a common resource that could be at the basis of more participatory urban design and management processes. Possibly also because of the novelty of the approach and of all the hype about the idea of smart cities, the project intrigued some city administrators and urban designers from all over Europe and the United States, who downloaded the platform and tried it. At first, when they were tinkering with the predefined algorithms, they all provided enthusiastic feedback. However, in the long run, they lost their interest when they realized that to perform analyses that more fully took into account the specificities of the context of application, they needed to heavily modify the platform and rewrite its core algorithms.

Open4Citizens

The EU-funded Open4Citizens project experimented with new forms of collaboration between citizens, public authorities, interest groups, local businesses, and IT experts with the aim of (1) aggregating communities around open data, (2) developing a set of practices and infrastructures for using such data, and (3) generating new public and private services. The project focused on open data as a new shared resource and aimed to generate the conditions for this resource to be used and managed as a commons (Bollier, 2014; Ostrom, 1990).

Open4Citizens was articulated in five pilots in Barcelona, Copenhagen, Karlstad, Milan and Rotterdam. Each of these pilots worked on different challenges ranging from migration to tourism, within a shared framework of design processes, tools, and methods to increase citizens' awareness of open data and to engage them in the creation of new solutions for their everyday problems.

The citizens' participation in the pilots was mostly supported through the organization of hackathons (Concilio et al., 2017), which would bring together

TABLE 1 **Pilots run during Open4Citizens.**

Pilot	Thematic areas	Participants	Stakeholders
Copenhagen (DK)	Smart tourism	60, mostly students	Municipalities, data owners, innovation network Infinit
Karlstad (SE)	Healthcare, quantified self	30 citizens	Municipality, data owners
Rotterdam (NL)	Parks and common spaces	35 activists	Local communities, municipality
Barcelona (ES)	Urban challenges, culture	110 citizens and programmers	Municipality, City Council's culture innovation grant programs, data owners
Milan (IT)	Integration	85 citizens and programmers	Municipality, NGOs, data owners

a local community with shared interests in specific problematic areas, such as public roadworks in urban contexts or issues related to healthcare, integration, pollen pollution, or regeneration of parks.

Furthermore, every pilot managed to involve a different kind of participants and stakeholders, as it is summarized in Table 1.

The convergence of different stakeholders around shared challenges would create an ecosystem to generate or consolidate the demand for open data or to organize crowdsourced gathering of new open data. For a common good to benefit the members of a community, there is a need for a shared understanding of the value of this resource and clear guidelines for its use (Bollier, 2014). The Open4Citizens hackathon approach focused on placing ordinary but engaged citizens at the center of the development of open data-driven solutions to address societal challenges they would consider relevant and important. This approach was oriented toward (a) facilitating the collaboration between hackathon participants with different skill sets and interests and (b) putting all the participants in the condition to tinker and work with such data and to understand the value of such data.

The participation in the hackathon was fostered by a starter kit that included tools that supported citizens in the various steps of the process while guiding them progressively from inspired ideas to concrete solutions (Fig. 2). Such tools were also collected into a virtual Open4Citizens software platform freely

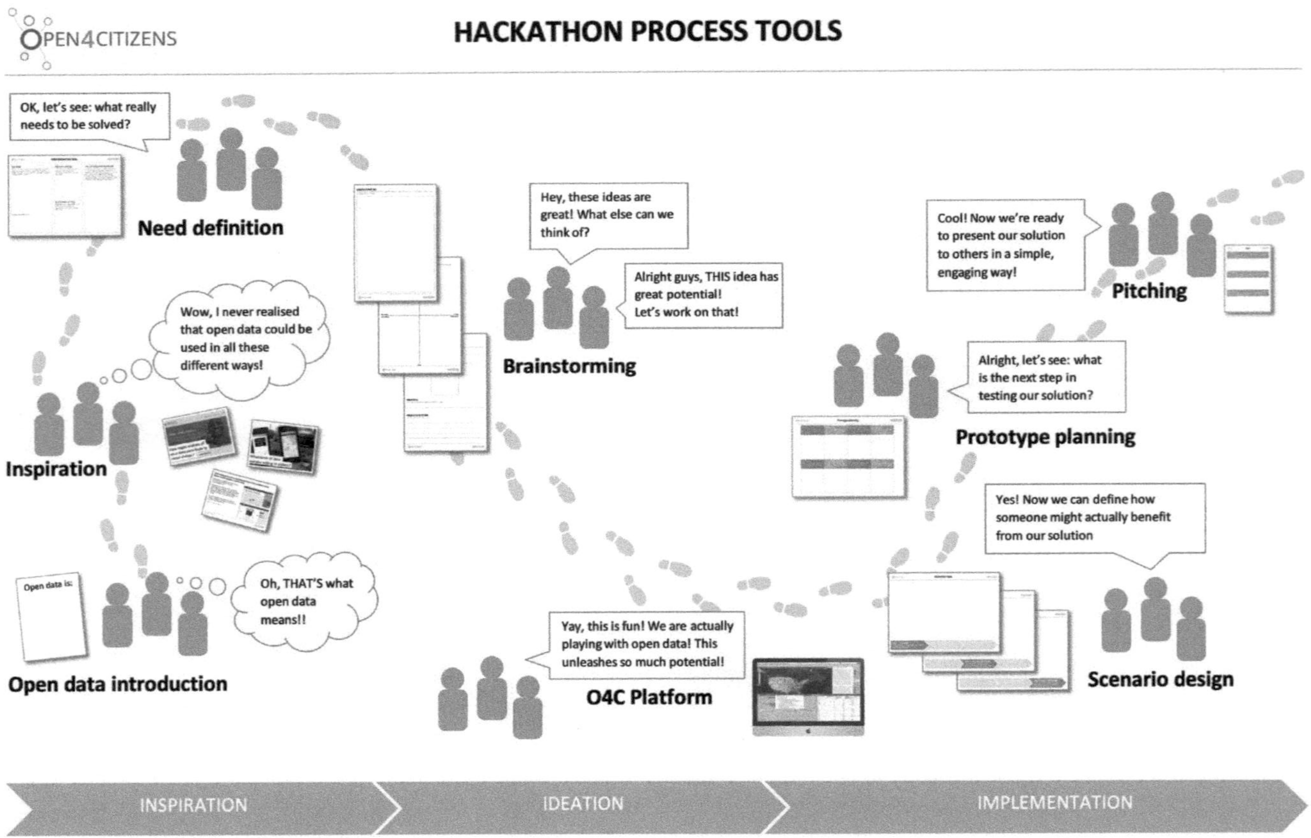

FIG. 2 The Open4Citizens starter kit.*(Credits: Open4Citizens project.)*

accessible to the public. Each hackathon was structured around three types of activities: (1) "inspiration," that is, understanding the opportunities at the intersection of citizens' needs and the current or potential availability of open data; (2) "ideation," that is, creatively addressing these opportunities; and (3) "implementation," that is, translating the ideas into concrete apps or service concepts and prototypes.

In each pilot, the experiments on open data had to be articulated along three main dimensions: the availability of data, the nature of the societal challenge(s) to be addressed and the creation of a community, that is, people with different interests, motivations, skills, and knowledge who could jointly experiment on such open data, possibly also beyond the duration of a hackathon. Some of the solutions emerged during the hackathons included mobile apps, online portals and platforms, data visualizations, and new or improved public services and their level of development varied from early concepts to more advanced prototypes. Some of these solutions were more original or innovative than others, but, in most cases, innovation emerged as a social process in which bottom-up and locally grounded dynamics played a primary role.

The Open4Citizens project also worked to consolidate all the knowledge and practices developed during the (temporary) hackathons into some "OpenDataLabs," that is, more permanent structured physical infrastructures. The concept of OpenDataLabs is inspired by other experiences of collaborative physical spaces, such as makerspaces or FabLabs, which generally aim at a strict integration in local urban contexts and in which emerging knowledge is exchanged around concrete ideas and opportunities. OpenDataLabs intend to be "public innovation places" (Manzini and Staszowski, 2013; Tassinari, 2013) where knowledge and practice about open data can be exchanged among people with different background and skills. The idea is that the OpenDataLabs can be a reference place for all individuals and private and public organizations that want to develop initiatives on open data; OpenDataLabs can offer workstations equipped with software applications to explore and tinker with data sets and can host educational activities and events, including hackathons. At the end of the EU-funding phase, Open4Citizens managed to prepare a detailed blueprint with guidelines on how to establish OpenDataLabs. To date, though, the creation of physical OpenDataLabs is still in progress: efforts are being made to secure funding, suitable physical spaces, and the continued interest of local communities.

Discussions

Different leverage points

The two projects adopted a different strategic approach as regards to where to direct their efforts.

Fig. 3 shows the strategy of Urban Sensing in relation to its focus on a single main leverage point.

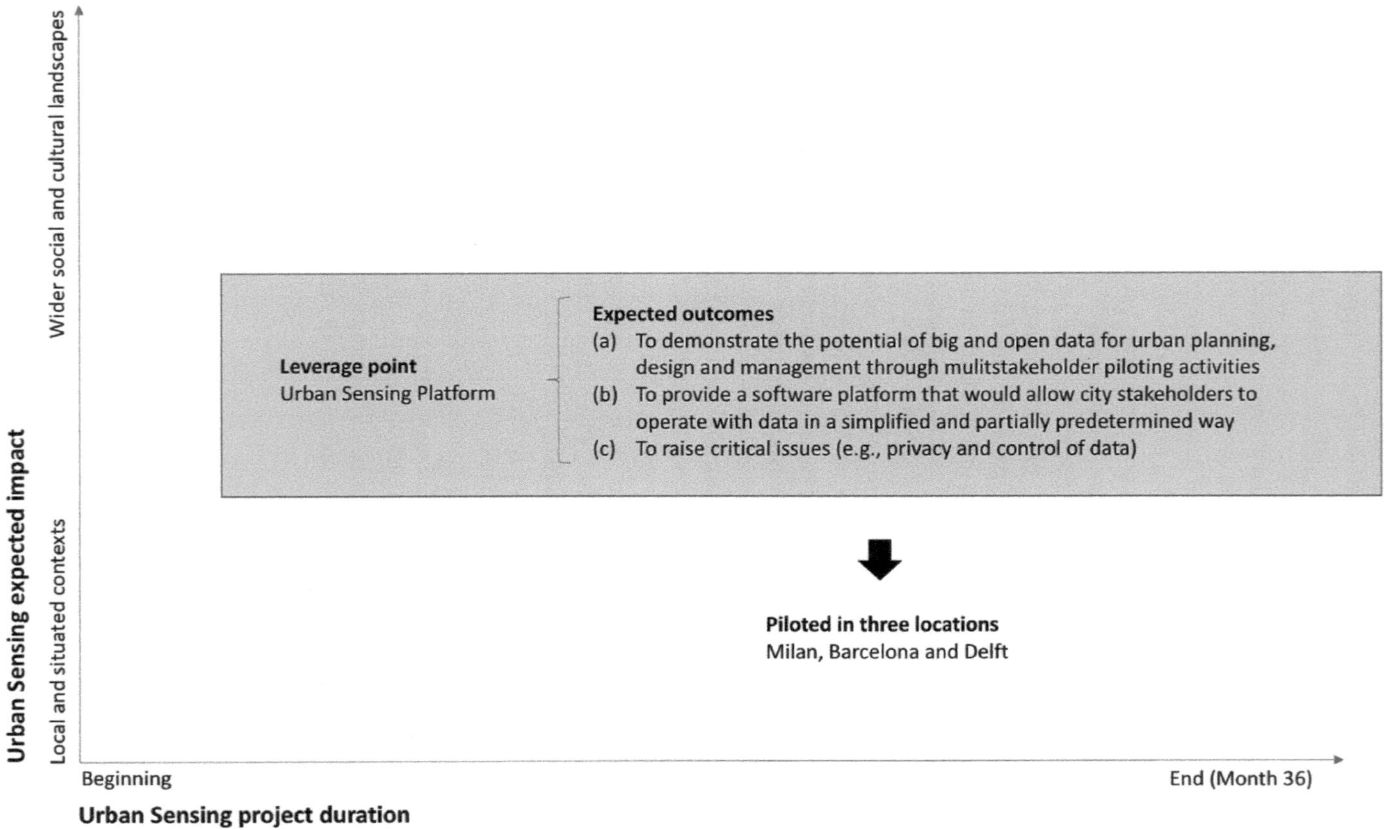

FIG. 3 Single-point strategy for Urban Sensing.

Urban Sensing focused all its efforts toward an advanced software platform that was initially codified by design experts but that could later make life easier for citizens and other city stakeholders because it would automatize various processes for data analysis and visualization. Although the design and development of this software platform was a participatory design process, this process was still quite technocentric and with a strong top-down approach. Ultimately, experts in data analysis and visualization were in charge of implementing the core algorithms and components of the Urban Sensing platform. All the resources of the project were directed toward a single leverage point.

Open4Citizens, conversely, adopted a more bottom-up approach by focusing on the design of an organizational infrastructure (a series of hackathons and the idea of OpenDataLabs) that is needed to aggregate a community around open data and by developing a set of practices and tools that facilitate the access to this resource and its usage. Some citizens were invited to a codesign process (the hackathon) through which they defined their needs and learnt how to address them through open data. Fig. 4 illustrates the multilayered strategy of Open4Citizens with respect to the three leverage points chosen to address the challenges behind the project.

Open4Citizens decided to focus on the leverage point 1 and 2 in the initial part of the project. Leverage point 1 had to do with the organization of the campaign needed to advertise the whole project and the single hackathons and to build relevant local communities of participants with different skills and knowledge about their urban contexts. Leverage point 2 had to do with the organization of the structured and sequential hackathon events. At a later stage, effort and resources were also directed toward leverage point 3 to consolidate the interventions carried out in the first phase of the project. The black arrows in Fig. 4 indicate that the leverage points supported each other. All the work behind leverage point 1 ignited and mobilized stakeholders in leverage point 2 and, at a later stage, would be crucial for the creation of OpenDataLabs (leverage point 3). Simultaneously, early work on the OpenDataLabs supported some of the hackathons in the second phase of the project. In the diagram the leverage point 3 extends also after the end of the project, as, ideally, the OpenDataLabs will become a permanent urban infrastructure.

This multilayered strategy was also aimed at mitigating the risks associated with the Open4Citizens ambition to work with social and cultural change. In a preliminary phase, by directing the efforts toward leverage point 1 and 2, Open4Citizens decided to work in specific geographic areas and contexts of action (e.g., the areas along the new metro line in Milan and a large green area in Rotterdam) where the possibility to successfully involve local stakeholders was deemed realistic. Hackathons were iteratively carried out in such locations through an incremental process that led to better results over time. In this way, Open4Citizens secured the first positive results while at the same time making the foundations for the riskier and more difficult leverage point 3. While the local dimension explored through the hackathons allowed the consortium

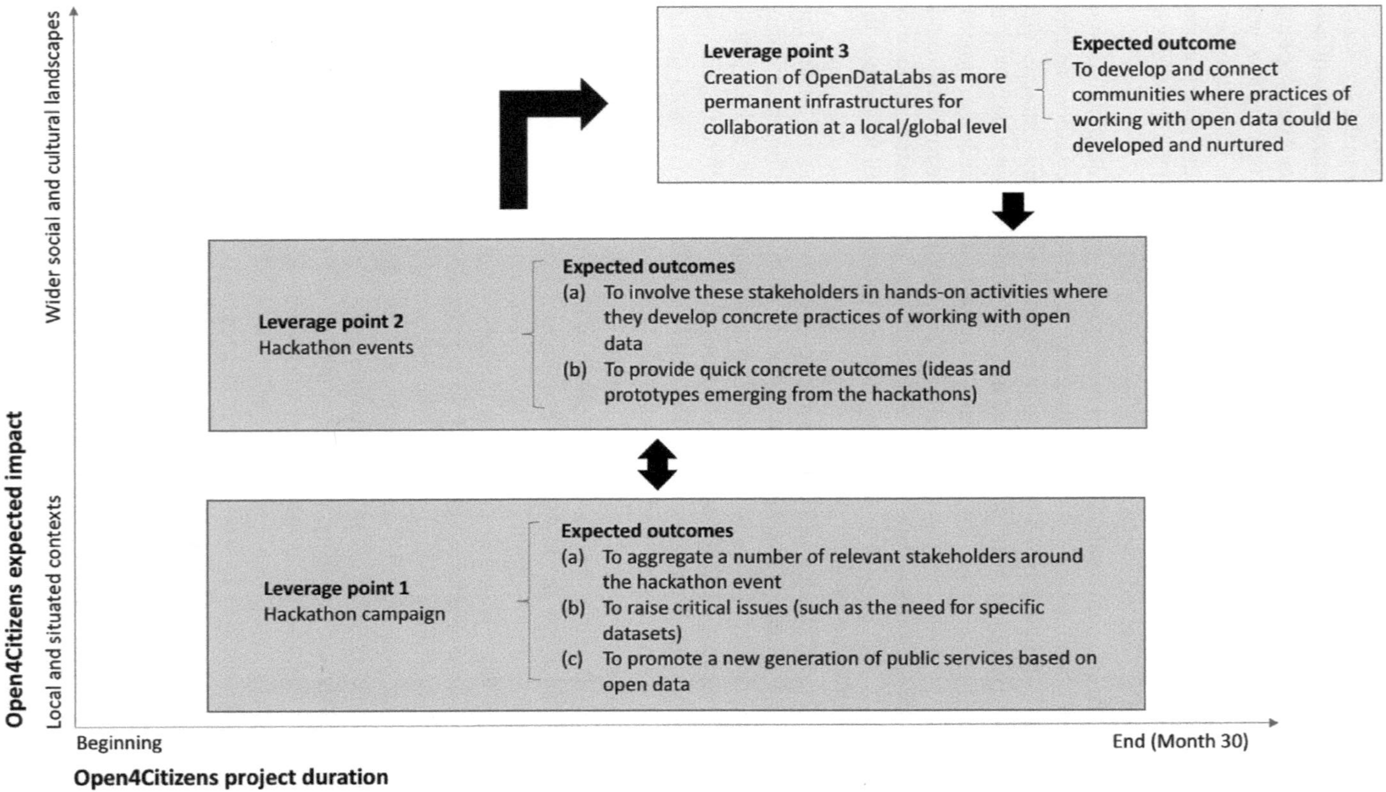

FIG. 4 Multilayered strategy of Open4Citizens with respect to leverage points and expected outcomes.

to establish initial local networks around specific challenges, it was also important to connect such local experiences, so that different communities could learn from each other's experiences and share tools, methods, and strategies. During the project, for instance, the challenges of migrants' integration have been investigated first in Copenhagen and then in Milan: despite the different contexts, the experience of the first pilot informed the process in the second one that could then get more tangible and durable results. While being deeply rooted in a specific place, these local networks should not stay isolated but rather become nodes of a large and open network, according to what Ezio Manzini defines a "cosmopolitan localism capable of generating a new sense of place" (Manzini, 2015, p. 25). For this purpose, toward the end of the EU-funding phase, Open4Citizens adopted a more top-down approach and pushed to create a first network of OpenDataLabs in Europe that would connect local communities operating with open data. To date, this process is still ongoing, and a network of interested communities and research centers across Europe is currently trying to secure some funding. Proper financial resources are needed not only to support the creation of local labs but also to consolidate collaboration flows among all these different organizations. Hopefully the first generation of OpenDataLabs will be established in the next years, but the outcome of such process is still uncertain.

Different ways to engage the stakeholders

Both projects relied upon design methods and processes to support the engagement of city stakeholders. Overall, Urban Sensing adopted a more vertical and tech-driven approach led by expert visual and interaction designers in collaboration with data scientists. The idea that city stakeholders often engaging for the first time with open data could get to a level of data literacy that could lead to interesting analyses was discarded as not realistic. Rather the approach was to create a software platform that would only require a minimum level of data literacy and that would be easy to use for a wide variety of city stakeholders. The challenge, here, was to create a software platform that would show such interesting analyses and visualizations that the city stakeholders would understand the crucial role that open data could have in urban studies, design, planning and management. In terms of expected impact, the project aimed at raising the level of awareness among key city stakeholders (including city managers) that could also become advocates for open data. In this sense, the project managed to ignite the initial interest of some city administrators and urban planners, perhaps also because it came at a time when this kind of urban analyses was not mainstream. As such, the project was featured in a variety of outlets (Lupi et al., 2012; Simeone, 2015; Simeone and Patelli, 2016). However, the interest that the project sparked was not enough to support the creation of a structured community that would outlive the duration of Urban Sensing and would keep advocating for open data. Surely, some of those urban designers, planners, managers, and researchers who used Urban Sensing are now much more aware of the potential of open data, but they mostly keep working at an individual level or, at best, at a city level. As a software platform, Urban Sensing was designed as to be somewhat "context neutral" as it could be easily plugged and configured to be used in different cities and urban contexts. This was possible thanks to some predefined algorithms to analyze and visualize data sets. However, this has proven to be also a limitation of the platform as city stakeholders realized that to produce more fine-grained and context-specific analyses, the platform needed to be heavily modified. This, perhaps, was one of the reasons why Urban Sensing did not directly contribute to the creation of a long-lasting and geographically distributed community concerned with open data.

In Open4Citizens, design processes and methods were used in a quite different perspective. Here the idea was to engage city stakeholders through a more bottom-up process relying on a set of instruments (e.g., the data starter kits and the Open4Citizens software platform) and processes (hackathons). Open data were not preprocessed by expert algorithms like in Urban Sensing but used as a starting material to be explored collaboratively in hackathons across Europe. The expected impact of this approach is to increase public awareness and understanding of the potential of open data, to provide opportunities for citizens to manipulate this new resource, to foster a stronger sense of ownership on open data, and to support social innovation processes. The limitations of this approach are in the readiness of data and their availability in the right format. It is often hard to make sure that sufficient and meaningful sets of data are available to be used in hackathons. On the one hand, the demand for public and open data is still weak and unable to generate standard formats and consolidated practices, on the other, often, when the data are available, there is a lack of awareness even among those civil servants that are the owners, if not the unconscious producers, of those datasets. This implies a complex work of standardization and preparation of data before they can be used in the hackathons and the need for the creation of better practices and new roles inside public organizations, such as some sort of "data intermediaries" that could constantly advocate for public and private organizations to release better data sets. These were some of the reasons why in Open4Citizens the organization of hackathons required a big preparatory effort to find the missing intermediaries, the data owners, and the stakeholders that could have been interested in the outcome of such bottom-up process. A big effort was spent in managing and keeping the stakeholders involved and securing their contribution for the hackathon. Another aspect that limits the impact of the Open4Citizens' strategy is its relevance in the political and social arena. The larger parts of the data published at present are touching politically neutral aspects of public domains, whereas the data owners are often reluctant to publish more sensitive datasets, although concerning the public domain, in fear of possible misuse or misinterpretation of the data. This is a clear limitation to the availability of meaningful datasets and, therefore, to the possibility to support the consolidation of common practices around open data.

Conclusions

In summary and also with regard to some possible practical implications emerging from our analysis, the two approaches led to interesting outcomes and had an impact on the engagement of city stakeholders. Each approach also had limitations and had to face challenges (Table 2).

Possibly a combination of the two approaches—and their technocentric versus socially emerging innovation perspectives—would have led to a wider impact and longer lasting results. Previous literature examined how similar participatory processes to the ones adopted by Urban Sensing and Open4Citizens (e.g., hackathons) tried out two different approaches, one in which tinkering with technology has a greater centrality and another one in which issues of local concern are put in the foreground and the technology used to address these issues remains in the background (Lodato and DiSalvo, 2016; Porter et al., 2017; Taylor and Clarke, 2018; Trainer et al., 2016). Open4Citizens promoted "civic hacking" (Tauberer, 2014) by trying to build social ties and social relations and, at the same time, by articulating local and social issues in relation to data-material production. In this context, data literacy proved to be the key factor to achieve a successful articulation of the issues. Possibly the Urban Sensing platform could have had a central role in this regard since one of the limitations of the Open4Citizens project was precisely the lack of a common technological framework to be adopted during the hackathons. The participants used their own different digital tools to explore the datasets, and this led to very diverse outcomes and the difficulty to share outcomes and results. In an ideal world the Urban Sensing platform could have served as a common framework to inspire city stakeholders and show them the full potential of advanced data analyses and visualization, while the Open4Citizens engagement process could have supported broader interaction with local communities.

The two projects told two different stories in relation to their potential to respond to local contexts and places. Urban Sensing tried to deliberately develop a platform that could be context neutral or, more precisely, easily plugged into different contexts through some preset functionalities to be quickly configured by local teams. On the one hand, this made the platform an object that could easily circulate across different contexts and sparkle interest among city stakeholders thanks to its polished and shiny analyses and visualizations precreated by professional designers, programmers, and data scientists. On the other, this one-size-fits-all platform showed its limitations when some users—some city administrators and urban planners—realized that the preset functionalities did not allow a fine-grained response to the local contexts of intervention. Open4Citizens, conversely, was more strictly anchored to the specific contexts in which the various hackathons were organized. The problem, here, was that engaging the local stakeholders and organizing the hackathons was a painstaking activity, which needed to be carried out for any hackathon with few economies of scale. To support wide and bottom-up participatory processes, a great deal of

TABLE 2 Key elements of differentiation between Urban Sensing and Open4Citizens.

	Urban Sensing	Open4Citizens
How the project used design to engage citizens	The whole platform was created through a somewhat collaborative process in which various local stakeholders—architects, urban planners, and city administrators—were involved, especially in the phases of user requirements and testing. However, most of the design and development process of the platform was in the hands of a small group of expert and professional designers who defined an initial number of data sets, specific processes to analyze and visualize them and the whole user interaction model. This software platform allowed end users to (individually) interact with data and perform some analyses at an urban scale. The platform was designed as to be configurable and easily instantiated in different contexts of use	Design is considered as a "diffuse capability" of citizens (Manzini, 2015), which can be supported and stimulated through the involvement of citizens in creative events (hackathons) that bring together people with different competences. Processes of data analyses are here conducted in groups
Anchorage to local conditions	Although, initially, Urban Sensing sought the collaboration of local stakeholders, in the long run, it did not manage to secure their interest over time and beyond the duration of the EU funding phase. Urban Sensing became a software platform freely accessible via web from all around the world but not specifically anchored to any local context	Open4Citizens had a firm anchorage to local contexts, both through the hackathon events organized in the different pilots and through the hackathon campaigns. Every hackathon was organized around some specific local and societal concerns that got investigated with local relevant stakeholders and the citizens themselves
Achievements	The software platform provided a simplified access to open data to a wide audience without specific data literacy. Such platform demonstrated and disseminated the potential of open data for urban analysis, design, and planning	Open4Citizens successfully leveraged broad citizen participation and raised awareness on the potential of open data

Limitations and challenges	• Technocentric approach • Data already come as preprocessed and the end users employ preprogrammed algorithms to analyze the data. This limits the potential of the instrument • Urban Sensing ignited interest, but did not foster long-term sustained engagement • This approach did not develop deep technological competences on how to operate with data	• The Open4Citizens hackathons are complex events that need to be carefully organized. The heterogeneity of the participants requires a preliminary work on the organization of the hackathon campaign, which would aggregate the whole social ecosystem participating in the hackathon. It is also critical to maintain a balance between people with high IT skills (e.g., software developers or coders) and other stakeholders with a lower level data literacy • Other challenges come from the capability to find adequate datasets to make the hackathon relevant, especially for coders, who expect to contribute with their data-related skills • OpenDataLabs are considered as a key component to support the integration of local communities and secure their interest over time but their success depends on the possibility to integrate OpenDataLabs into other urban infrastructures on the basis of a sustainable business model

centralized organization on behalf of the core Open4Citizens consortium was needed. The two projects highlighted a common dilemma of design-based strategies for the engagement of city stakeholders (Ehn et al., 2014; Seravalli and Simeone, 2016), which is to what extent it is possible to rely on self-organizing and emergent bottom-up participation and to what extent, instead, a certain level of top-down organization is needed to provide indications and directions and to activate and sustain local and wider conversations.

The chapter looked at these two projects as a way to develop some critical reflections on at least two aspects: (1) the current hype for big and open data and (2) the somewhat diffused idea that design can work as a quick magic gesture to fix complex and wicked problems by involving a variety of participants in collaborative processes (Badke-Schaub et al., 2010; Johansson-Sköldberg et al., 2013; Nussbaum, 2011). The analysis of Urban Sensing and Open4Citizens showed all the challenges emerging while design-based strategies try to ignite and sustain such collaborative processes. As such the potential of design-based strategies for the engagement of city stakeholders with open data needs to be weighed against careful considerations on the shortcomings of such approaches.

Acknowledgments

The research leading to this contribution has received funding from the European Commission under Grant Agreements No. ICT-687818 and No. 314887. This article reflects the authors' views. The European Commission is not liable for any use that may be made of the information contained therein.

References

Badke-Schaub, P., Roozenberg, N., Cardoso, C., 2010. Design thinking: a paradigm on its way from dilution to meaninglessness? In: Proceedings of the 8th Design Thinking Research Symposium (DTRS8). Presented at the Sidney. Sidney.

Barney, J., 1991. Firm resources and sustained competitive advantage. J. Manag. 17 (1), 99–120. https://doi.org/10.1177/014920639101700108.

Barney, J., 2001. Resource-based theories of competitive advantage: a ten-year retrospective on the resource-based view. J. Manag. 27, 643–650.

Bijker, W.E., 1995. Of Bicycles, Bakelites, and Bulbs: Toward a Theory of Sociotechnical Change. MIT Press, Cambridge, MA.

Bijker, W.E., Hughes, T.P., Pinch, T.J. (Eds.), 1987. The Social Construction of Technological Systems: New Directions in the Sociology and History of Technology. MIT Press, Cambridge, MA.

Björgvinsson, E., Ehn, P., Hillgren, P.-A., 2010. Participatory design and democratizing innovation. In: Proceedings of the 11th Biennial Participatory Design Conference. ACM, Sidney, Australia, pp. 41–50.

Bollier, D., 2014. Think Like a Commoner: A Short Introduction to the Life of the Commons. New Society Publishers, Gabriola, British Columbia.

Castells, M., 1996. The Rise of the Network Society. Volume 1 of the Information Age: Economy, Society and Culture. Blackwell, Oxford; Malden, MA.

Choi, J., Tausczik, Y., 2017. Characteristics of collaboration in the emerging practice of open data analysis. In: Proceedings of the 2017 ACM Conference on Computer Supported Cooperative Work and Social Computing (CSCW '17), pp. 835–846, https://doi.org/10.1145/2998181.2998265.

Ciuccarelli, P., Lupi, G., Simeone, L., 2014. Visualizing the Data City—Social Media as a Source of Knowledge for Urban Planning and Management. Springer, Milan, Heidelberg, New York, Dordrecht, London.

Concilio, G., Tosoni, I., 2019. Innovation Capacity and the City: The Enabling Role of Design. Springer.

Concilio, G., Molinari, F., Morelli, N., 2017. Empowering citizens with open data by urban hackathons. In: Proceedings of Conference for E-Democracy and Open Government, pp. 125–134, https://doi.org/10.1109/CeDEM.2017.28.

Davies, T., 2010. Open Data, Democracy and Public Sector Reform: A Look at Open Government Data Use from Data.gov.uk. Retrieved from Practical Participation website http://www.opendataimpacts.net/report/wp-content/uploads/2010/08/How-is-open-government-data-being-used-in-practice.pdf.

Davies, T., 2012. The promises and perils of open government data (OGD). Autori 8 (2), 1–6.

de Götzen, A., Kun, P., Simeone, L., Morelli, N., 2018. Making sense of data in a Service Design education. In: ServDes Linköping Electronic Conference Proceedings. Presented at the ServDes.

Dosi, G., 1982. Technological paradigms and technological trajectories: a suggested interpretation of the determinants and direction of technical change. Res. Policy 11 (3), 147–162.

Dul, J., Neumann, W.P., 2009. Ergonomics contributions to company strategies. Appl. Ergon. 40 (4), 745–752. https://doi.org/10.1016/j.apergo.2008.07.001.

Echevarria, A.J., 2017. Military Strategy: A Very Short Introduction. Oxford University Press, Oxford; New York.

Ehn, P., Nilsson, E.M., Topgaard, R., 2014. Making Futures: Marginal Notes on Innovation, Design, and Democracy. MIT Press, Cambridge, MA and London.

Eisenhardt, K.M., 1989. Building theories from case study research. Acad. Manag. Rev. 14 (4), 532–550.

Eisenhardt, K.M., Graebner, M.E., 2007. Theory building from cases: opportunities and challenges. Acad. Manag. J. 50 (1), 25–32.

Eisenhardt, K.M., Martin, J.A., 2000. Dynamic capabilities: what are they? Strateg. Manag. J. 21 (10–11), 1105–1121. https://doi.org/10.1002/1097-0266(200010/11)21:10/11<1105::AID-SMJ133>3.0.CO;2-E.

Freedman, L., 2013. Strategy: A History. Oxford University Press, Oxford; New York.

Freeman, C., 1987. Technology Policy and Economic Performance: Lessons from Japan. Pinter Pub, London; New York.

Freeman, C., Perez, C., 1988. Structural crises of adjustment, business cycles and investment behaviour. In: Dosi, G. (Ed.), Technical Change and Economic Theory. Francis Pinter, London, pp. 38–66.

Fréry, F., 2006. The fundamental dimensions of strategy. MIT Sloan Manag. Rev. Retrieved from https://sloanreview.mit.edu/article/the-fundamental-dimensions-of-strategy/.

Hill, D., 2012. Dark Matter and Trojan Horses. A Strategic Design Vocabulary. Strelka Press.

Holt, K., 1991. The impact of technology strategy on the engineering design process. Des. Stud. 12 (2), 90–95. https://doi.org/10.1016/0142-694X(91)90050-7.

Hsu, Y., 2006. Comparative study of product design strategy and related design issues. J. Eng. Des. 17 (4), 357–370. https://doi.org/10.1080/09544820600664655.

Hsu, Y., 2009. Exploring design innovation and performance: the roles of issue related to design strategy. J. Eng. Des. 20 (6), 555–569. https://doi.org/10.1080/09544820802043609.

Hyysalo, V., Hyysalo, S., 2018. The mundane and strategic work in collaborative design. Des. Issues 34 (3), 42–58. https://doi.org/10.1162/desi_a_00496.

Janssen, M., Charalabidis, Y., Zuiderwijk, A., 2012. Benefits, adoption barriers and myths of open data and open government. Inf. Syst. Manag. 29 (4), 258–268. https://doi.org/10.1080/10580530.2012.716740.

Johansson-Sköldberg, U., Woodilla, J., Çetinkaya, M., 2013. Design thinking: past, present and possible futures. Creat. Innov. Manag. 22 (2), 121–146. https://doi.org/10.1111/caim.12023.

Julien, P.-A., 2007. A Theory of Local Entrepreneurship in the Knowledge Economy. Edward Elgar Publishing, Cheltenham, UK; Northampton, MA.

Kalampokis, c., Tambouris, E., Tarabanis, K., 2013. Linked open government data analytics. In: International Conference on Electronic Government. Springer, pp. 99–110.

Kapoor, S., Mojsilovic, A., Strattner, J.N., Varshney, K.R., 2015. From open data ecosystems to systems of innovation: a journey to realize the promise of open data. In: Presented at the Bloomberg Data for Good Exchange Conference, New York.

Karasti, H., 2014. Infrastructuring in participatory design. In: Participatory Design Conference, Windhoek, Namibia. vol. 14, pp. 141–150.

Kempenaar, A., van den Brink, A., 2018. Regional designing: a strategic design approach in landscape architecture. Des. Stud. 54, 80–95. https://doi.org/10.1016/j.destud.2017.10.006.

Kuhn, T.S., 1962. The Structure of Scientific Revolutions. University of Chicago Press, Chicago, IL.

Lockwood, T., Walton, T., 2008. Building Design Strategy. Allworth Press, New York.

Lodato, T.J., DiSalvo, C., 2016. Issue-oriented hackathons as material participation. New Media Soc. 18 (4), 539–557. https://doi.org/10.1177/1461444816629467.

Lupi, G., Patelli, P., Simeone, L., Iaconesi, S., 2012. Maps of babel. Urban sensing through user generated content. In: Proceedings of Human Cities Symposium. Presented at the Brussels (BE). Brussels (BE).

Manzini, E., 2015. Design, when Everybody Designs: An Introduction to Design for Social Innovation. The MIT Press, Cambridge, MA.

Manzini, E., Staszowski, E., 2013. Public and Collaborative: Exploring the Intersection of Design. Social Innovation and Public Policy. Retrieved from http://www.designagainstcrime.com/files/publications/pub_2013_public_and_collaborative.pdf.

Meadows, D., 1999. Leverage Points. Places to Intervene in a System. The Sustainability Institute, Hartland, VT USA.

Morelli, N., de Götzen, A., Simeone, L., 2018. A system of innovation to activate practices on open data: The Open4Citizens project. In: Proceedings of 3rd International Conference on Smart Learning Ecosystem and Regional Development - The Interplay of Data, Technology, Place and People. Presented at the Aalborg, Denmark. Aalborg, Denmark.

Nelson, R.R., Winter, S.G., 1982. An Evolutionary Theory of Economic Change. Belknap Press of Harvard University Press, Cambridge, MA.

Niinimäki, K., Hassi, L., 2011. Emerging design strategies in sustainable production and consumption of textiles and clothing. J. Clean. Prod. 19 (16), 1876–1883. https://doi.org/10.1016/j.jclepro.2011.04.020.

Nixon, N.W. (Ed.), 2016. Strategic Design Thinking. Bloomsbury, New York; London.

Nussbaum, B., 2011. Design Thinking Is a Failed Experiment. So What's Next? May 4. Retrieved from Fast Company/Co.Design website http://www.fastcodesign.com/1663558/design-thinking-is-a-failed-experiment-so-whats-next. (Accessed 26 February 2016).

O'Neil, C., 2016. Weapons of Math Destruction: How Big Data Increases Inequality and Threatens Democracy. Crown, New York.

Ostrom, E., 1990. Governing the Commons: The Evolution of Institutions for Collective Action. Cambridge University Press, Cambridge.

Paré, G., 2004. Investigating information systems with positivist case research. Commun. Assoc. Inf. Syst. 13, 1. Retrieved from http://aisel.aisnet.org/cais/vol13/iss1/18.

Pettigrew, A.M., 1990. Longitudinal field research on change: theory and practice. Organ. Sci. 1 (3), 267–292.

Porter, E., Bopp, C., Gerber, E., Voida, A., 2017. Reappropriating hackathons: the production work of the CHI4Good day of service. In: Proceedings of the 2017 CHI Conference on Human Factors in Computing Systems (CHI '17), pp. 810–814, https://doi.org/10.1145/3025453.3025637.

Ratti, C., Claudel, M., 2016. The City of Tomorrow: Sensors, Networks, Hackers, and the Future of Urban Life. Yale University Press, New Haven; London.

Ratti, C., Townsend, A., 2011. Harnessing residents' electronic devices will yield truly smart cities. Sci. Am. 305 (3), 42–48.

Rumelt, R., 2011. Good Strategy, Bad Strategy. Crown Business, New York.

Schumpeter, J.A., 2003. Capitalism, Socialism and Democracy. Routledge, London; New York (Original work published 1942).

Schumpeter, J.A., 2011. In: Becker, M.C., Knudsen, T., Swedberg, R. (Eds.), The Entrepreneur. Stanford University Press, Stanford, CA.

Seravalli, A., Simeone, L., 2016. Performing Hackathons as a way of positioning boundary organizations. J. Organ. Chang. Manag. 29 (3), 326–343.

Shepard, M. (Ed.), 2011. Sentient City: Ubiquitous Computing, Architecture, and the Future of Urban Space. MIT Press, Cambridge, MA.

Simeone, L., 2015. Missing intentionality: The limitations of social media analysis for participatory urban design. In: Eric, G., Mihailidis, P. (Eds.), Civic Media Project. Retrieved from www.civicmediaproject.org.

Simeone, L., Patelli, P., 2016. Urban sensing: potential and limitations of social network analysis and data visualization as research methods in urban studies. In: Kubitschko, S., Kaun, A. (Eds.), Innovative Methods in Media and Communication Research. Palgrave Macmillan, London.

Sung, T.-J., Lu, Y.-T., Ho, S.-S., 2010. Time-based strategy and business performance under environmental uncertainty: an empirical study of design firms in Taiwan. Int. J. Des. 4 (3), 29–42.

Tassinari, V., 2013. From welfare state to partner state: the case of welcome to saint-Gilles. In: Public and Collaborative. Exploring the Intersection of Design, Social Innovation and Public Policy. Desis Network, pp. 39–50.

Tauberer, J., 2014. Open Government Data: The Book. Lulu.com.

Taylor, N., Clarke, L., 2018. Everybody's hacking: participation and the mainstreaming of Hackathons. In: Proceedings of the 2018 CHI Conference on Human Factors in Computing Systems (CHI '18), pp. 1–12, https://doi.org/10.1145/3173574.3173746.

Teece, D.J., Pisano, G., Shuen, A., 1997. Dynamic capabilities and strategic management. Strateg. Manag. J. 18 (7), 509–533.

Townsend, A.M., 2013. Smart Cities: Big Data, Civic Hackers, and the Quest for a New Utopia. W. W. Norton & Company, New York, NY.

Trainer, E.H., Kalyanasundaram, A., Chaihirunkarn, C., Herbsleb, J.D., 2016. How to Hackathon: socio-technical tradeoffs in brief, intensive collocation. In: Proceedings of the 19th ACM Conference on Computer-Supported Cooperative Work & Social Computing (CSCW '16), pp. 1116–1128, https://doi.org/10.1145/2818048.2819946.

Turki, S., Martin, S., Renault, S., 2017. How open data ecosystems are stimulated? In: Proceedings of the Internationsl Conference on Electronic Governance and Open Society Challenges in Eurasia (EGose '17), pp. 179–187, https://doi.org/10.1145/3129757.3129787.

Verganti, R., 2009. Design Driven Innovation: Changing the Rules of Competition by Radically Innovating What Things Mean. Harvard Business Review Press, Cambridge, MA.

von Clausewitz, C., 1984. On War. Princeton University Press, Princeton, NJ (Original work published 1832).

West, P.C., Gerber, J.S., Engstrom, P.M., Mueller, N.D., Brauman, K.A., Carlson, K.M., et al., 2014. Leverage points for improving global food security and the environment. Science 345 (6194), 325–328. https://doi.org/10.1126/science.1246067.
Willis, K.S., Aurigi, A., 2017. Digital and Smart Cities. Routledge, New York.
Yin, R.K., 2009. Case Study Research: Design and Methods, Fourth Edition. Sage, Thousand Oaks.

Author biographies

Luca Simeone is Associate Professor and a member of the Service Design Lab at Aalborg University. He is particularly interested in the strategic, organizational, economic, and financial dimensions of design and on how design can support entrepreneurship, innovation, and managerial processes. Simeone has conducted research and teaching activities in various international centers (Harvard, MIT, Polytechnic University of Milan, Malmö University, and University of the Arts London), (co)authoring 4 books and some 20 publications for design and management journals.

Nicola Morelli is Professor in Service Design and Director of the Service Design Lab at Aalborg University. He has a Bachelor's in Architecture (University of Naples) and a PhD in Industrial Design (Politecnico di Milano). His research focuses on methodological issues related to service design and on strategic issues related to the role of service design in social innovation, sustainability, and sociotechnical transitions.

Amalia De Götzen is Associate Professor in Service Design and a member of the Service Design Lab at Aalborg University. She graduated in Electronic Engineering at the University of Padova and was awarded a PhD in Computer Science from the University of Verona. Her research activity focuses on digital social innovation, in particular, on the intersection between interaction design and service design. She is interested in tools and methods that bridge the analog and digital world of services.

Index

Note: Page numbers followed by *f* indicate figures and *t* indicate tables.

C

F

G

H

I

J

K

L

M

S

T

FSC
www.fsc.org
MIX
Papier | Fördert
gute Waldnutzung
FSC® C083411

Kolibri 360 GmbH
Ferdinand-Jühlke-Straße 7
99095 Erfurt, Deutschland
produktsicherheit@kolibri360.de